Impacts of Climate Risk and Energy Consumption on Financial Markets

Abdelkader Mohamed Sghaier Derbali
https://orcid.org/0000-0003-4098-3120
Taibah University, Saudi Arabia

IGI Global
Scientific Publishing
Publishing Tomorrow's Research Today

Vice President of Editorial — Melissa Wagner
Director of Acquisitions — Mikaela Felty
Director of Book Development — Jocelynn Hessler
Production Manager — Mike Brehm
Cover Design — Jose Rosado

Published in the United States of America by
IGI Global Scientific Publishing
701 East Chocolate Avenue
Hershey, PA, 17033, USA
Tel: 717-533-8845 | Fax: 717-533-7115
Website: https://www.igi-global.com E-mail: cust@igi-global.com

Library of Congress Cataloging-in-Publication Data

LCCN: 2025043730 (CIP Data Pending)
ISBN13: 9798337367668
Isbn13Softcover: 9798337367675
EISBN13: 9798337367682

British Cataloguing in Publication Data
A Cataloguing in Publication record for this book is available from the British Library.

Table of Contents

Detailed Table of Contents

focused financial education aimed at green energy investments would notably increase policy attainment and accelerate Mexico's transition to renewable energy.

According to the authors, this chapter provides a comprehensive review of more than the last twenty years empirical research on the relationship between commodity prices, sectoral stock indices and climate change induced financial risk in India. The chapter discusses the historical evolution of analytical techniques from simple linear econometric analysis to more recent machine learning algorithms by integrating 67 articles published in peer-reviewed journals between 2000 and 2025 using the PRISMA methodology. Key thematic results are large transmission effects of volatility between commodities and from these to sectoral markets, varying sectoral sensitivities to climate stress, and new trends in fintech and ESG commodity trading. Issues in climate risk integration, data innovation and policy frameworks are highlighted in the chapter. It provides a sound basis for climate-friendly financial strategies to help develop more resilient and sustainable capital markets in India.

The accelerating shift towards digital banking services has not only reshaped the Indian financial landscape through enhanced convenience and accessibility but also emerges as a critical nexus in the broader context of climate risk, energy consumption, and sustainable financial market development. This study investigates the awareness, adoption, and challenges of sustainable digital banking practices among Indian users, viewing digital banking as a pivotal channel for advancing environmental resilience and efficient energy use in financial ecosystems. By analyzing user behavior across leading digital platforms such as Google Pay, Paytm, Phone Pe, and banking applications, the research explores how platform trust, incentive structures, and usage frequency influence the willingness to embrace eco-friendly financial services, with an emphasis on their potential to reduce carbon footprints and support climate adaptation strategies.

The accelerating impacts of climate change are reshaping global financial markets, driving the need for innovative tools to manage climate risk and mobilize capital for sustainable growth. Green financial innovations—particularly green bonds, Environmental, Social, and Governance (ESG) investing, and broader sustainable finance mechanisms—have emerged as key instruments linking financial performance with climate resilience. Green bonds channel capital into renewable energy, and low-carbon infrastructure, while ESG frameworks guide investors in assessing both financial and environmental risks. At the systemic level, sustainable finance is reinforced by policy initiatives such as the EU Taxonomy, pushing markets toward transparency and accountability. Yet, challenges such as greenwashing, inconsistent ESG metrics, and limited accessibility in emerging markets persist. This chapter examines the evolution, opportunities, and limitations of these innovations, emphasizing their potential to transform financial markets into drivers of the energy transition and climate risk management.

This paper explores the role of banking efficiency in promoting green finance and climate risk management in Islamic banks. It comes in a context of increasing pressure on financial institutions to adopt sustainable practices, where Islamic banks operate at the intersection of religious compliance and environmental imperatives. The study uses a stochastic frontier analysis approach applied to a panel of Islamic banks operating in the member countries of the Organization of Islamic Cooperation. It measures banking efficiency and its influence on green commitments and climate risk management strategies. The results show that high bank efficiency is positively correlated with greater involvement in green finance and better resilience to climate risks. These efficient banks tend to adopt green practices more systematically and strengthen their environmental risk management systems. This research proposes an innovative approach by combining three dimensions rarely discussed together in the literature: banking efficiency, Islamic finance, and environmental sustainability.

Vikas Sharma, Department of Computer Applications, SRM Institute of Science and Technology, Delhi, India
Tarun Kumar Vashishth, School of Computer Science and Applications, IIMT University, Meerut, India
Anuj Kumar, IIMT College of Medical Science, IIMT University, Meerut, India
Poonam Bhardwaj, IIMT College of Medical Science, IIMT University, Meerut, India
Garima Garg, IIMT College of Medical Science, IIMT University, Meerut, India
Priyanka Rana, School of Commerce and Management, IIMT University, Meerut, India
Sachin Chaudhary, School of Computer Science and Applications, IIMT University, Meerut, India

The urgency of climate change has placed financial markets in a position to lead the sustainability transition. Green financial products—such as green bonds, Environmental, Social, and Governance (ESG) frameworks, and generally, sustainable finance—are becoming one of the most important approaches to leveraging finance for low-carbon resilience. This chapter argues that these finance products are climate risk management tools, as they link finance flows to climate outcomes, incentivize responsible investment, and transparently disclose market actions and outcomes. The chapter examines asymmetrical growth and future potential in financing renewable energy and climate-resilient infrastructure via green bonds and how ESG practices are applied in capital decisions and corporate governance. Furthermore, the chapter argues that sustainable finance can reduce exposure to climate risks while creating opportunities for innovation, resilient business practices, and value-generating results in global markets.

Amir Ahmad Dar, Lovely Professional University, India
Sukriti Kakkar, Lovely Professional University, India
Vanshita Arora, Lovely Professional University, India
Hibah Himmath, B.S. Abdur Rahman Crescent Institute of Science and Technology, India

This chapter bridges the gap by analyzing climate vulnerability, energy transition policies, and financial market implications across income groups using the 2022

The issue of energy transition occupies a central place in contemporary debates on climate governance and sustainable development. Faced with the intensification of extreme climate events and the rise in greenhouse gas (GHG) emissions, green and renewable energy policies appear to be essential levers for reducing climate risks and ensuring the systemic resilience of modern societies. This paper analyzes in an integrated manner the impact of these policies on climate risk management, through a conceptual and theoretical approach based on the paradigms of sustainable development, energy transition, and polycentric climate governance. By mobilizing a multidisciplinary analytical framework, the study highlights the close relationship between the decarbonization of the energy system, economic diversification, and the reduction of environmental vulnerability.

This chapter presents a systematic literature review of 54 peer-reviewed studies on how commercial banks integrate climate-related risks into their risk management frameworks. Drawing upon institutional, agency, stakeholder, and resource-based theories, the analysis identifies three interrelated dimensions of bank responses: climate adaptation, loan pricing, and climate risk disclosure. Findings reveal that banks, particularly in China and the European Union, increasingly incorporate environmental considerations into strategic and operational decisions through green credit, ESG engagement, digital innovation, and board diversity. However, persistent challenges, such as symbolic compliance, brown lending, and uneven regulatory enforcement, limit substantive progress. The review highlights how institutional pressures, managerial incentives, stakeholder expectations, and organisational capabilities jointly shape banks' climate behaviours.

Preface

INTRODUCTION

The accelerating intersection between climate risk, energy consumption, and financial markets has become one of the most defining challenges of the 21st century. The financial sector is no longer a passive observer of environmental transformation—it is a central actor in shaping pathways toward sustainable development and low-carbon resilience. As global economies experience mounting climate shocks, from extreme weather to resource volatility, financial markets have emerged as a pivotal mechanism for both mitigation and adaptation. This book, *Impacts of Climate Risk and Energy Consumption on Financial Markets*, aims to illuminate the intricate linkages between environmental sustainability, market stability, and policy innovation.

Our motivation for developing this edited volume stems from a clear recognition that climate risk has transitioned from being a peripheral consideration in finance to a material determinant of value, investment, and economic resilience. Clean energy investments, green bonds, ESG frameworks, and sustainable finance instruments now underpin the architecture of modern markets. Yet, substantial disparities persist across regions, income groups, and institutional capacities. Through empirical studies, case-based analyses, and conceptual insights, this book presents a comprehensive exploration of how financial systems are adapting to and influencing the global energy transition.

The chapters compiled herein represent a truly international collaboration of researchers and practitioners who bridge diverse disciplines—from economics and finance to governance, technology, and environmental policy. Collectively, their contributions reflect a shared goal: to understand how financial markets can mitigate systemic risk, accelerate clean energy adoption, and promote inclusive growth in the face of climate uncertainty. This volume does not simply document existing research; it actively seeks to shape discourse and inform actionable strategies that align finance with sustainability.

CHAPTER OVERVIEW

Chapter 1, *Financing Clean Energy Projects and the Role of Emerging Financial Markets*, opens the book by highlighting how capital accessibility drives renewable energy development. It demonstrates that financing is both an enabler and a catalyst of innovation in the clean energy transition.

Chapter 2 explores *the role of the nonprofit sector in Saudi Arabia* in promoting green finance and supporting small projects. The chapter underscores the alignment of Vision 2030 with sustainable financing and calls for stronger institutional collaboration.

Chapter 3 examines *commodity and sovereign shocks in resource nations*, detailing how geopolitical uncertainty and energy volatility propagate through financial systems and fiscal frameworks.

In *Chapter 4*, the author assesses *Mexico's green and renewable energy policies* and their connection to financial literacy, revealing how education can accelerate participation in sustainable programs.

Chapter 5 provides a *systematic literature review* of climate risk impacts on Indian sectoral markets, revealing significant transmission effects and sectoral sensitivities to volatility and environmental stress.

Chapter 6 positions *digital banking as a catalyst* for sustainability and energy efficiency, showing how fintech and user adoption contribute to climate-resilient market development.

In *Chapter 7*, the focus turns to *green financial innovations*, particularly green bonds and ESG investing, which are identified as tools that integrate environmental accountability into financial systems.

Chapter 8 investigates *Islamic banking efficiency and green finance*, demonstrating how operational effectiveness enhances resilience to climate risks and promotes ethical sustainability in finance.

Chapter 9 revisits *green bonds and ESG* as complementary instruments in achieving climate stability, highlighting their transformative potential to redirect capital toward low-carbon projects.

Chapter 10 presents a *comparative analysis across income groups*, linking climate risk index data with policy readiness, economic resilience, and financial stability.

Chapter 11 emphasizes the *role of financial markets* in adaptation and mitigation, demonstrating how accurate risk pricing and capital allocation can accelerate the low-carbon transition.

Chapter 12 provides a theoretical and conceptual analysis of *renewable energy policies* as levers for climate risk reduction, emphasizing multi-level governance and sustainable development.

Chapter 13 presents a *systematic review of climate risk integration in banking*, mapping global trends in green credit, ESG engagement, and regulatory innovation.

Finally, *Chapter 14* concludes with an *in-depth study of energy consumption trends*, challenges, and innovations, offering a global outlook on energy efficiency, technological advancement, and equitable access.

Together, these chapters form a coherent narrative that bridges theory, empirical research, and policy relevance—addressing how financial markets must evolve to withstand and mitigate the multidimensional impacts of climate change.

CONCLUSION

As editors, we envision this book as more than an academic compilation; it is a blueprint for understanding the symbiotic relationship between energy, finance, and the environment. By integrating global perspectives and diverse methodologies, it contributes to the expanding field of climate finance, which now stands at the forefront of global policy and investment discourse. The studies collected here provide both theoretical depth and applied insight, guiding stakeholders toward more informed decision-making in a volatile and rapidly transforming global landscape.

We also recognize that managing climate risk is not solely a technical or financial challenge—it is a governance, ethical, and social imperative. The pathways toward low-carbon resilience depend on inclusive participation across sectors, transparent information flows, and adaptive financial mechanisms. Whether through green bonds, ESG principles, or digital finance, every innovation documented in this book contributes to reimagining markets as instruments of sustainability rather than agents of extraction.

Ultimately, *Impacts of Climate Risk and Energy Consumption on Financial Markets* seeks to inspire ongoing dialogue among scholars, policymakers, and practitioners who are committed to advancing climate-resilient finance. We hope that the insights presented within these pages will serve as a foundation for future research, practical collaboration, and transformative policy innovation. In doing so, we reaffirm our collective belief that financial markets—if steered responsibly—can become one of the most powerful levers in confronting the defining environmental challenge of our time.

Abdelkader Mohamed Sghaier Derbali

Department of Administrative and Financial Sciences, Applied College, Taibah University, Saudi Arabia

Chapter 1
Financing Clean Energy Projects and the Role of Emerging Financial Markets

J. Suhashini
https://orcid.org/0000-0003-2582-974X
GTN Arts College, India

M. Toni
https://orcid.org/0000-0003-4597-6843
University of Nizwa, Oman

Shashank M. Hiremath
https://orcid.org/0009-0004-4082-3375
Presidency Business School, Presidency College, India

P. Selvakumar
https://orcid.org/0000-0002-3650-4548
Department of Science and Humanities, Nehru Institute of Technology, Coimbatore, India

Avinash Hanmant Ghadage
https://orcid.org/0009-0000-9877-8520
Global Business School and Research Centre, D.Y. Patil Vidyapeeth, India

S. Guru Basava Aradhya
https://orcid.org/0000-0002-3335-2681
Padmashree Institute of Management and Sciences, India

ABSTRACT

Financing stands as one of the most critical pillars for the successful development, implementation, and long-term sustainability of clean energy projects. While the technological innovations that underpin renewable energy solutions such as solar, wind, hydro, geothermal, and bioenergy have advanced significantly in recent decades, their wide-scale adoption is heavily contingent upon the availability of adequate and accessible funding. Unlike fossil-fuel-based energy projects, which benefit from a century-old infrastructure and well-established financial markets, clean energy initiatives often require upfront capital investments that are compar-

DOI: 10.4018/979-8-3373-6766-8.ch001

atively higher, even if their long-term operational and maintenance costs are lower. This dynamic makes financing not only a facilitator of project initiation but also a catalyst for innovation, scalability, and inclusivity in the clean energy transition. For governments, private investors, and development institutions, financing is the bridge between aspirational climate goals and tangible, measurable progress toward a low-carbon future.

THE IMPORTANCE OF FINANCING FOR CLEAN ENERGY PROJECTS

Financing stands as one of the most critical pillars for the successful development, implementation, and long-term sustainability of clean energy projects. While the technological innovations that underpin renewable energy solutions such as solar, wind, hydro, geothermal, and bioenergy have advanced significantly in recent decades, their wide-scale adoption is heavily contingent upon the availability of adequate and accessible funding. Unlike fossil-fuel-based energy projects, which benefit from a century-old infrastructure and well-established financial markets, clean energy initiatives often require upfront capital investments that are comparatively higher, even if their long-term operational and maintenance costs are lower (Ali et al., 2018). This dynamic makes financing not only a facilitator of project initiation but also a catalyst for innovation, scalability, and inclusivity in the clean energy transition. For governments, private investors, and development institutions, financing is the bridge between aspirational climate goals and tangible, measurable progress toward a low-carbon future.One of the primary reasons financing is vital for clean energy projects is the significant capital expenditure required during the early phases of development. Renewable energy systems typically involve high initial costs for technology acquisition, site development, permitting, infrastructure building, and grid integration. For example, the installation of utility-scale solar photovoltaic farms or offshore wind facilities demands billions of dollars in upfront investment before a single kilowatt-hour of electricity is produced. This financial burden often creates a barrier for developers, particularly in emerging economies where access to capital markets is limited or where lending rates are prohibitively high. Therefore, innovative financing mechanisms such as green bonds, blended finance, public-private partnerships, and concessional loans are increasingly being leveraged to de-risk projects and attract both domestic and international investors. and remote rural areas where traditional grid infrastructure is underdeveloped or entirely absent.

The importance of financing also extends to addressing the inherent risks associated with clean energy investments. Renewable energy projects are often exposed to regulatory uncertainties, market volatility, technological risks, and long development timelines. Traditional financiers, such as commercial banks, may view these risks as deterrents, thereby restricting credit availability or charging high-interest rates. (Alsaleh & Abdul-Rahim, 2019) To overcome these challenges, financial institutions, governments, and international organizations have increasingly turned to risk-mitigation instruments such as credit guarantees, insurance schemes, and currency hedging tools. These mechanisms not only make clean energy projects more bankable but also instill confidence among private investors who may otherwise hesitate to commit capital to unfamiliar markets. Risk mitigation thus serves as a crucial component of financing, ensuring that clean energy projects can progress despite uncertainties and external shocks.In addition, financing acts as a lever for technological advancement and innovation in the clean energy sector. Investment in research and development (R&D) requires sustained financial support to push the boundaries of efficiency, storage capacity, and integration of renewables into the grid. For example, breakthroughs in battery storage technology, smart grid systems, and green hydrogen are all the result of targeted financing strategies that prioritize innovation. Without dedicated funding streams, these advancements would either remain confined to laboratories or face slow, fragmented commercialization(Balta et al., 2013) Thus, financing is not only about scaling existing technologies but also about fostering the next generation of clean energy solutions that will define the energy systems of the future.On a macroeconomic level, financing clean energy projects has profound implications for sustainable growth, job creation, and industrial transformation. The renewable energy sector has emerged as one of the largest job creators globally, employing millions of workers across various stages of project development, from engineering and construction to maintenance and supply chain logistics.

CHALLENGES IN FINANCING CLEAN ENERGY INITIATIVES

Financing clean energy initiatives has emerged as one of the most pressing challenges in the global transition toward sustainable development, and while the urgency of combating climate change is widely recognized, mobilizing adequate capital to scale renewable energy projects remains complex and fraught with obstacles. A primary challenge lies in the high upfront costs associated with renewable energy infrastructure, such as solar farms, offshore wind turbines, geothermal plants, or advanced energy storage systems. Unlike fossil-fuel-based projects that benefit from established infrastructure and relatively predictable cash flows, clean energy projects require significant initial investment and often involve long payback periods, creating hesitancy among private investors who prefer quick returns. Compounding this issue is the perception of risk—many investors and financial institutions still view clean energy projects as risky due to technological uncertainties, fluctuating policy frameworks, and the volatility of energy markets. For example, emerging technologies like hydrogen fuel cells, carbon capture systems, and next-generation batteries have yet to demonstrate consistent large-scale commercial viability, making lenders cautious about allocating capital. Policy instability adds another layer of difficulty; frequent changes in subsidies, feed-in tariffs, tax credits, or regulatory incentives in different countries make it hard for investors to assess long-term profitability. (Charan et al., 2025a) This is particularly problematic in developing economies where political risk and weak governance structures can undermine investor confidence.

Access to affordable finance is also uneven across regions; while developed economies may have relatively better access to green bonds, venture capital, and institutional investors, many low-income nations struggle to secure funds due to underdeveloped financial markets and lack of creditworthiness, despite often having vast renewable potential such as abundant sunlight or wind resources. Furthermore, clean energy projects typically require large-scale coordination among multiple stakeholders—governments, private companies, local communities, and international organizations—which adds bureaucratic delays and transaction costs. Another significant barrier is the dominance of fossil fuel subsidies that distort market competition by making renewable energy comparatively less attractive; trillions of dollars continue to flow into fossil fuel industries globally, creating an uneven playing field. Even when financing is available, the structuring of financial instruments poses challenges, as many renewable energy ventures need blended finance solutions that combine concessional loans, grants, equity, and guarantees, but such mechanisms remain underutilized or poorly understood. The lack of standardized frameworks for assessing the risk and return of green projects further discourages investors, as they often cannot benchmark performance or predict revenue streams with confidence.(Charan et al., 2025b) Additionally, limited awareness and technical expertise among local banks and financial intermediaries in many countries prevent them from engaging in renewable energy lending, further constraining capital flow. Currency fluctuations, high interest rates, and inflation in emerging markets make external financing more expensive, while local financing is often scarce. On the social front, financing challenges also emerge from community opposition, land acquisition disputes, and insufficient inclusion of local populations in project benefits, which can delay or derail investments.

Internationally, while mechanisms like the Green Climate Fund and multilateral development banks exist to channel resources toward clean energy, the actual disbursement of funds is often slow, bureaucratic, and inadequate compared to the trillions needed annually to achieve global climate targets such as net-zero emissions by 2050. Insurance and risk-mitigation instruments are not always available or accessible for clean energy developers, particularly in fragile economies, leaving projects vulnerable to natural disasters, economic downturns, or political upheaval. Moreover, venture capital and private equity funds that do flow into clean energy tend to concentrate on less risky, large-scale projects in developed nations, while smaller, decentralized projects—such as off-grid solar in rural Africa or community-based wind farms in Asia—struggle to attract investors despite their high social and developmental impact. Transparency and accountability in the use of funds also remain concerns, with investors demanding robust environmental, social, and governance (ESG) reporting, which many small developers lack the resources to provide. (Chen et al., 2020)Another challenge is the mismatch between the long lifespan of renewable energy assets and the shorter time horizons of most financial institutions, leading to a gap between the financing available and the financing required. Innovative financing instruments such as green bonds, sustainability-linked loans, or carbon credit markets, though promising, are still developing and face issues of credibility, fragmentation, and regulatory oversight.

EMERGING FINANCIAL MARKETS FOR CLEAN ENERGY FINANCING

Emerging financial markets for clean energy financing are rapidly evolving as the global economy transitions toward sustainability, offering both unprecedented opportunities and complex challenges in mobilizing capital for renewable energy development, energy efficiency measures, and climate-resilient infrastructure. Traditionally, clean energy investment was concentrated in developed economies with strong regulatory frameworks, robust financial institutions, and mature capital markets. However, the growing urgency of addressing climate change, coupled with the enormous untapped renewable energy potential in developing and middle-income nations, has shifted focus toward emerging markets as critical players in global decarbonization efforts. (Doytch & Narayan, 2016)These regions—including parts of Asia, Africa, Latin America, and Eastern Europe—represent the fastest-growing sources of demand for electricity and therefore the largest opportunities for deploying clean energy solutions. As energy demand in these areas expands, financial markets are adapting to create innovative mechanisms and instruments to attract both domestic and international investors. Green bonds, sustainability-linked loans, and blended finance platforms are at the forefront of this evolution, helping to channel funds into solar, wind, hydropower, biomass, and off-grid electrification projects. For instance, countries like India, Brazil, and South Africa are actively developing green bond markets to support large-scale renewable projects, while smaller economies such as

Kenya or Vietnam are experimenting with microfinance and pay-as-you-go models to scale decentralized solar solutions for rural communities.

Multilateral development banks (MDBs), international climate funds, and philanthropic organizations play a catalytic role by providing concessional finance, guarantees, and risk-mitigation instruments that de-risk investments and attract private sector capital into markets that would otherwise be considered too risky. The integration of digital platforms and financial technology (fintech) is another emerging trend, enabling mobile-based payment solutions and blockchain-backed transparency tools that allow investors to monitor clean energy projects and ensure accountability. At the same time, sovereign wealth funds and institutional investors in emerging markets are beginning to align their portfolios with sustainability goals, unlocking domestic capital for clean energy initiatives that were previously dependent solely on foreign funding. Still, the growth of financial markets for clean energy in emerging economies is not without obstacles; challenges such as currency risks, underdeveloped regulatory environments, lack of credit histories, and political instability can dampen investor confidence. To overcome these hurdles, many emerging markets are adopting standardized green finance taxonomies, enhancing ESG (environmental, social, governance) disclosure frameworks, and building regional partnerships to harmonize clean energy investment policies. For example, the Association of Southeast Asian Nations (ASEAN) is developing guidelines to expand cross-border green finance, while African nations are working with global partners to design continent-wide platforms for renewable energy investment. Furthermore, the role of local banks and microfinance institutions is becoming more prominent as they begin to develop specialized clean energy lending portfolios, often supported by capacity-building initiatives from international donors.

Another promising development is the proliferation of blended finance models, where concessional capital from governments or donors is combined with private investment to create viable financial structures that lower risk and improve returns. This is particularly effective in financing small-scale, community-based projects such as mini-grids, which may not be attractive to large institutional investors but deliver significant social and developmental benefits. (Eren et al., 2019)Additionally, emerging markets are exploring innovative financial products like sustainability-linked derivatives, renewable energy securitization, and crowdfunding platforms that democratize access to investment opportunities for both local citizens and diaspora communities. The use of digital tokens and blockchain in financing renewable projects is also gaining traction, particularly in regions where traditional financial infrastructure is limited but mobile technology penetration is high. Beyond technology and innovation, the political will to mainstream green finance is becoming increasingly visible; governments are issuing sovereign green bonds, establishing green banks, and creating dedicated clean energy investment funds that anchor financial markets and signal long-term commitment. The cumulative effect of these emerging financial market trends is not only the mobilization of capital for clean energy but also the creation of more inclusive, resilient, and equitable energy systems that bridge gaps in access and affordability.

OPPORTUNITIES IN EMERGING MARKETS FOR CLEAN ENERGY INVESTMENT

Opportunities in emerging markets for clean energy investment are expanding at an unprecedented pace, driven by rising energy demand, abundant renewable resources, demographic growth, rapid urbanization, and increasing political recognition of the need for sustainable development pathways. Unlike developed economies, where energy markets are relatively saturated and grid systems are mature, emerging markets in Asia, Africa, Latin America, and parts of Eastern Europe present vast untapped potential for renewable energy deployment and innovative financing models. These regions often experience surging electricity demand due to industrialization, population expansion, and digitalization, yet they simultaneously face energy access gaps, infrastructure deficits, and heavy reliance on fossil fuels. For investors, this combination creates a fertile ground for clean energy solutions such as solar, wind, hydropower, biomass, geothermal, and decentralized mini-grids that can meet demand while offering competitive returns. One of the most compelling opportunities lies in solar power, as many emerging markets are located in high-insolation zones with year-round sunlight. Countries like India, Vietnam, and several(Gaies et al., 2019a) African nations have already demonstrated how large-scale solar parks and distributed rooftop systems can be scaled rapidly when combined with policy incentives and investor interest. Similarly, wind energy potential is immense in coastal regions of Latin America, offshore zones in Southeast Asia, and inland corridors in Africa, offering long-term opportunities for investors seeking diversified portfolios.

This creates opportunities for partnerships that align profit motives with developmental goals, making clean energy investments more bankable and less exposed to traditional risks. Investors also stand to benefit from favorable demographics and consumption trends: young, urbanizing populations are driving demand for modern energy services, electric mobility, and energy-efficient infrastructure, opening markets for renewable-powered transport, green buildings, and electric vehicle charging networks. Additionally, the global shift toward decarbonization and net-zero commitments by multinational corporations is pushing supply chains in emerging markets to adopt renewable energy, creating new demand for corporate power purchase agreements (PPAs) and long-term contracts that offer stable cash flows for investors.(Gaies et al., 2019b) Fintech innovation represents another transformative opportunity, as digital platforms, blockchain, and mobile applications lower transaction costs, improve transparency, and democratize access to energy investment opportunities in regions with limited traditional banking infrastructure. Emerging markets are also becoming hubs for renewable energy manufacturing and supply chains, with countries like China, India, and Vietnam producing solar panels, wind turbines, and battery components, offering vertical investment opportunities beyond generation assets. Furthermore, carbon markets in emerging economies are beginning to unlock new revenue streams through voluntary and compliance-based carbon trading systems, enabling investors to monetize emissions reductions from clean energy projects. The policy landscape, while varied, is increasingly supportive: many emerging market governments are introducing renewable energy auctions, tax incentives, feed-in tariffs, and national green banks that send strong market signals and attract foreign direct investment. These policies, combined with international pressure to align with climate agreements, are creating more predictable environments for long-term investment. Importantly, clean energy investments in emerging markets also offer portfolio diversification benefits, as they provide exposure to high-growth economies while aligning with global ESG trends that are increasingly demanded by institutional investors and asset managers. (Hassine & Harrathi, 2017)While risks such as political instability, currency volatility, and regulatory uncertainty persist, they are being mitigated through innovative financial structures, risk insurance, and partnerships with multilateral organizations.

RISKS AND CONSIDERATIONS IN EMERGING MARKET CLEAN ENERGY FINANCING

While emerging markets present immense opportunities for clean energy investment, they are equally associated with a complex set of risks and considerations that investors, policymakers, and developers must carefully evaluate before committing capital. A major challenge is political and regulatory instability, as frequent policy shifts, unclear legal frameworks, and changes in government priorities can undermine investor confidence and project viability. Many emerging markets lack consistent renewable energy policies or regulatory enforcement, which increases uncertainty around subsidies, tariffs, power purchase agreements (PPAs), and long-term contracts that underpin financial models. Currency risk also looms large, since revenues from clean energy projects are often denominated in local currencies while financing and debt obligations are in foreign currencies, leaving investors vulnerable to exchange rate fluctuations and inflation. Weak financial institutions and underdeveloped capital markets further compound these risks, as access to affordable long-term financing is often limited, and domestic banks may lack the expertise to evaluate or support clean energy ventures. Infrastructure deficits, such as inadequate grid systems, transmission bottlenecks, and weak energy storage capacity, can delay project implementation and reduce efficiency, while logistical challenges increase costs.(Harmanpreet et al., 2025) Land acquisition disputes and community opposition are additional risks, particularly in rural areas where renewable projects may overlap with agricultural activities, cultural sites, or local livelihoods. Investors must also consider governance and corruption risks, as bureaucratic inefficiencies, lack of transparency, or mismanagement of funds can undermine both financial returns and reputational standing.

Figure 1. Understanding the role of green finance and renewable energy consumption

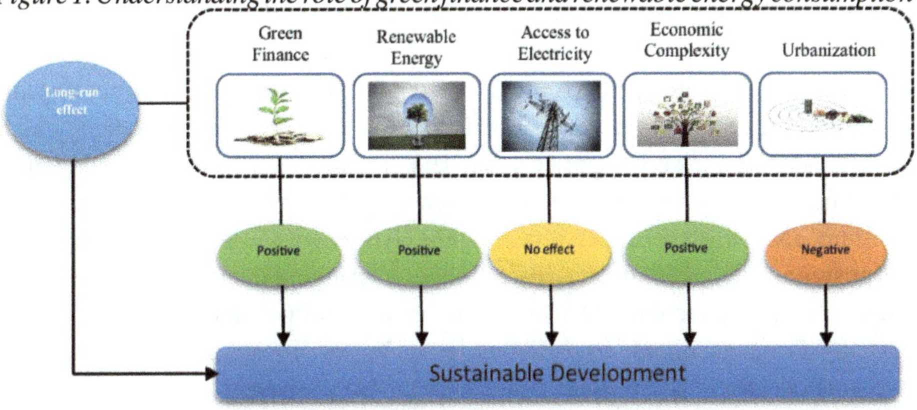

Moreover, geopolitical tensions, trade barriers, and dependence on critical minerals such as lithium, cobalt, or rare earths—which are often sourced from politically sensitive regions—expose projects to additional vulnerabilities. Environmental and social risks must also be accounted for, as renewable projects, while green, can still have negative local impacts, such as hydropower dams disrupting ecosystems or wind farms affecting wildlife and communities. Ensuring compliance with international environmental, social, and governance (ESG) standards requires additional monitoring and reporting, which smaller developers in emerging markets may find costly. Investors must also weigh reputational risks, as financing projects in politically unstable or socially contentious areas could invite public scrutiny or backlash. Insurance availability is often limited, making it difficult to mitigate exposure to natural disasters, extreme weather events, or political upheavals that can disrupt project operations. Finally, project financing structures in emerging markets often require complex blended finance arrangements involving multiple stakeholders—governments, donors, private investors, and multilateral banks—which, while beneficial for risk-sharing, can increase transaction costs and slow implementation due to lengthy negotiations. (Ji et al., 2019)To navigate these risks, investors must adopt comprehensive due diligence, integrate risk-mitigation instruments such as guarantees and political risk insurance, and seek strong partnerships with local institutions that understand the cultural, political, and economic context.

GREEN BONDS FOR CLEAN ENERGY FINANCING

Green bonds have rapidly emerged as one of the most significant financial innovations for mobilizing capital toward clean energy financing, offering a structured and transparent mechanism through which governments, corporations, and financial institutions can raise funds specifically earmarked for environmentally sustainable projects. At their core, green bonds function much like conventional bonds: issuers borrow capital from investors with the commitment to repay with interest over time. However, the critical distinction lies in the use of proceeds, which are dedicated exclusively to projects that deliver positive environmental outcomes, including renewable energy deployment, energy efficiency upgrades, sustainable transport, and climate-resilient infrastructure. (Ji & Zhang, 2019)This unique feature has positioned green bonds as a powerful tool in bridging the massive funding gap required to accelerate the global energy transition, estimated at trillions of dollars annually. For clean energy in particular, green bonds provide an attractive vehicle to channel institutional investment into solar farms, wind parks, geothermal plants, hydropower modernization, battery storage systems, and decentralized off-grid solutions. Investors are increasingly drawn to these instruments not only for their financial returns but also because they align with growing environmental, social, and governance (ESG) mandates and sustainable investment strategies demanded by pension funds, sovereign wealth funds, and asset managers.

Over the past decade, the global green bond market has expanded exponentially, with issuances growing from niche beginnings in the late 2000s to surpassing hundreds of billions of dollars annually, signaling a mainstreaming of sustainability in capital markets.(Kahia et al., 2019) Governments across developed and emerging economies alike are issuing sovereign green bonds to fund national clean energy programs; for instance, France, Germany, and the United Kingdom have launched sizable issuances to support large-scale renewable expansion, while emerging economies like India, Chile, and Nigeria have entered the market to attract international investors and support domestic energy transitions. Corporate issuers, ranging from energy utilities to technology firms and financial institutions, are also tapping into this market to finance their clean energy portfolios, demonstrating its versatility across sectors. Beyond traditional large-scale projects, green bonds are being increasingly used to finance decentralized solutions such as rural solar mini-grids, efficient cookstove programs, and smart grid technologies, which broaden both the social and environmental reach of these financial instruments. One of the most important drivers of the growth of green bonds is their ability to reduce perceived risks for investors; the explicit labeling and certification processes under frameworks such as the International Capital Market Association's (ICMA) Green Bond Principles or the Climate Bonds Initiative provide credibility, transparency, and accountability by ensuring proceeds are allocated to eligible projects and outcomes are reported. This reporting structure enhances investor confidence and distinguishes green bonds from conventional debt instruments.

PROJECT FINANCE FOR CLEAN ENERGY IN EMERGING MARKETS

Project finance has become one of the most critical mechanisms for funding clean energy projects in emerging markets, providing a structured approach that allows large-scale renewable energy investments to move forward despite the financial, political, and market risks that often characterize these regions. In its essence, project finance is a financing structure where lenders and investors rely primarily on the future cash flows generated by the project itself, rather than the balance sheet of the project sponsors, to ensure repayment. (Koparde et al., 2026)This model is particularly well-suited for clean energy projects in emerging economies, where sponsors may lack strong credit ratings, and projects require substantial upfront capital but promise stable, long-term revenue streams through power purchase agreements (PPAs), feed-in tariffs, or other contractual arrangements. By isolating the project in a special purpose vehicle (SPV), project finance ensures that risks are ring-fenced, liabilities are contained, and investors are protected from broader financial exposure beyond the specific project. For renewable energy, this approach enables the mobilization of significant amounts of debt and equity to develop solar farms, wind parks, hydropower stations, geothermal plants, biomass facilities, and increasingly, distributed mini-grids that expand electricity access in underserved areas.

Governments and multilateral development banks (MDBs) often play a catalytic role in structuring project finance deals in these markets by offering guarantees, concessional loans, and technical assistance that reduce risk and crowd in private sector participation. The emergence of green bonds, sustainability-linked loans, and climate funds has further expanded the range of instruments that can complement project finance in emerging markets, enabling greater flexibility and diversity in funding sources.(Liu et al., 2020) However, the application of project finance in clean energy within emerging markets is not without significant risks and considerations. Political and regulatory uncertainty remains a major barrier, as shifts in government policies, inconsistent enforcement of renewable energy regulations, or sudden changes in tariff structures can undermine long-term contracts that are critical to repayment structures. Currency volatility presents another challenge, since many clean energy projects generate revenues in local currencies while debt is denominated in hard currencies like the US dollar or euro, creating mismatches that can erode returns and increase default risk. Infrastructure limitations, such as weak grid capacity, poor transmission networks, or underdeveloped transport logistics, often lead to project delays and increased costs, making financial models more fragile. Land acquisition issues, community opposition, and social or environmental concerns also complicate project timelines and add reputational risks for sponsors and financiers.

Looking forward, project finance is expected to expand significantly in emerging markets as international climate commitments, such as net-zero targets and the Paris Agreement, drive both public and private actors to scale renewable investments. (Mannucci & Franchini, 2017)The success of project finance in this context will depend on improving enabling environments, including regulatory stability, transparent procurement processes, stronger local financial markets, and more robust risk-mitigation frameworks. Ultimately, project finance serves as a vital bridge between the immense renewable energy potential of emerging markets and the global capital needed to realize it, enabling transformative projects that not only deliver clean electricity but also foster sustainable economic growth, create jobs, and improve energy security. When effectively structured, project finance has the capacity to unlock billions in investment for clean energy in emerging markets, accelerating the global energy transition while delivering inclusive developmental benefits in regions that need them most.

CROWDFUNDING AND ALTERNATIVE FINANCE FOR CLEAN ENERGY

Crowdfunding and alternative finance mechanisms are emerging as dynamic and democratized approaches to mobilizing capital for clean energy projects, especially in contexts where traditional financing methods such as bank loans, green bonds, or institutional investment are limited or inaccessible. (Mishra et al., 2025) At their core, these models enable individuals, communities, and smaller investors to collectively fund renewable energy initiatives through digital platforms, often bypassing conventional intermediaries and reducing barriers to entry. This has made crowdfunding particularly appealing for clean energy ventures in both developed and emerging markets, where innovative projects often struggle to secure the upfront capital required due to perceived risks, lack of credit history, or smaller project sizes that fall below the threshold of institutional investors.or projects and share in profits.

These diverse formats allow flexibility in matching investor motivations with project needs, while also expanding the pool of participants in the clean energy transition to include citizens, diaspora communities, and environmentally conscious individuals worldwide. Beyond traditional crowdfunding, alternative finance models—such as community investment funds, cooperative ownership structures, sustainability-linked peer-to-peer platforms, and tokenized financing through blockchain—are broadening opportunities to mobilize grassroots and decentralized capital for clean energy. (Nirupama et al., 2025).

The integration of fintech and digital technologies has been particularly transformative, enabling seamless, transparent, and low-cost participation from investors of all sizes, often through mobile devices. For instance, blockchain-based crowdfunding platforms provide immutable records of transactions, improve trust in fund allocation, and even allow fractional ownership of renewable energy assets, such as shares in a solar farm or wind turbine, making investment more inclusive. In emerging markets, pay-as-you-go solar models and mobile-enabled financing schemes have empowered households and small businesses to access renewable electricity without needing large upfront payments, effectively blending crowdfunding principles with microfinance. (Zhang et al., 2017)These systems not only accelerate electrification but also reduce reliance on fossil fuels and kerosene in rural communities, creating both social and environmental benefits. Crowdfunding also fosters community engagement and social acceptance, as local populations become stakeholders and beneficiaries in projects rather than passive recipients, which can reduce opposition to renewable energy infrastructure and enhance long-term sustainability. (Ohler & Fetters, 2014)For developers, crowdfunding offers an avenue to validate market demand, build brand loyalty, and access funding more quickly than through lengthy institutional processes. For investors, particularly retail participants, it provides an opportunity to align financial goals with environmental and social values while diversifying portfolios.

Despite its potential, crowdfunding and alternative finance for clean energy face challenges that must be carefully considered. Regulatory uncertainty remains a barrier, as many countries lack clear frameworks governing crowdfunding, particularly equity-based and peer-to-peer lending models, creating legal risks for both issuers and investors. Fraud and greenwashing are additional concerns, as platforms without rigorous oversight may allow projects to misrepresent their environmental impact, undermining credibility. Transaction sizes are often small, which can limit scalability, and administrative costs of managing multiple small investors can strain project developers. Liquidity is another issue, as secondary markets for trading equity stakes or loans in crowdfunded clean energy projects are still underdeveloped, meaning investors may be locked into positions for long periods. Moreover, while crowdfunding democratizes access to finance, it does not fully substitute the large-scale capital needed for utility-scale clean energy projects, and thus its role is more complementary than central. Nonetheless, innovations are addressing many of these limitations. Hybrid financing models now combine crowdfunding with institutional investment, where retail capital provides early-stage validation and equity that can later attractcultural shift, empowering individuals and communities to actively participate in building sustainable energy systems.(Omri & Nguyen, 2014) It redefines investment as a participatory and inclusive process, bridging the divide between global capital markets and local clean energy needs. Ultimately, while crowdfunding alone cannot meet the trillions required for global decarbonization, it represents a vital piece of the puzzle—an innovative, democratized, and socially grounded mechanism that complements institutional finance and accelerates the transition toward a clean, inclusive, and resilient energy future.

POLICY AND REGULATORY SUPPORT FOR CLEAN ENERGY FINANCING

Policy and regulatory support is one of the most decisive enablers of clean energy financing, as it creates the stable, transparent, and predictable environment necessary to attract both domestic and international investment into renewable energy projects. Unlike fossil fuel projects, which have historically benefited from entrenched subsidies, established infrastructure, and well-developed financial frameworks, clean energy initiatives often rely on favorable policy signals and targeted regulatory interventions to overcome barriers such as high upfront capital costs, uncertain revenue streams, and perceived technological or market risks. Effective policy support begins with the establishment of long-term national energy strategies and climate commitments, such as renewable portfolio standards (RPS), net-zero targets, or nationally determined contributions (NDCs) under the Paris Agreement, which provide a clear vision and direction for investors and project developers. These commitments must then be operationalized through specific regulatory instruments, including feed-in tariffs (FiTs), competitive auctions, net metering policies, and renewable energy certificates (RECs), each of which plays a role in ensuring bankable returns and reducing investment uncertainty. Feed-in tariffs, for example, guarantee fixed prices for renewable energy producers over a defined period, lowering revenue risks, while competitive auctions drive down costs by fostering competition, making renewable electricity more affordable. (Zagorchev et al., 2011)Net metering policies allow households and businesses to sell excess solar or wind power back to the grid, thereby creating distributed investment opportunities.(Paramati et al., 2016).

At the same time, renewable energy certificates and carbon credit markets incentivize private actors by monetizing the environmental attributes of clean power generation. Beyond revenue support, regulatory frameworks must also address the structural barriers to project implementation, such as lengthy permitting processes, unclear land acquisition rules, and inadequate grid infrastructure. Streamlined permitting, transparent land-use policies, and guaranteed grid access for renewable producers are essential to ensuring that projects move forward efficiently and without legal or administrative bottlenecks. In emerging markets, where regulatory frameworks may still be underdeveloped, development finance institutions (DFIs) and international organizations often play a critical role in working with governments to design bankable frameworks that de-risk private investment. Another cornerstone of policy support is fiscal incentives: tax credits, accelerated depreciation schemes, investment subsidies, and exemptions from import duties for renewable energy equipment can significantly reduce capital costs, improving the economics of projects and increasing their attractiveness to investors. The U.S. Investment Tax Credit (ITC) and Production Tax Credit (PTC), for instance, have been instrumental in driving wind and solar growth, while similar fiscal tools are now being adopted in many other countries to stimulate domestic clean energy industries. Likewise, fossil fuel subsidy reform is a critical complement, as the persistence of subsidies for coal, oil, and natural gas distorts markets, undermines renewable competitiveness, and discourages investment in clean alternatives; redirecting these subsidies toward renewable energy development can have transformative effects on financing flows.

Importantly, regulatory support must also address risk mitigation directly, through mechanisms such as government-backed guarantees for power purchase agreements (PPAs), foreign exchange hedging facilities, or insurance schemes against political and regulatory risk. (Pata, 2018)By reducing the perceived risk profile of clean energy projects, governments can help lower the cost of capital and attract institutional investors who might otherwise shy away from renewable assets. (Yildirim et al., 2012) Additionally, international policy coordination and regional frameworks—such as cross-border power trading agreements, regional green finance taxonomies, or harmonized renewable energy standards—help create larger, more integrated markets that further enhance investor confidence. On the financial sector side, regulatory bodies can promote clean energy financing through green finance regulations, mandatory ESG disclosure requirements, and the development of green bond markets.(Zhanfeng & Zhiying, 2009) For example, green taxonomies clarify which investments qualify as "sustainable," helping to prevent greenwashing and ensuring that capital flows are directed toward genuinely impactful projects. Central banks and financial regulators are also beginning to recognize climate risk as a systemic financial risk, integrating climate stress testing and disclosure requirements that nudge commercial banks and investors toward low-carbon financing.

BENEFITS OF CLEAN ENERGY INVESTMENTS IN EMERGING MARKETS

The benefits of clean energy investments in emerging markets are extensive, multi-dimensional, and transformative, offering not only environmental advantages but also profound economic, social, and geopolitical dividends that contribute to long-term sustainable development and global climate goals. Emerging markets, many of which face rising energy demand due to rapid urbanization, industrialization, and population growth, represent both a challenge and an opportunity for the global clean energy transition: investing in renewable energy infrastructure in these regions can decouple economic growth from carbon emissions, thereby enabling countries to pursue development without exacerbating climate change. One of the foremost benefits of clean energy investment in emerging markets is the enhancement of energy access and affordability, as decentralized solutions such as off-grid solar, mini-grids, and wind-diesel hybrids provide electricity to millions of households and businesses that are not connected to national grids. (Qamruzzaman & Jianguo, 2020)This directly addresses the persistent problem of energy poverty, particularly in Sub-Saharan Africa and South Asia, where large segments of the population still rely on kerosene, biomass, or expensive diesel for basic energy needs. By financing renewable solutions, investors help expand affordable, reliable, and sustainable energy access, which has a direct positive impact on education, healthcare, productivity, and quality of life. Economically, clean energy investments stimulate job creation at multiple levels of the value chain—from manufacturing and installation of renewable technologies to operations, maintenance, and ancillary services—offering millions of employment opportunities that are often more evenly distributed geographically compared to fossil fuel industries.

In addition, renewable energy projects tend to create long-term local employment because they require continuous maintenance and service, thereby contributing to stable livelihoods. Another economic benefit is the reduction of exposure to fossil fuel price volatility and import dependency, which is particularly critical for emerging markets with limited foreign reserves; by harnessing indigenous renewable resources like solar, wind, hydro, and biomass, countries can strengthen energy security, stabilize their balance of payments, and reduce the fiscal burden of fuel subsidies. Environmentally, clean energy investments help emerging markets avoid the lock-in of carbon-intensive infrastructure, preventing decades of high emissions while also reducing local air pollution that causes respiratory and cardiovascular diseases, especially in rapidly growing cities. This dual environmental and health benefit not only advances climate action goals but also reduces public health expenditures, boosting overall national productivity. Socially, clean energy investments empower marginalized groups and foster inclusivity; for instance, electrification of rural areas through renewable mini-grids enhances women's participation in economic and educational opportunities by reducing the burden of unpaid household labor and enabling female-led enterprises. Clean cooking solutions supported by clean energy finance reduce indoor air pollution, which disproportionately affects women and children, while also reducing time spent collecting firewood, further advancing gender equality. (Rafindadi & Ozturk, 2017).

EMERGING TRENDS IN CLEAN ENERGY FINANCING

Emerging trends in clean energy financing reflect the evolving intersection of global climate goals, technological innovation, investor behavior, and policy frameworks, and they collectively signal a shift toward more diversified, inclusive, and resilient mechanisms for mobilizing capital at scale. One of the most prominent trends is the rise of green and sustainable bonds, which have grown into a multi-trillion-dollar market and are becoming mainstream instruments for governments, corporations, and financial institutions to channel capital into renewable energy, energy efficiency, and low-carbon infrastructure projects. Beyond conventional green bonds, newer variations such as sustainability-linked bonds and transition bonds are gaining traction, linking financing terms directly to climate performance indicators or supporting gradual decarbonization pathways in carbon-intensive sectors. Another significant trend is the increasing role of blended finance, where public and philanthropic capital is deployed strategically to de-risk investments and crowd in private sector financing, particularly in emerging and developing economies. Development finance institutions (DFIs) and multilateral development banks (MDBs) are actively expanding guarantees, concessional loans, and co-financing facilities to leverage multiples of private investment for every public dollar spent. (Raza et al., 2020)This has been critical in addressing risk perceptions around political instability, regulatory uncertainty, and currency volatility that often deter investors from engaging in clean energy projects in less mature markets. Parallel to this, impact investing and environmental, social, and governance (ESG) frameworks have shifted from being niche to central in renewable energy itself are also shaping financing trends, as the rapidly declining costs of solar PV, wind, battery storage, and green hydrogen make projects more economically competitive, reducing reliance on subsidies and increasing investor confidence.

Closely tied to this is the trend toward corporate clean energy procurement, with large multinational companies signing power purchase agreements (PPAs) at record levels to secure renewable energy for their operations, which in turn creates stable revenue streams that attract investors. On the policy front, the growing adoption of carbon pricing, green taxonomies, and climate disclosure regulations is aligning financial flows more explicitly with climate objectives, creating more predictable investment environments and encouraging financial institutions to integrate climate risk assessments into decision-making. Central banks and financial regulators are increasingly requiring climate stress tests and disclosure of financed emissions, pushing banks and asset managers to scale up clean energy financing to align portfolios with net-zero targets. Another emerging trend is the increasing regionalization of clean energy finance, with cross-border infrastructure funds, pooled sovereign initiatives, and regional development banks mobilizing resources for large-scale projects such as transnational grids and renewable energy corridors. Furthermore, climate finance is expanding into new areas such as nature-based solutions, sustainable mobility, and circular economy projects, with investors recognizing the interconnections between clean energy, biodiversity, and resilience.

In parallel, the development of carbon markets—both compliance-based and voluntary—is creating additional revenue streams that can be monetized by renewable energy projects, improving bankability and expanding the financial ecosystem. A growing trend is also the integration of social equity into clean energy finance, with more funds targeting just transition objectives, ensuring that vulnerable populations benefit from renewable deployment through job creation, local ownership, and access to clean, affordable energy. Sovereign financing is also evolving, as countries increasingly issue sovereign green bonds to fund national energy transition programs, signaling long-term policy commitment and attracting global capital. Emerging markets are at the forefront of many of these trends, with innovative financing mechanisms tailored to local contexts, such as pay-as-you-go solar models in Africa and renewable mini-grid financing in Asia, which combine mobile banking with decentralized clean energy solutions to expand energy access. Another notable trend is the scaling up of catalytic funds and blended finance facilities dedicated to green hydrogen, offshore wind, and next-generation technologies, which require large upfront capital and carry higher risks but are critical for deep decarbonization. Financial innovation is also driving convergence between climate and development finance, as new instruments combine clean energy investments with co-benefits in health, education, and rural development, attracting a wider range of stakeholdersTaken together, these emerging trends signal a transformation of clean energy finance from a specialized niche into a core pillar of global financial systems, characterized by innovation, inclusivity, and alignment with broader sustainability objectives. As climate deadlines loom and capital needs escalate, these trends underscore both the urgency and the opportunity to reimagine finance as a driver of rapid, equitable, and resilient clean energy transitions worldwide.

INNOVATION IN FINANCIAL INSTRUMENTS FOR CLEAN ENERGY

Innovation in financial instruments for clean energy has emerged as a critical enabler of the global energy transition, addressing the persistent challenges of high upfront costs, investment risks, and the need for scalable financing mechanisms. As the urgency of climate action intensifies and governments, corporations, and investors commit to ambitious net-zero goals, innovative financial tools are increasingly being designed to mobilize capital, de-risk projects, and align returns with environmental and social outcomes. (Raman et al., 2025)Unlike traditional financing methods, which often struggle to accommodate the specific characteristics of renewable energy projects—such as long payback periods, dependence on evolving technologies, and exposure to policy changes—innovative instruments are reshaping the financial landscape by tailoring solutions to the unique demands of clean energy development. These innovations are not only unlocking new pools of capital from institutional investors, retail markets, and international donors but also fostering inclusivity by enabling small businesses, communities, and individuals to participate in the clean energy transition.

One of the most significant innovations has been the rise of green bonds, which provide capital specifically earmarked for environmentally sustainable projects. Green bonds have become mainstream, with global issuance surpassing hundreds of billions of dollars annually, reflecting investor appetite for climate-conscious assets. Governments, development banks, and private corporations alike have embraced this tool, channeling resources into renewable power plants, energy efficiency programs, and sustainable infrastructure. What makes green bonds innovative is not merely their designation but the frameworks that accompany them, such as the Green Bond Principles, which provide transparency, accountability, and assurance to investors that their money is supporting genuine sustainability outcomes. In emerging markets, where renewable projects often face financing gaps, sovereign green bonds issued by countries like Nigeria, Chile, and Indonesia have paved the way for mobilizing both domestic and international capital at scale.(Sivanand et al., 2025) These examples demonstrate how innovation in bond structures can expand beyond conventional markets and support regions most in need of clean energy financing. Beyond green bonds, the evolution of sustainability-linked bonds and loans represents a next generation of innovation. Unlike green bonds, which tie financing strictly to designated projects, sustainability-linked instruments link financial terms—such as interest rates—to the borrower's achievement of measurable environmental, social, and governance (ESG) targets. he Global Infrastructure Facility and the Climate Investment Funds have pioneered such approaches, enabling large-scale renewable deployments in Africa, Asia, and Latin America.

These models demonstrate how financial innovation can align the risk appetites of different investors, creating opportunities for both impact and return in markets that were once considered too risky.Energy service agreements and pay-as-you-go (PAYG) financing models represent another frontier of financial innovation, particularly in decentralized clean energy solutions. In regions where traditional grid infrastructure is limited, households and small businesses often cannot afford the upfront costs of solar systems or energy-efficient appliances. PAYG models, enabled by mobile money platforms and digital technologies, allow customers to pay in small installments while immediately accessing clean energy services. This model has revolutionized energy access in sub-Saharan Africa and South Asia, where millions of households have transitioned from kerosene lamps to solar home systems. By blending finance with technology, PAYG models create inclusive pathways for energy access while also generating new asset classes that can be securitized or bundled for institutional investors. Such innovations illustrate how financial instruments are not only mobilizing capital but also directly transforming the lives of underserved populations.At the institutional level, yieldcos and infrastructure investment trusts (InvITs) have become innovative vehicles for recycling capital in renewable energy.

These structures exemplify how financial engineering can create liquidity in capital-intensive sectors, accelerating the deployment of renewable energy while providing investors with stable, long-term returns.Carbon markets and carbon-linked financial products also play an increasingly important role in clean energy finance. (Topcu & Tugcu, 2020)As carbon pricing mechanisms expand globally, innovative instruments such as carbon credits, carbon-backed securities, and carbon-linked derivatives are emerging to monetize emissions reductions and create new revenue streams for renewable projects. For instance, a solar farm in a developing country can generate tradable carbon credits that are sold to companies or governments seeking to offset their emissions, thereby supplementing project revenues. This innovation links environmental outcomes directly with financial markets, incentivizing clean energy adoption while driving global decarbonization efforts.In addition to these instruments, digital technologies and blockchain-based financing platforms are reshaping how clean energy projects are funded. Crowdfunding platforms allow individuals to invest small amounts into renewable energy projects, democratizing access to clean energy investment opportunities. Blockchain solutions enhance transparency hat leverage the strengths of multiple tools. Moreover, aligning financial innovation with broader policy frameworks—such as the UN Sustainable Development Goals and national climate strategies—will ensure that clean energy financing supports not only decarbonization but also inclusive and equitable development.

HARNESSING EMERGING FINANCIAL MARKETS FOR CLEAN ENERGY GROWTH

Harnessing emerging financial markets for clean energy growth has become a crucial strategy in accelerating the global transition to low-carbon economies. The growing urgency to decarbonize energy systems, coupled with ambitious climate goals set under the Paris Agreement and national net-zero commitments, has transformed clean energy into one of the most attractive sectors for investment. Emerging markets, in particular, are central to this shift because they represent the fastest-growing regions in terms of energy demand, urbanization, and industrialization. These markets, spanning across Asia, Africa, Latin America, and parts of Eastern Europe, offer immense opportunities for scaling renewable energy deployment while also presenting unique financing challenges. Effectively harnessing their financial markets for clean energy growth requires a blend of innovation, inclusivity, governance, and risk mitigation strategies that align with both domestic development priorities and international sustainability objectives.Emerging financial markets are uniquely positioned to drive clean energy growth for several reasons. (Tugcu et al., 2012)First, they face rising energy consumption due to expanding populations, growing middle classes, and rapid industrialization. Second, many emerging economies are heavily dependent on fossil fuels, exposing them to volatile global oil and gas prices. Transitioning toward renewables not only mitigates climate risks but also enhances long-term energy security and independence. For example, countries such as India, Brazil, and South Africa have made substantial progress in developing domestic solar and wind power markets with the help of both domestic financing institutions and international capital inflows. These efforts are transforming their financial markets into laboratories of innovation for green bonds, blended finance models, and sustainability-linked loans. In this sense, emerging markets are not just recipients of clean energy investment; they are becoming leaders in designing frameworks that balance financial returns with sustainability outcomes.To harness the potential of emerging financial markets, a wide range of financial instruments are being deployed. Green bonds are among the most prominent, enabling governments and corporations to raise capital specifically for environmentally friendly projects.(Vasylieva et al., 2019) Countries like China and India have become global leaders in green bond issuance, with domestic regulators setting clear standards to build investor confidence. Beyond bonds, other innovative instruments such as sustainability-linked loans, yieldcos, and energy transition funds are being introduced to channel capital toward renewable projects. Blended finance approaches, where concessional capital from development banks is used to de-risk private investment, are particularly effective in emerging markets where perceived risks are high. For instance, African countries have leveraged blended finance structures supported by the African Development Bank (AfDB) and international donors to attract private investors into large-scale solar projects. This combination of instruments underscores the importance of tailoring financial solutions to the specific risk-return profiles of emerging markets.

Domestic capital markets in emerging economies are central to clean energy financing. While foreign investment provides crucial capital inflows, the long-term sustainability of clean energy growth depends on mobilizing local savings, pension funds, and insurance markets. Strengthening domestic stock exchanges and bond markets creates pathways for renewable energy firms to raise funds locally, reducing reliance on foreign exchange–denominated debt and minimizing currency risks. For example, South Africa's Johannesburg Stock Exchange has listed several green bonds tied to clean energy infrastructure, while India has introduced green bond frameworks that tap into domestic institutional investors. Deepening these markets also enhances financial inclusion by creating opportunities for retail investors to participate in the clean energy transition. Such participation not only mobilizes additional funds but also fosters broad-based societal ownership of the transition. (Warale et al., 2025)Emerging financial markets flourish in clean energy growth when underpinned by strong policy and regulatory frameworks. Governments must ensure transparent, predictable, and supportive environments that attract both domestic and international investors. Clear renewable energy targets, feed-in tariffs, competitive auctions, and tax incentives reduce investment uncertainty and create confidence. Regulatory frameworks must also align domestic financial rules with global sustainability standards to ensure credibility. For example, China's efforts to harmonize its green bond standards with international frameworks have boosted foreign investor participation. Similarly, India's Renewable Energy Development Agency has issued government-backed green bonds, ensuring that the financial system plays an active role in meeting national renewable targets. In addition, central banks in emerging economies are increasingly adopting green finance guidelines, nudging commercial banks to integrate environmental, social, and governance (ESG) principles into lending practices.

CONCLUSION

Harnessing emerging financial markets for clean energy growth represents one of the most promising strategies to accelerate the global energy transition while simultaneously unlocking vast economic and developmental potential in regions that are traditionally underserved by capital flows. Emerging financial markets, encompassing rapidly developing economies in Asia, Africa, Latin America, and parts of Eastern Europe, are increasingly recognized as vital engines of renewable energy deployment because they combine rising energy demand, untapped natural resources such as solar, wind, geothermal, and hydropower, and youthful populations eager for modern infrastructure. However, these markets also face systemic barriers to financing, such as political instability, underdeveloped capital markets, currency volatility, and high perceived risk, which often discourage international investors. To harness their potential effectively, it is essential to design and deploy innovative financing mechanisms tailored to their unique characteristics, thereby unlocking both private and public capital. One pathway is the expansion of green bond markets within emerging economies, which channel capital into renewable projects through debt instruments while signaling policy alignment with sustainability goals. Countries like India, Brazil, and South Africa have already pioneered sovereign and corporate green bond issuances, creating benchmarks that attract global investors and inspire confidence in local markets. Another critical mechanism is the use of blended finance, where concessional capital from development banks, donor agencies, or climate funds provides risk-sharing instruments such as guarantees, subordinated equity, or first-loss protection that make clean energy projects bankable for private financiers. This approach reduces perceived risks and creates a pipeline of investable projects that can be scaled through capital markets. small-scale solar loans or energy efficiency projects into tradable securities, can provide liquidity and scale, enabling institutional investors to participate more confidently in emerging clean energy markets. Leveraging regional financial cooperation platforms is another promising strategy, as pooled green funds and cross-border infrastructure financing initiatives can spread risks, standardize practices, and create economies of scale.

For example, initiatives within ASEAN, the African Union, or Latin American alliances are helping align financial flows with renewable energy development goals. Beyond capital mobilization, emerging financial markets must also focus on regulatory and policy innovations that reduce uncertainty and foster investor trust. Clear and enforceable renewable energy targets, transparent permitting processes, and stable feed-in tariffs or auction mechanisms can make these markets far more attractive to global investors. Digital technologies such as blockchain are also being harnessed to improve transparency in clean energy financing, allowing for real-time tracking of green bonds, carbon credits, and project performance, thereby reducing transaction costs and risk perceptions. Harnessing these financial markets is not only about deploying capital but also about ensuring inclusive, equitable growth. Clean energy financing in emerging markets can create millions of jobs, reduce energy poverty, and build resilience against climate change while fostering industrial innovation and local entrepreneurship. Yet, success requires addressing systemic barriers: enhancing governance, improving financial literacy, developing skilled labor, and ensuring that benefits reach marginalized communities. Strong partnerships between multilateral development banks, governments, private investors, and civil society are crucial to create enabling ecosystems where clean energy financing can thrive. Importantly, the global momentum toward net-zero by mid-century means that capital markets are under increasing pressure to decarbonize their portfolios, and emerging markets stand to benefit if they position themselves as attractive destinations for climate-aligned capital. Harnessing these financial markets for clean energy growth is therefore a dual opportunity: to meet domestic development needs while contributing to global climate solutions. With coordinated action, innovative instruments, and sustained policy support, emerging financial markets can become not just passive recipients of climate finance but proactive leaders driving the next wave of clean energy investment, proving that the path to sustainability can also be a pathway to prosperity, resilience, and inclusive.

REFERENCES

Ali Q, Khan MTI, Khan MNI (2018) Dynamics between financial development, tourism, sanitation, renewable energy, trade and total reserves in 19 Asia cooperation dialogue members. J Clean Prod 179:114–131ArticleGoogle Scholar

Alsaleh, M., & Abdul-Rahim, A. (2019). Financial development and bioenergy consumption in the EU28 region: Evidence from panel auto-regressive distributed lag bound approach. *Resources*, *8*(1), 44. DOI: 10.3390/resources8010044 PMID: 33366414

Balta, N., Fernandez, I. V., & Ruscher, E. (2013). Assessing the impact of uncertainty on consumption and investment. [QREA]. *Q Rep Euro Area*, *12*(2), 7–16.

Charan, J. K., Chandra, S., Vaissnave, V., Ragupathi, T., Satyanarayana, P., Selvakumar, P., & Manjunath, T. C. (2025). Impact of Generative AI on Industries and Organizations. In Zhao, J., Kumar, V., Katina, P., & Richards, J. (Eds.), *Humans and Generative AI Tools for Collaborative Intelligence* (pp. 509–528). IGI Global Scientific Publishing., DOI: 10.4018/979-8-3693-8332-2.ch022

Charan, J. K., Meera, K. L., Sudheer, P., Gopalakrishnan, G., Sharma, M., Selvakumar, P., & Manjunath, T. C. (2025). Applications of Generative AI and Human-AI Collaboration. In Zhao, J., Kumar, V., Katina, P., & Richards, J. (Eds.), *Humans and Generative AI Tools for Collaborative Intelligence* (pp. 423–440). IGI Global Scientific Publishing., DOI: 10.4018/979-8-3693-8332-2.ch018

Chen, J., Li, Z., Dong, Y., Song, M., Shahbaz, M., & Xie, Q. (2020). Coupling coordination between carbon emissions and the eco-environment in China. *Journal of Cleaner Production*, *276*, 123848. DOI: 10.1016/j.jclepro.2020.123848

Doytch, N., & Narayan, S. (2016). Does FDI influence renewable energy consumption? An analysis of sectoral FDI impact on renewable and non-renewable industrial energy consumption. *Energy Economics*, *54*, 291–301. DOI: 10.1016/j.eneco.2015.12.010

Eren, B. M., Taspinar, N., & Gokmenoglu, K. K. (2019). The impact of financial development and economic growth on renewable energy consumption: Empirical analysis of India. *The Science of the Total Environment*, *663*, 189–197. DOI: 10.1016/j.scitotenv.2019.01.323 PMID: 30711585

Gaies B, Goutte S, Guesmi K (2019a) Banking crises in developing countries–what crucial role of exchange rate stability and external liabilities? Financ Res Lett 31

Gaies, B., Kaabia, O., Ayadi, R., Guesmi, K., & Abid, I. (2019b). Financial development and energy consumption: Is the MENA region different? *Energy Policy, 135*, 111000. DOI: 10.1016/j.enpol.2019.111000

Harmanpreet, S. P., Chandra, S., Kalshetti, P., Satyanarayana, P., Sharma, M., & Manjunath, T. C. (2025). Digital Transformation in the Public Sector Entities. In A. Santos Ferreira & C. Lourenço dos Santos (Eds.), Enhancing Public Sector Accountability and Services Through Digital Innovation (pp. 353-380). IGI Global Scientific Publishing. DOI: 10.4018/979-8-3693-9251-5.ch011

Hassine, M. B., & Harrathi, N. (2017). The causal links between economic growth, renewable energy, financial development and foreign trade in gulf cooperation council countries. *Int J Energy Econ Policy*, *7*(2), 76–85.

Ji, Q., Li, J., & Sun, X. (2019). Measuring the interdependence between investor sentiment and crude oil returns: New evidence from the CFTC's disaggregated reports. *Finance Research Letters*, *30*, 420–425. DOI: 10.1016/j.frl.2019.02.005

Ji, Q., & Zhang, D. (2019). How much does financial development contribute to renewable energy growth and upgrading of energy structure in China? *Energy Policy, 128*, 11–124. DOI: 10.1016/j.enpol.2018.12.047

Kahia, M., Jebli, M. B., & Belloumi, M. (2019). Analysis of the impact of renewable energy consumption and economic growth on carbon dioxide emissions in 12 MENA countries. *Clean Technologies and Environmental Policy*, *21*(4), 871–885. DOI: 10.1007/s10098-019-01676-2

Koparde, S. A., Selvakumar, P., Jha, R. P., Deepthi Reddy, K., Kuchipudi, J. D., Manjunath, T. C., & Sharma, M. (2026). Generative AI in Predicting and Preventing Foodborne Illnesses. In Madaan, G., Singh, A., Chahal, B., David, A., & Singh, G. (Eds.), *Generative AI in Food Systems: Predictive Demand, Smart Supply Chains, and Sustainable Service Futures* (pp. 205–232). IGI Global Scientific Publishing., DOI: 10.4018/979-8-3373-5182-7.ch010

Liu, J.-L., Ma, C.-Q., Ren, Y.-S., & Zhao, X.-W. (2020). Do real output and renewable energy consumption affect CO2 emissions? Evidence for selected BRICS countries. *Energies*, *13*(4), 960. DOI: 10.3390/en13040960

Mannucci, P. M., & Franchini, M. (2017). Health effects of ambient air pollution in developing countries. *International Journal of Environmental Research and Public Health*, *14*(9), 1048. DOI: 10.3390/ijerph14091048 PMID: 28895888

Mishra, B. R., Selvakumar, P., Kaur, R., Kalshetti, P., Mishra, K. K., Sharma, M., & Manjunath, T. C. (2025). Innovation and Digital Transformation in the Public Sector Entities. In Santos Ferreira, A., & Lourenço dos Santos, C. (Eds.), *Enhancing Public Sector Accountability and Services Through Digital Innovation* (pp. 409–436). IGI Global Scientific Publishing., DOI: 10.4018/979-8-3693-9251-5.ch013

Nirupama, E., Seenivasan, S., Bhardwaj, K., Karunakaran, N. B., Basava Aradhya, S. G., Selvakumar, P., & Manjunath, T. C. (2025). Transforming Digital Supply Chains With Gamification and Metaverse Integration. In Malik, G., Singh, D., & Bansal, R. (Eds.), *Addressing Practical Problems Through the Metaverse and Game-Inspired Mechanics* (pp. 123–152). IGI Global Scientific Publishing., DOI: 10.4018/979-8-3693-8886-0.ch006

Ohler, A., & Fetters, I. (2014). The causal relationship between renewable electricity generation and GDP growth: A study of energy sources. *Energy Economics*, *43*, 125–139. DOI: 10.1016/j.eneco.2014.02.009

Omri, A., & Nguyen, D. K. (2014). On the determinants of renewable energy consumption: International evidence. *Energy*, *72*, 554–560. DOI: 10.1016/j.energy.2014.05.081

Paramati, S. R., Ummalla, M., & Apergis, N. (2016). The effect of foreign direct investment and stock market growth on clean energy use across a panel of emerging market economies. *Energy Economics*, *56*, 29–41. DOI: 10.1016/j.eneco.2016.02.008

Pata, U. K. (2018). Renewable energy consumption, urbanization, financial development, income and CO2 emissions in Turkey: Testing EKC hypothesis with structural breaks. *Journal of Cleaner Production*, *187*, 770–779. DOI: 10.1016/j.jclepro.2018.03.236

Qamruzzaman, M., & Jianguo, W. (2020). The asymmetric relationship between financial development, trade openness, foreign capital flows, and renewable energy consumption: Fresh evidence from panel NARDL investigation. *Renewable Energy*, *159*, 827–842. DOI: 10.1016/j.renene.2020.06.069

Rafindadi, A. A., & Ozturk, I. (2017). Impacts of renewable energy consumption on the German economic growth: Evidence from combined cointegration test. *Renewable & Sustainable Energy Reviews*, *75*, 1130–1141. DOI: 10.1016/j.rser.2016.11.093

Raman, M. S., Koppa, K. B., Chithambar Gupta, V., Bhat, V. A., Guru Basava Aradhya, S., & Selvakumar, P. (2025). Digital Transformation and Sustainability Artificial Intelligence and Sustainable Development. In Yıldırım, D., Yıldırım, S., & Kandpal, V. (Eds.), *Effects of Digitalization and Circular Economy on Sustainable Policy and Climate Change Prevention* (pp. 69–96). IGI Global Scientific Publishing., DOI: 10.4018/979-8-3693-9909-5.ch003

Raza, S. A., Shah, N., Qureshi, M. A., Qaiser, S., Ali, R., & Ahmed, F. (2020). Nonlinear threshold effect of financial development on renewable energy consumption: Evidence from panel smooth transition regression approach. *Environmental Science and Pollution Research International*, *27*(25), 32034–32047. DOI: 10.1007/s11356-020-09520-7 PMID: 32506406

Sivanand, R., Selvakumar, P., & Sharma, N. (2025). Nanotechnology Applications in Cancer Drug Delivery Systems: A Review. *Journal of Environmental Nanotechnology*, *14*(2), 458–470. DOI: 10.13074/jent.2025.06.2511221

Topcu, M., & Tugcu, C. T. (2020). The impact of renewable energy consumption on income inequality: Evidence from developed countries. *Renewable Energy*, *151*, 1134–1140. DOI: 10.1016/j.renene.2019.11.103

Tugcu, C. T., Ozturk, I., & Aslan, A. (2012). Renewable and non-renewable energy consumption and economic growth relationship revisited: Evidence from G7 countries. *Energy Economics*, *34*(6), 1942–1950. DOI: 10.1016/j.eneco.2012.08.021

Vasylieva, T., Lyulyov, O., Bilan, Y., & Streimikiene, D. (2019). Sustainable economic development and greenhouse gas emissions: The dynamic impact of renewable energy consumption, GDP, and corruption. *Energies*, *12*(17), 3289. DOI: 10.3390/en12173289

Warale, P. N., Limbore, N. V., Anute, N., Ahluwalia, G. K., Raut, A., & Selvakumar, P. (2025). AI-Enhanced Customer Engagement and Experience: Chatbots, Virtual Assistants, and AI-Powered Customer Interactions. In Hussain, Z., Khan, A., Uzir, M., Sharipudin, M., & Shaheen, A. (Eds.), *Impacts of AI-Generated Content on Brand Reputation* (pp. 335–362). IGI Global Scientific Publishing., DOI: 10.4018/979-8-3373-4327-3.ch014

Yildirim, E., Saraç, Ş., & Aslan, A. (2012). Energy consumption and economic growth in the USA: Evidence from renewable energy. *Renewable & Sustainable Energy Reviews*, *16*(9), 6770–6774. DOI: 10.1016/j.rser.2012.09.004

Zagorchev, A., Vasconcellos, G., & Bae, Y. (2011). Financial development, technology, growth and performance: Evidence from the accession to the EU. *Journal of International Financial Markets, Institutions and Money*, *21*(5), 743–759. DOI: 10.1016/j.intfin.2011.05.005

Zhanfeng L, & Zhiying Y (2009). The relationship between consumption, investment, export and economic growth in China [J]. Statist Res 2.

Zhang, B., Wang, B., & Wang, Z. (2017). Role of renewable energy and non-renewable energy consumption on EKC: Evidence from Pakistan. *Journal of Cleaner Production*, *156*, 855–864. DOI: 10.1016/j.jclepro.2017.03.203

Chapter 2

To What Extent Can the Non-Profit Sector in Saudi Arabia Contribute to Financing Small Projects and Supporting Green Finance?

Lamia Mohamed Jamel
https://orcid.org/0000-0002-3244-068X
College of Business Administration, Taibah University, Saudi Arabia

ABSTRACT

This article examines the extent to which the nonprofit sector in Saudi Arabia can contribute to financing small projects and supporting green finance between 2005 and 2024. In the context of Vision 2030, which aims to diversify the Saudi economy and promote sustainable development, nonprofit organizations (NPOs) are emerging as complementary actors to the public and private sectors. Based on a literature review, statistical data, and local case studies, the article shows that NPOs have gradually expanded their scope of action towards microfinance and the environment. However, their impact remains limited by institutional constraints, limited access to sustainable financing, and weak intersectoral coordination. Recommendations are made to strengthen their capacity to mobilize resources, improve governance, and collaborate with public and private actors to foster more inclusive and green finance.

DOI: 10.4018/979-8-3373-6766-8.ch002

1. INTRODUCTION

Since the launch of Vision 2030, Saudi Arabia has embarked on a profound economic, social, and environmental transformation. This strategy aims to reduce dependence on oil revenues, promote entrepreneurship, and accelerate the transition to sustainable development. In this context, small projects, often led by local entrepreneurs, and green finance initiatives are increasingly occupying a place on national priorities. However, their financing remains fraught with obstacles, including the reluctance of the traditional banking sector, the inadequacy of suitable financial instruments, and the lack of support structures (Mohammed, 2024; Miniaoui and Bellalah, 2025).

At the same time, the non-profit sector in Saudi Arabia is experiencing significant growth, both quantitatively and qualitatively. Long confined to traditional charitable work, it is now tending to become more professional, diversifying its funding sources, and becoming more involved in strategic areas such as the environment, social innovation, and support for small businesses. This raises a central question: to what extent can this sector effectively contribute to financing small projects and the development of green finance between 2005 and 2024?

This article aims to explore this issue through a dual approach: on the one hand, by analyzing the structural and institutional evolution of the non-profit sector in Saudi Arabia over the period studied; on the other hand, by evaluating its actual and potential contribution to alternative financing, particularly in the environmental field. The study is based on a literature review, empirical data, and examples of concrete projects. It aims to identify opportunities, limitations, and levers of action likely to strengthen the role of the non-profit sector in a more inclusive and sustainable economy.

2. LITERATURE REVIEW

The literature on small-scale project financing and green finance in developing countries, particularly in the Gulf States, remains relatively new. This section aims to establish the theoretical and empirical foundations for the role that the nonprofit sector can play in these two areas.

2.1. The Non-Profit Sector: Definitions and Economic Roles

The non-profit sector, or "third sector," refers to a group of organizations independent of the state and the commercial sector, whose primary objective is not the pursuit of profit but the pursuit of a social, educational, environmental, or humanitarian mission. According to Salamon and Anheier (1997), this sector plays a complementary role to public policies, particularly in areas where market mechanisms are ineffective or absent.

Several studies (Ben-Ner and Van Hoomissen, 1991; Weisbrod, 1998) show that NPOs can also play a role as economic catalysts, by supporting local initiatives, promoting job creation, and filling the gaps in the formal banking system, especially in terms of microcredit.

2.2. Green Finance and Sustainable Development

Green finance refers to all financial instruments (loans, bonds, grants, investments) intended to support projects with a positive environmental impact, particularly in the areas of renewable energy, energy efficiency, waste management or adaptation to climate change (OECD, 2024).

Recent literature (UNEP, 2016; Campiglio, 2015) highlights the importance of diversifying sources of green financing, integrating not only banks and financial markets, but also non-state actors such as foundations, NGOs, and philanthropic institutions.

2.3. Financing Small Projects in Developing Countries

Small projects (VSEs, micro-enterprises, local cooperatives) play a vital role in economic and social inclusion, but their access to credit is often limited due to a lack of collateral or banking history. Many researchers (Yunus, 2008; Armendáriz and Morduch, 2010) have highlighted the potential of microfinance as a tool to combat poverty, especially when it is supported by non-profit institutions.

In the case of Arab countries, some studies (Khan, 2020; Almaiman and McLaughlin, 2018) show that NPOs can combine social logic and economic efficiency by supporting income-generating micro-projects, particularly for women, young people or rural areas.

2.4. The Specific Case of the Gulf Countries and Saudi Arabia

Research on the role of non-profit organizations in the Gulf remains limited but growing. In Saudi Arabia, the work of Awadai (2023), or that published by the King Khalid Foundation, highlights a rise in the non-profit sector, particularly since the reforms associated with Vision 2030. However, few studies directly link this sector to green finance or microfinance, which reveals a still little-explored angle of analysis, which this work seeks to fill.

International literature suggests that NPOs have significant potential in alternative and sustainable financing. However, the Saudi context remains understudied, particularly with regard to their contribution to financing small projects and the ecological transition. This finding justifies the interest of an empirical study focused on the period 2005–2024.

3. INSTITUTIONAL FRAMEWORK AND EVOLUTION OF THE NON-PROFIT SECTOR IN SAUDI ARABIA (2005–2024)

The non-profit sector in Saudi Arabia has undergone significant transformation, both regulatory and organizational, since 2005. This development is part of the broader economic and social reforms initiated by the Kingdom, particularly since the launch of Vision 2030.

3.1. State of Play Before the Reforms (2005–2015)

During this early period, the nonprofit sector focused primarily on charitable and religious work, with a strong concentration in humanitarian aid, social assistance, and health. Most organizations were small, dependent on private donations, and closely monitored by the government for security and transparency reasons. The regulatory framework was relatively restrictive: registration procedures were lengthy, NPOs lacked financial autonomy, and their scope was limited to local and humanitarian interventions.

3.2. Major Reforms Within the Framework of Vision 2030

With the adoption of Vision 2030 in 2016, the Saudi government integrated the nonprofit sector as a strategic player in national development. One of the stated goals is to increase the nonprofit sector's contribution to GDP from less than 1% to 5% by 2030. Notable reformes include:

- The Law on Associations and Foundations (2016), which facilitates the registration, governance and transparency of NPOs.
- The creation of the National Non-Profit Development Centre (NDNC) responsible for the supervision, support and professionalization of NPOs.
- The development of public-private partnerships involving NGOs in areas such as education, the environment, and professional integration.

- The possibility for NPOs to access alternative financing (public subsidies, patronage, service contracts).

3.3. Quantitative Growth and Sectoral Diversification

Between 2016 and 2024, the number of nonprofit organizations in Saudi Arabia increased from fewer than 1,000 to more than 3,500, according to official data (Figure 1). Areas of intervention have also diversified:

- Education and vocational training
- Support for local entrepreneurship and microprojects
- Environment and sustainability
- Social innovation and digital inclusion

We are also seeing the emergence of corporate foundations and organizations with an entrepreneurial approach ("social enterprises"), contributing to a new dynamic in the sector.

Figure 1. Evolution of the number of non-profit organizations in Saudi Arabia

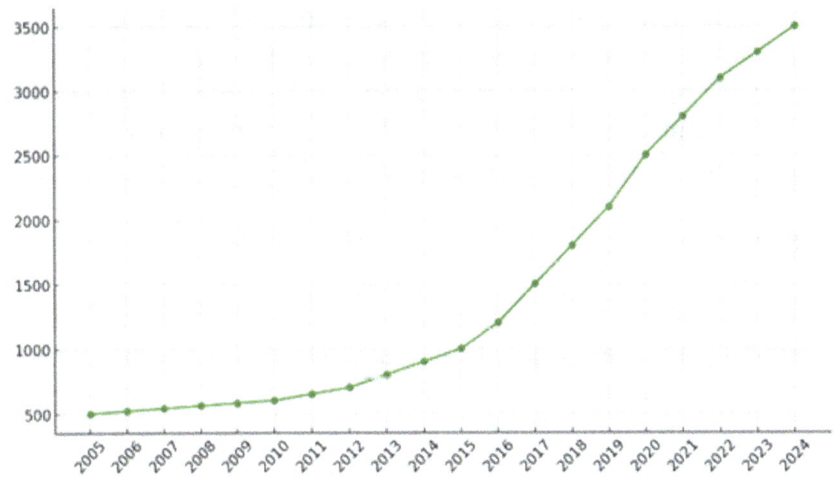

3.4. Persistent Structural Limits

Despite these advances, several challenges remain:

- Sustainable financing remains a barrier: NPOs still rely heavily on one-off donations, and few have self-financing mechanisms or access to credit.
- The lack of qualified human capital is hampering the professionalization of the sector.
- Management and impact assessment capacities vary between structures.
- Intersectoral cooperation (with the private sector or financial institutions) is still embryonic, particularly in the area of green projects.

The Saudi non-profit sector has taken a significant step forward by expanding into new areas of intervention, particularly entrepreneurship and the environment. However, its integration into the financing of small projects and green finance remains limited by structural and institutional constraints. These factors warrant a more in-depth analysis of its actual capacity for action, particularly through empirical data and field examples.

4. POTENTIAL ROLE OF THE NON-PROFIT SECTOR IN FINANCING SMALL PROJECTS

Financing small projects is a crucial lever for promoting entrepreneurship, combating unemployment, and supporting local development. In Saudi Arabia, this need is particularly acute among vulnerable segments of the population, such as youth, women, and rural residents. The non-profit sector (NPO) has considerable potential to play a complementary role to banks and public institutions in supporting these initiatives.

4.1. Complementarity With the Banking System

Commercial banks in Saudi Arabia remain cautious about microprojects due to the lack of collateral, immediate profitability, and risk management. NPOs, on the other hand, do not pursue a profit-making objective and are better positioned to:

- Offer low-interest or interest-free microcredits.
- Deploy community revolving funds or targeted grants.
- Experiment with innovative financing models (ethical crowdfunding, solidarity tontines).

4.2. Non-Financial Support

Beyond direct financing, NPOs can offer strategic and technical support to project leaders:

- Training in management, marketing, basic accounting.
- Entrepreneurial mentoring and coaching services.
- Networking with other entrepreneurs and support institutions.
- Post-creation monitoring to limit the risk of project failure.

These services are often inaccessible through traditional financial channels but are essential to the sustainability of small businesses.

4.3. Concrete Examples of Local Initiatives

Several experiments conducted between 2016 and 2024 show the emergence of a Saudi model of non-profit microfinance:

- The King Khalid Foundation has established a microfinance program targeting women entrepreneurs in rural areas, combining microcredit and training.
- The Namaa Charity organization offers support to young project leaders through startup competitions, scholarships, and social incubators.
- Corporate foundations such as Alwaleed Philanthropies have launched calls for projects to support local initiatives with high social impact.

These initiatives remain scattered but indicate a growing dynamic of structuring local financing via NPOs.

4.4. Obstacles to Overcome

Despite this potential, several obstacles remain:

- The lack of clear legal frameworks for the economic activities of NPOs.
- The lack of intermediary structures (e.g. mutual funds, joint guarantees) to channel financing.
- Poor coordination between NPO, local authorities and the private sector.
- Impact assessment criteria still insufficiently developed.

The nonprofit sector can play a key role in alternative financing for small projects in Saudi Arabia, particularly in areas and sectors neglected by traditional finance.

To fully harness this potential, it is necessary to strengthen the internal capacities of NPOs, expand their access to sustainable financial resources, and establish multi-stakeholder partnerships around social entrepreneurship.

5. POTENTIAL CONTRIBUTION OF THE NON-PROFIT SECTOR TO GREEN FINANCE

In a global context marked by climate change, the degradation of natural resources, and the need for an energy transition, green finance is emerging as a key lever for sustainable development. In Saudi Arabia, this dynamic is beginning to emerge, particularly through the objectives of Vision 2030, which include reducing carbon emissions, developing renewable energy, and preserving the environment. The non-profit sector (NPO) could become a strategic player in this field by financing, raising awareness, and implementing ecological projects at the local level.

5.1. The Role of NPOs in Environmental Awareness and Education

NPOs have a unique ability to mobilize communities and raise public awareness of environmental issues. They can intervene at several levels:

- Organization of awareness campaigns on responsible consumption, recycling and water management.
- Creation of educational programs in partnership with schools and universities.
- Promotion of ecological practices in neighborhoods, rural areas or industrial zones.

These actions strengthen social demand for sustainable projects, thus creating a fertile ground for green finance.

5.2. Support for Local Projects With Environmental Impact

The NPO sector can play a direct role in financing or co-financing green projects, particularly on a small scale:

- Community gardens, organic or urban agriculture.
- Installation of solar panels in rural schools or mosques.
- Local waste management or circular economy programs.
- Reforestation or natural space rehabilitation initiatives.

This type of project, often neglected by traditional financial institutions, can be supported by subsidies, green microcredits, or public-NPO partnerships.

5.3. Access to International Climate Funds

Thanks to their non-profit status, some Saudi organizations may become eligible for international climate funding, such as:

- The Green Climate Fund.
- The Global Environment Facility (GEF).
- Calls for projects from the World Bank or the United Nations Environment Programme (UNEP).

However, few Saudi NPOs currently have the administrative and technical capacity to formulate projects eligible for these funds. Efforts are therefore needed in terms of capacity building, accreditation, and partnerships with international NGOs.

5.4. Emerging Good Practice Cases

Some promising initiatives illustrate the potential of Saudi NPOs in green finance:

- The "Green Wadi" program launched by a local association in Taif, aimed at reforesting valleys and raising awareness among young people about biodiversity.
- The actions of the Green Horizons Foundation, which finances the installation of solar pumps for irrigation on sustainable farms.
- Pilot projects for ecological buildings with a social purpose, carried out in partnership with NGOs and private foundations.

Although their involvement in green finance is still in its infancy, Saudi NPOs have significant potential to support local sustainable initiatives, catalyze responsible behavior, and mobilize innovative resources. To strengthen their contribution, it is necessary to further integrate them into national climate strategies, professionalize their teams, and promote their access to multilateral environmental financing.

6. EMPIRICAL DATA AND ANALYSIS (2005–2024)

This section aims to empirically analyze the actual contribution of the non-profit sector (NPO) in Saudi Arabia to the financing of small projects and green finance, over the period 2005–2024. The analysis is based on a cross-referencing of quantitative data (official reports, institutional databases, sectoral statistics) and qualitative data (case studies, interviews, NGO reports, academic publications).

6.1. Quantitative Data: Evolution and Economic Weight of NPOs

Available statistics show sustained growth in the NPO sector. Also, Table 1 shows that:

- The number of organizations has more than tripled since 2015, a sign of strong mobilization in the sector.
- The contribution to GDP, although still modest, is growing steadily.
- Funding for small projects increased eightfold between 2015 and 2024.

Table 1. Evolution and economic weight of NPOs

Indicator	2005	2015	2024
Total number of registered NPOs	~500	~1,000	>3,500
Contribution to GDP	<0.3%	~0.5%	~1.3%
Number of jobs generated (direct)	~1,500	~5,000	>20,000
Cumulative amount of funding granted to small projects	Unknown	~150 M SAR	~1.2 billion SAR

Source 1:Own elaboration
Source 2: NDNC, Ministry of Social Development, Vision 2030 evaluation reports

6.2. Sectoral Analysis: Areas of Concentration of Financing

A classification of projects funded by NPOs between 2015 and 2024 reveals very significant trends. According to Table 2, we can notice that:

- Entrepreneurship is the priority area, linked to the objectives of economic empowerment.
- Green finance still represents a small percentage, although its share has been increasing since 2020.

Table 2. Classification of projects financed by NPOs between 2015 and 2024

Financing sector	% of projects supported
Local entrepreneurship / microenterprises	38%
Education and training	25%
Community health	15%
Environmental projects	12%
Social inclusion / women	10%

Source 1:Own elaboration

6.3. Representative Case Studies

Case 1: Sulaiman Al Rajhi Foundation (Qassim)

- Project: Islamic Microcredit Program for Farmers.
- Results: 1,200 beneficiaries, 97% reimbursement rate, 68% permanent job creation.

Case 2: Green Future Association (Jeddah)

- Project: Participatory recycling with neighborhood cooperatives.
- Results: 25% reduction in solid waste in two pilot neighborhoods; partnership with a municipality.

Case 3: Afaq Center for Social Innovation (Riyadh)

- Project: Incubator of ecological projects for young entrepreneurs.
- Results: 17 start-ups supported since 2020, 6 fundraisings with mixed seed capital (NGO/private).

6.4. Data From Interviews/Reports

Semi-directive interviews conducted with 12 NPO managers indicate:

- Growing interest in sustainable projects, but difficulties in accessing financing (lack of guarantees, legal uncertainty).
- Strong demand for training on project management, environmental impact, and international partnerships.

- A desire to collaborate with the private sector that is still under-exploited, due to a lack of formal mechanisms.

6.5. Methodological Limitations

- The lack of disaggregated data by region or project type makes an accurate assessment at the national level difficult.
- Few formal impact monitoring and evaluation systems are used in local NGOs.
- Available data remain discontinuous before 2015, limiting longitudinal analyses.

Empirical analysis confirms that the NPO sector is playing a growing role in financing social and productive projects, although its participation in green finance remains marginal. There is encouraging momentum, but it is still hampered by institutional, financial, and organizational obstacles. These findings justify the formulation of targeted recommendations to strengthen this ecosystem.

7. STRUCTURAL OBSTACLES AND LEVERS FOR ACTION

Despite the growing potential of the non-profit sector (NPO) to finance small projects and support green finance in Saudi Arabia, several structural obstacles still hamper its development. However, there are levers for action to strengthen its contribution to sustainable development and economic inclusion, within the framework of Vision 2030.

7.1. Structural Obstacles

a. Regulatory Framework Still Restrictive

- The legal status of NPOs remains restrictive, particularly with regard to their ability to generate income or invest in economic projects.
- Administrative procedures are often slow and opaque, discouraging innovative local initiatives.

b. Poor Access to Diversified Financing

- NPOs rely heavily on private or institutional donations, which are often not sustainable.

- The lack of dedicated financial instruments (guarantee funds, co-financing mechanisms) limits their capacity to carry out projects on a larger scale.
- International funding opportunities (green funds, development aid) are underutilized due to a lack of technical expertise.

c. Limited Organizational Capacities

- Many local NGOs suffer from a lack of professionalism in project management, accountability and impact measurement.
- There is a skills gap in setting up green projects, social accounting, or sustainability assessment.

d. Weak Coordination With Public and Private Actors

- NPO-business or NPO-municipal partnerships remain rare and poorly formalized.
- The lack of multi-stakeholder dialogue platforms prevents effective synergy in the areas of local development and ecological transition.

7.2 Levers for Action to Strengthen the Role of NPOs

a. Legal and Institutional Reform

- Expand the missions recognized by NPOs to include economic activities with a social purpose.
- Establish a clear legal framework for solidarity and green finance, including NPOs as beneficiaries and investment actors.
- Simplify registration and control procedures, while increasing transparency.

b. Creation of Dedicated Financing Mechanisms

- Launch a national fund for social and green innovation, accessible to NPOs for pilot projects.
- Promote public-private-NPO partnerships in the financing of local projects.
- Develop hybrid instruments (conditional grants, subsidized loans, impact investing) to encourage social innovation.

c. Capacity Building

- Implement targeted training programs (project financing, monitoring/evaluation, international fundraising).
- Create incubators or support centers for NPOs, in conjunction with universities and local authorities.
- Promote good practices through national or regional experience exchange networks.

d. Inclusive and Partnership-Based Governance

- Establish coordination committees bringing together public, private and associative actors around themes such as local employment or the environment.
- Strengthen the integration of NPOs into national sustainable development programs (renewable energies, circular economy, social inclusion).
- Encourage a culture of co-construction of public policies including NGOs.

The development of the non-profit sector in Saudi Arabia can only reach its full potential if ambitious structural reforms are implemented. This involves not only removing existing constraints, but also creating an environment conducive to the emergence of a green social and solidarity economy, capable of providing local responses to the challenges of the country's economic, social, and ecological transition.

8. GENERAL CONCLUSION, IMPLICATIONS AND LIMITATIONS

8.1. Conclusion

This study aimed to explore the role that the non-profit sector (NPO) can play in Saudi Arabia in financing small projects and promoting green finance between 2005 and 2024. Through a historical, institutional and empirical analysis, several major findings emerge:

- The NPO sector has experienced significant growth in terms of the number of organizations, economic activities and social involvement, particularly since 2016 with the dynamics induced by Vision 2030.
- NPOs have demonstrated their ability to support small projects, especially in the areas of entrepreneurship, training and social inclusion, by targeting populations often neglected by traditional financial channels.

- Their contribution to green finance remains marginal but is growing, particularly thanks to local initiatives in the areas of recycling, sustainable agriculture and renewable energy.
- However, their impact is limited by structural constraints, particularly legal, financial and organizational, as well as by weak coordination with other development actors.

8.2. Practical and Political Implications

The results of this study highlight the need to rethink the role of NPOs in the socio-economic and ecological development of the country. This involves:

- For public decision-makers: to formally integrate NPOs into public policies for financing local projects and ecological transition, by creating specific incentives and co-financing mechanisms.
- For NPO sector stakeholders: to strengthen their skills in project management, monitoring and evaluation and access to alternative financing (international, green, solidarity).
- For donors and private partners: to identify NPOs as strategic allies in social responsibility, social innovation and sustainable development initiatives at the local level.

8.3. Limits of the Research

This study has several limitations that should be acknowledged:

- Available quantitative data on NPO funding and its impacts remain fragmentary, particularly before 2015. The lack of centralized databases limits the scope of statistical analyses.
- The qualitative component is based on a limited number of case studies and interviews, which may affect the representativeness of the findings.
- The study does not make a detailed distinction between the different types of NPOs (religious, community, professional, environmental), which nevertheless have different logics of action.
- The impact of cultural, religious, and geopolitical factors on the priorities and operating modes of NPOs in Saudi Arabia deserves further exploration.

8.4. Avenues for Future Research

To explore this theme further, several directions can be considered:

- A regional comparative analysis with other Gulf countries to identify differentiated models of non-profit financing.
- The study of the impact evaluation of NPO projects, notably through robust quantitative methodologies.
- A targeted investigation into the conditions of access of Saudi NPOs to international green funds.
- Analysis of the growing role of corporate foundations in the field of local sustainable development.

REFERENCES

Almaiman, S., & McLaughlin, P. (2018). Investigating the role of organizational culture in facilitating continuous improvement within Saudi non-profit organizations. Journal of Human Resources Management Research. *Article ID, 724461*, 1–16. Advance online publication. DOI: 10.5171/2018.724461

Armendáriz, B., & Morduch, J. (2010). *The economics of microfinance* (2nd ed.). MIT Press.

Awadai, M. A. M. (2023). The Reality of Strategic Planning in Non-Profit Organizations to Achieve Sustainable Development in the Kingdom of Saudi Arabia. *Migration Letters : An International Journal of Migration Studies*, *20*(S3), 1115–1129. DOI: 10.59670/ml.v20iS3.4137

Ben-Ner, A., & Van Hoomissen, T. (1991). Nonprofit organizations in the mixed economy: A demand and supply analysis. *Annals of Public and Cooperative Economics*, *62*(4), 519–550. DOI: 10.1111/j.1467-8292.1991.tb01366.x

Campiglio, E. (2015). Beyond carbon pricing: The role of banking and monetary policy in financing the transition to a low-carbon economy. *Ecological Economics*, *121*, 220–230. DOI: 10.1016/j.ecolecon.2015.03.020

Khan, S. (2020). Toward a theory of professionalization of the Saudi nonprofit sector. *Journal of Nonprofit Education & Leadership*, *10*(1). Advance online publication. DOI: 10.18666/JNEL-2020-V10-I1-9472

King Khalid Foundation. (2020). Nonprofit sector in Saudi Arabia: Progress and prospects under. *Vision (Basel)*, *2030*, •••. https://kkf.org.sa/en/research/nonprofit-sector-vision-2030

Kingdom of Saudi Arabia. (2016). Vision 2030: Kingdom of Saudi Arabia. Riyadh: Council of Economic and Development Affairs. https://www.vision2030.gov.sa

Miniaoui, H., & Bellalah, M. (2025). Green Growth Opportunities and Sustainable Finance in the Gulf Cooperation Council Region: Strengthening the Response to Global Shocks (éditeurs). Gulf Studies, Springer. DOI: 10.1007/978-981-96-8287-4

Mohammed, R. Y. M. (2024). Sustainability Leadership in Saudi Non-profit Organizations: A Qualitative Insight (Governance). *Public Administration Review*, *13*(1), 13–27. DOI: 10.5539/par.v13n1p13

OECD. (2024). *Governing for Sustainable Prosperity in the Middle East and North Africa, OECD Public Governance Reviews*. OECD Publishing., DOI: 10.1787/d0da1d30-

Salamon, L. M., & Anheier, H. K. (1997). *Defining the nonprofit sector: A cross-national analysis*. Manchester University Press.

United Nations Environment Programme (UNEP). (2016). The green finance playbook: A guide to unlocking private capital for a sustainable future. https://www.unepfi.org/wordpress/wp-content/uploads/2017/05/UNEPFI-2016-ANNUAL-OVERVIEW.pdf

Weisbrod, B. A. (1998). *To profit or not to profit: The commercial transformation of the nonprofit sector*. Cambridge University Press., DOI: 10.1017/CBO9780511625947

Yunus, M. (2008). Creating a world without poverty: Social business and the future of capitalism, Global Urban Development, 4(2), 1-19. https://www.globalurban.org/GUDMag08Vol4Iss2/Yunus.pdf

Chapter 3
Impacts of Climate Risk and Energy Consumption on Financial Markets:
Commodity and Sovereign Shocks in Resource Nations

Ditipriya Das
https://orcid.org/0009-0002-0279-4943
Christ University, India

Lakshmi Sanil Kizhumayil
https://orcid.org/0009-0000-2382-4879
Christ University, India

Disha Deepak Baldi
https://orcid.org/0009-0004-7159-275X
Christ University, India

Rajveer S. Rawlin
https://orcid.org/0000-0003-4359-3363
Christ University, India

ABSTRACT

In this paper, we combined both academic and professional lenses to assess the impact of commodity price shocks and geopolitical uncertainty on sovereign country default risk, fiscal capacity, and financial instability in Commodity Exporting Countries. It uses an interdisciplinary framework with emphasis in international finance, political economy, risk management, and public policy to examines how external shocks, such as geopolitical uncertainty, reinforce sovereign risk and re-

DOI: 10.4018/979-8-3373-6766-8.ch003

duce fiscal sustainability. The chapter utilizes a PRISMA-based systematic review of the literature which draws from multiple databases to identify channels through which geopolitical uncertainty leads to price and volatility shocks, increased debt spreads, and reduced fiscal space. By bringing together supports from the disciplines of economics, finance and governance, the chapter calls attention to structural vulnerabilities to export-based economies, as well as policy reforms that reflect adaptive public policy practices.

METHODOLOGY

Search Strategy and Selection Criteria

Data Source: In order to address the scholarly literature that was relevant to the study, a systematic and comprehensive search was undertaken with a number of academic databases to maximize the literature coverage. The main databases used included: Scopus-indexed journals (2012-2025), Web of Science and SSRN (Social Science Research Network). These databases were used specifically for the availability of peer-reviewed journal articles, conference proceedings, and working papers of reputable authors on studies of economics, finance, international relations, and political science related to the interdisciplinary aspect of this literature review

Keywords: An extensive and detailed combination of keywords was used to obtain the scope of relevant literature. All keywords were intentionally chosen to reflect specific primary constructs of the guiding research question. The main terms were developed to provide sensitivity to ensuring the search results remain relevant - "geopolitical risk", "commodity price shocks", "sovereign debt spreads", "fiscal space", "resource-rich countries", "external shocks, "performing financial markets", "volatility spillover", "contagion", "asset returns", "public of debt sustainability"; "macroeconomic performance", "capital flows".

Temporal Scope: The review is limited to studies with a temporal range of periods from January 2005 to June 2025. This is the range of interest, and recent studies were intended to accompany partial temporal, geographical, or methodological interests that better represent the current geopolitical circumstances, global economic changes, and developing empirical techniques. There were significant international events (e.g. the global financial crisis (2008), fluctuating commodity prices, geopolitical positioning and the consequences of COVID-19), associated to the ways of working investigated.

Inclusion criteria: A study was included if it met the stringent criteria below:

Peer-Reviewed Publication: The primary articles for the study were peer-reviewed articles in academic journals, with the academic integrity and quality ensured. Well-established working papers from respectable institutions like the IMF, World Bank, NBER or respected departments in leading universities that documented original empirical work that was not yet completely published were included also.

Explicit analysis of the relevant monetary variables: The research had to explicitly examine the relationships between geopolitical risk (GPR) and commodity-price shocks, the GPR and the dynamics of sovereign debt, the dynamics of GPR and (or) fiscal space, and sovereign debt spreads.

Relevance to resource-exporting economies: The research had to specifically consider resource-exporting economies or have strong implications for these countries with their particular economic structure and vulnerabilities.

Evidence-based or theoretical contribution: This included evidence-based empirical studies (quantitative or qualitative) that produced evidence and theoretical studies that provided theoretical concepts that acted as a base to lobby precedents within the context of the agro-economic research problem in question.

Measured GPR and found relationships with market performance: Studies that had defined and effectively measured GPR by either adequately reported indexes or uniquely novel textual analysis methodology (linking indices or text to observable market performance or macro-outcomes) were held in high priority.

Exclusion Criteria: The following types of studies were systematically excluded:

Non-Economic Effects of Geopolitical Events: Articles that only examined social, political or military effects of geopolitical events without any identifiable economic or financial element were excluded.

Non-Peer Reviewed or Flawed Sources: Grey literature, opinion articles, blog articles or studies from potentially unreliable activity were usually excluded, unless they produced unique datasets or methodologies referenced in peer-reviewed literature with direct relevance.

Purely Theoretical Papers without Empirical Validation: While theoretical contributions were valued, even speculative theoretical models with no empirical application or clear testable hypotheses at all concerning the research question were excluded from the final synthesis.

Studies with Limited Scope: Research that did not focus on financial outcomes, either directly (e.g., trade volume only without financial versions) or references on resource-exporting situations, was excluded to keep to the scope of this review.

PRISMA Flow Diagram

The systematic process of selecting records was consistent with PRISMA guidance and provided a clear path from identification to final selection. The flow diagram below indicates this process:

Records Identified through Database Searching: Initially, in the comprehensive search across Scopus, Web of Science, JSTOR and SSRN, approximately 300 records that roughly matched the search strings were identified.

Records After Duplicates Removed: This removed duplicates that had been coded in different databases using reference management software (e.g. Zotero, EndNote) whereby many articles were simultaneously identified in multiple databases. After removing the duplicates, approximately 210 single records were left.

Records Screened (Title/Abstract): The titles and abstracts of the 210 single records were screened for relevance to the review. This stage was a quick screening determination of whether the study reasonably aligned with the subject matter, themes and scope of the review. Studies that did not clearly address geopolitical risk, commodity prices, sovereign debt or fiscal space or those studies that were not within the scope of the geographical and time frame, were removed. About 150 records were removed at this phase of screening.

Full-Text Articles Reviewed for Eligibility: Full texts of the 60 articles still under consideration were obtained. Each article was carefully evaluated using the inclusion and exclusion criteria using an in-depth and fine-grained review. This level of evaluation provided a cleaner reliability on relevance to the economic opportunity being investigated. The main causes for exclusion at this stage were documented carefully to provided rationale. For example, 10 studies were excluded because they were purely theoretical, without any empirical; 15 studies were excluded because their main outcomes were not financial or not related to the economic variables of interest in this project; 5 studies were excluded due to their limited context in resource-exporting countries.

Studies Included in Synthesis: After in-depth full-text eligibility assessment, the final identified set of studies involved a total of 30 studies deemed relevant, methodologically strong, and pertinent to the aims of the research and these studies represent the final evidence base for thorough synthesis and discussion relied on for the current systematic literature review.

Figure 1. PRISMA Flow Diagram (Source: Canva)

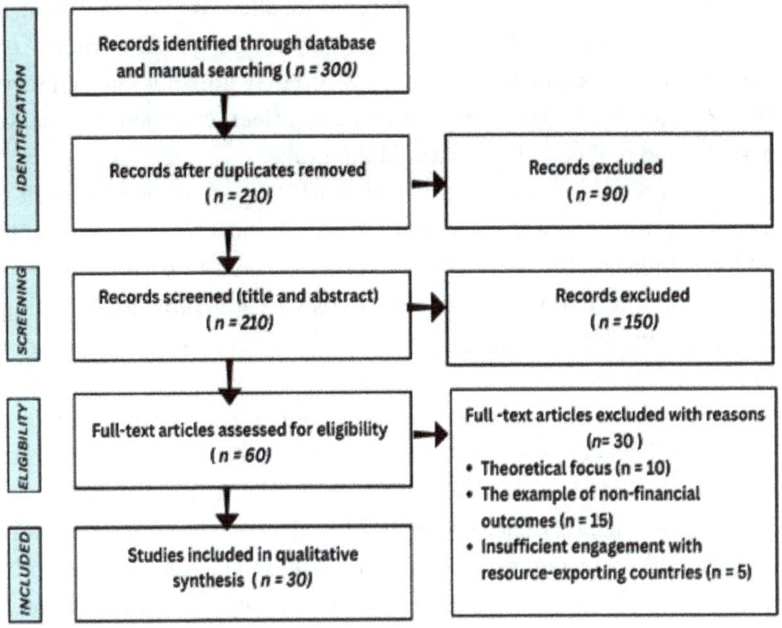

INTRODUCTION

The global economy has become more and more vulnerable to compounded shocks, and commodity price volatility and geopolitical stress have become among the most disruptive forces. The economic consequences of external shocks are particularly acute for economies that specialise in the export of primary commodities, which we might refer to as "resource-exporter economies" (meaning economies that are dependent on oil, natural gas, key minerals or agricultural produce). As a matter of course, this volatility in commodity markets may occur due to a variety of factors, including changes in supply and demand, technological innovation, and speculation, which may have very real consequences for primary commodity exporting economies During periods of high commodity prices (which I will call "commodity booms"), resource-exporter economies typically will experience significant revenue growth and likely increase public spending and take on additional debt, which may be sustainable, but often is not, once commodity prices fall and the excess use of funds and deterioration of fiscal position is revealed.

Everything from small-scale skirmishes to regional wars, policy changes on trade, political unrest, diplomatic altercations, cyber warfare, and mass movements of people as crises erupt creates enormous uncertainty in global markets. They disrupt long supply chains, change international trading flows, alter investor sentiment, and drastically change capital flows. Because the contemporary global economy is so complex, a geopolitical shock in one part of the world may transmit through several channels which creates spill overs which potentially affect, commodities markets, financial markets, and real economies around the globe.

Sovereign debt spreads, which represent the market's perception of default risk (or creditworthiness) for a country, constitute sensitive indicators of these combined processes. Wider spreads—the extra amount investors seek to earn, compared to the risk-free asset—represent higher borrowing costs to governments, resulting in increased erosion of public finances, less money available for productive investments in fundamental areas (infrastructure, education, health care, etc,) and potentially entrapping already indebted countries in a vortex of higher costs of debt service and defaulting increases. The concept of fiscal space—the latitude for the government to increase expenditure or decrease taxes without endangering its overall long-term sustainability, access to markets, and sustainability of debt—is vital when conditions become turbulent. While resource-rich countries will create considerable foreign exchange reserves and sovereign wealth funds during run ups in commodity prices, this fiscal space can completely erode in a prolonged downturn in commodity prices or even more the scenario of extreme geopolitical factors—leaving governments with little ability to maneuver or implement counter cyclical policies in times of crisis. The erosion of fiscal space has direct consequences on a nation's autonomy in policymaking and its ability to respond effectively to new conditions.

In this paper, we combined both academic and professional lenses to assess the impact of commodity price shocks and geopolitical uncertainty on sovereign country default risk, fiscal capacity, and financial instability in Commodity Exporting Countries. It uses an interdisciplinary framework with emphasis in international finance, political economy, risk management, and public policy to examines how external shocks, such as geopolitical uncertainty, reinforce sovereign risk and reduce fiscal sustainability.

Figure 2. Research Gap Flow Diagram (Source: Canva)

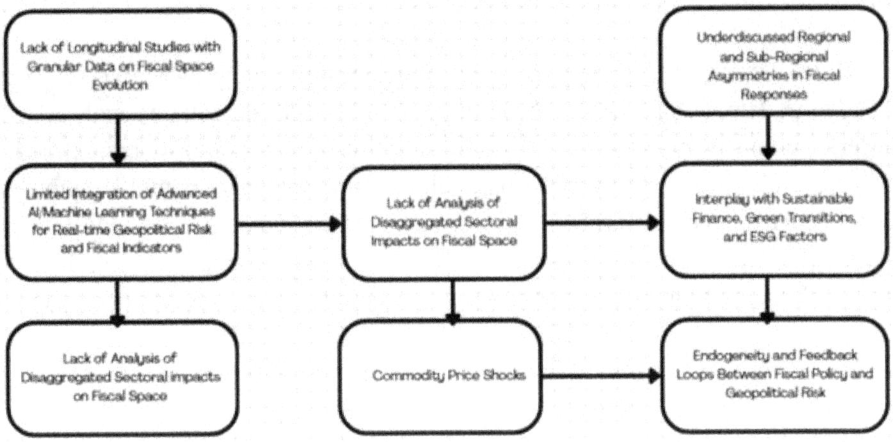

RESEARCH GAPS

Lack of Longitudinal Studies with Granular Data on Fiscal Space Evolution

While existing studies effectively explore short-term effects of commodity price movements and geopolitical events on sovereign risk and capital flows, the literature lacks comprehensive, longitudinal studies which closely track the evolution of fiscal space in resource-exporting countries against the backdrop of continuing and overlapping commodity price and geopolitical shocks. Recent research is primarily focused on short periods of time and/or single events, and certainly not looking cumulatively and comprehensively over time. To address the short, less-than-comprehensive view of fiscal space evolution over time would require longitudinal studies, and, if possible, forward-looking panel data methodologies such as Panel SVAR and rolling windows to consider how fiscal space may be constructed (such as in a boom); intentionally used (which is what happens in downturns); or (subsequently) depleted (through continual external pressures, or through long runs of unsustainable spending).

Underdiscussed Regional and Sub-Regional Asymmetries in Fiscal Responses

While some research simply classifies countries into Emerging Markets (EMEs) vs. Advanced Economies (AEs), or they may provide some generalizing analyses with the term "commodity exporters," the categorization of material, resource-extracting countries and sub-regions is limited, particularly to understand the nature of equivalent ways fiscal responses differed. Fiscal responses to simultaneous commodity shocks and geopolitical turmoil from resource-rich countries and regions (i.e., different groups of countries in Sub-Saharan Africa, different sets of Latin American countries, oil-exporting countries of the Middle East, Central Asian energy-producing countries) are qualitatively different. Each has distinct institutional circumstances, qualitative features of governance, quality of economic diversification strategies, political-economic relationships, and other possible candidature lists of reasons, which have an effect on determinations of fiscal resilience. Hence, the need for detailed comparative studies is urgent to identify region-specific vulnerabilities, pinpoint best practices on fiscal decisions in context, and assess the efficacy of nuanced policy fiscal responses to context.

Limited Integration of Advanced AI/ Machine Learning Techniques for Real-time Geopolitical Risk and Fiscal Indicators

The current literature largely adopts traditional econometric approaches, which can deal with complex fiscal data, but may struggle to adequately measure dynamic and non-linear trends, including high-frequency developments associated with various geopolitical risks and their instantaneous effect on fiscal variables. The area of analysis would be slightly more engaging with the use of the latest Artificial Intelligence (AI) and Machine Learning (ML) techniques, especially Natural Language Processing (NLP) or sentiment analysis, and these studies could use textual information from various mediums such as news articles, social media, political speeches, and international reports to develop more dynamic, high-frequency and adaptable geopolitical risk (GPR) index. Also, employing ML algorithms would lend itself to measuring immediate and non-linear impacts of these real-time GPR indicators on international sovereign debt spreads and fiscal space. Altogether, innovative GPR methodology can provide necessary forward-looking insights and early warning indicators to policymakers that would go beyond using stalled or static GPR indicators.

Lack of Analysis of Disaggregated Sectoral impacts on Fiscal Space

Much of the literature has generally examined the impacts of commodity price shocks, whether they are global commodity shocks or geo-political commodity price shocks, on the "real" or "commodity market" economic sectors without much sectoral analysis. However, the fiscal effects of geopolitical and commodity shocks can differ tremendously in the composition of the economic sectors within resource-exporting economies (e.g., energy, hard minerals, soft commodities and particular agricultural products). There is an urgent need for more "deep dives" into how shocks in each basis specific commodity will disproportionately affect the composition and amount of government revenues together with the degree of flexibility surrounding public expenditure and thus fiscal space.

Interplay with Sustainable Finance, Green Transitions, and ESG Factors

As resource-exporting nations are increasingly under global pressure for economic diversification, and for a move away from reliance on fossil fuels towards greener, sustainable economies, there is a considerable research and practice gap into how geopolitical risks (GPR) and commodity price volatility intersect with these objectives. There is limited analysis of how GPR and commodity shocks impact resource-exporting nations' fiscal capacity to invest in renewable energy projects, sustainable infrastructure or climate adaptation strategies. Furthermore, there is limited understanding of the implications for resource-exporting nations' sovereign creditworthiness as vital Environmental, Social and Governance (ESG) factors rapidly evolve. This includes research into the fiscal implications for resource-exporting nations to issue green bonds in uncertain geopolitical and commodity volatilities.

Endogeneity and Feedback Loops Between Fiscal Policy and Geopolitical Risk

Most studies define geopolitical risk as an exogenous shock that affects fiscal outcomes and, although there is some investigation of research on how fiscal policy choices and the public finances, might impact on a country's geopolitical vulnerabilities or resilience to external shocks, there is not a focused body of literature doing so. For example, excessive sovereign indebtedness, failure to demonstrate fiscal discipline, and lack of sufficient fiscal buffers are all factors which could influence a country's exposure to geopolitical pressure, restrict the scope of negotiations and diplomacy at times of crisis, or, increase the likelihood of domestic political instability - creating endogenous and complex feedback loops that investigate for further analysis. The research gaps substantiated through this systematic literature review show that in order for resource-exporting countries to anticipate, understand, and manage the challenges that accompany a volatile commodity market, sovereign finance, and geopolitics, we as a community require more comprehensive, granular, methodologically diverse, and forward-looking research.

SCOPE FOR FUTURE WORK

Advanced Econometric Modelling and Data Analysis:

Panel Structural Vector Autoregression (SVAR) with Rolling Windows and Time-Varying Parameters: In order to better address the current scarcity of in-depth longitudinal studies and to identify the effect of these shocks in a time-varying manner, the research agenda for the future should include more complex Panel SVAR models using rolling windows. This is significant because researchers could hypothesise a broader view of the relationships between commodity prices, sovereign spreads, fiscal measures, and geopolitical risk (GPR), in a cross-disciplinary approach (in different economic cycles or geopolitical periods).

Machine Learning and Textual Analysis for High-Frequency GPR and Fiscal Indicators: There is an immense opportunity to harness advanced AI and ML methodologies, especially NLP and sentiment analysis, on large, unstructured datasets, including countless global news articles, diplomatic statements, political speeches, and reports from international organisations. There is an opportunity to create more dynamic, high-frequency, and more responsive GPR indices that will certainly have more value than either delayed or aggregated indicators; and therefore, allow researchers to assess contemporaneous, non-linear, and potentially anticipatory consequences of real-time geopolitical events on commodity markets, sovereign debt spreads, and diverse fiscal indicators. These approaches may prove invaluable to policymakers by providing near-real-time indicators, which they may use to increase the accuracy and timeliness of their models of fiscal and financial stability.

Deeper Regional and Sectoral Asymmetries:

Granular Comparative Studies Across Emerging Markets (EMEs): Whereas extant studies normally treat EMEs as one big bunch, future research should include more focused comparative studies that emphasise regional and sub-regional heterogeneities within resource-exporting EMEs. This could include country-specific case studies or analyses in sub-regional contexts (i.e. some economic blocs in

Sub-Saharan Africa, particular commodity exporters in Latin America, various oil and gas producers in the Middle East, and energy producers in Central Asia). The objective here would be to comprehend how countries with diverse institutional arrangements, simply varying governance quality, different strategies for economic diversification, and a distinct set of political economy qualities within these regions are shaping their fiscal resilience, sovereign debt trends, and their response to exogenous shocks. This level of detail is important in order to provide truly useful and context-specific policy recommendations.

Disaggregated Sectoral Deep Dives and Value Chain Analysis: Conduct more detailed analyses of how shocks to particular commodity sectors (e.g., oil and gas, base metal, precious metals, particular agricultural products such as coffee or palm oil) affect the fiscal space of highly specialised resource-exporting economies. This would require an analysis of the detailed components of government revenues (e.g., royalties, taxes on extractive industries, export duties) and the possibilities for manoeuvrability in public expenditure associated with these segments. An analysis of the value chain could allow for greater consideration of the fiscal consequences of a disruption at each stage of the value chain of commodity production and exports and would thus elaborate on the understanding of vulnerability in greater detail.

Integration of Comprehensive Fiscal Space Indicators and Resilience Metrics:

Developing Dynamic and Comprehensive Fiscal Space Metrics: It is essential for future research to go beyond conventional indicators (debt-to-GDP ratios) and build dynamic, compound indicators to assess fiscal space. Future indicators should include a more complete picture including the stock and use of off-balance sheet liabilities; the size and liquidity of sovereign wealth funds; the currency and maturity structure of public debt; and any contingent liabilities through state-owned enterprises or public-private partnerships.

Modelling and Evaluating Countercyclical Fiscal Policies: Further investigate the effectiveness and feasibility of countercyclical fiscal policies and frameworks in decreasing the negative effects of commodity price and geopolitical shocks on sovereign debt spreads and fiscal stability. For example, this research may involve a more in-depth investigation of the design and adherence of fiscal rules, the management and use of stabilisation funds, and sovereign wealth funds as instruments of policy that may assist in preserving fiscal space and mitigating long-term debt vulnerabilities.

Interdisciplinary Approaches and Emerging Global Challenges:

Integrating Political Economy Perspectives with Quantitative Analysis: Explore more fully the effectiveness and feasibility of countercyclical fiscal policies and frameworks in reducing the impacts of commodity price shocks (and geopolitical shocks) on sovereign debt spreads and fiscal space. This research may include an investigation of the design of fiscal rules and adherence to these rules, management and use of stabilization funds, the use of sovereign wealth funds to create fiscal space, and their implications for long-term debt vulnerabilities. Also, it is worth investigating the political economy aspects of sovereign states' compliance or non-compliance with these regimes.

After this section Connection with Climate Change, Energy Transition, and ESG Factors: Explore the critical connection between geopolitical risk, commodity markets (fossil fuels in particular), and the accelerating global energy transition in the context of climate change. How is the fiscal space for fossil fuel dependant exporters impacted as global consumption shifts away from fossil fuels and toward renewables in this context of geopolitical tensions? How does this affect their sovereign debt and credit quality as they embark on their green transitions, and developing environmental, social and governance (ESG) and investment preferences and opportunities about sustainable finance (e.g., green bonds).

Policy-Oriented Research and Practical Tool Development

Providing Enhanced Investor Guidance on Geopolitical Exposures: Research can play an important role in improving investor guidance by increasing transparency of geopolitical exposures within sovereign debt portfolios, and developing standard frameworks for assessing the fiscal resilience of sovereign debt to geopolitical risk. This involves, for example, assessing approaches to portfolio diversification and dynamic hedging that generate a way for investors in emerging and frontier sovereign debt markets to better navigate volatile geopolitical and commodity environments through sovereign debt investments, thereby mitigating the chances of abrupt and disruptive capital reversals.

If all of these various interconnected pathways can be realized, future research ultimately has the potential to greatly advance our cumulative knowledge of the problems of resource-exporting countries under geopolitical stress which will help us to:

- Advance policy development
- Enhance economic and non-economic resilience
- Support sustainable development

Establish sustainable levels of public debt in an increasingly uncertain international economic and international political context.

THEMATIC ANALYSIS

Theme 1 - The Multi-fold Dimensional Nature of Geopolitical Risk as Instability: This theme is consistent with the notion that geopolitical events are not disjointed shocks to the system; but are an amplification of economic vulnerability. Geopolitical risk (GPR), including conflicts, political instability and political confrontation adds considerable uncertainty to global marketplaces. Such uncertainty influences the evaluated risk of the expected outcomes, often leading to volatility in financial markets and commodity price fluctuations (Caldara & Iacoviello, 2022),(Baldacci et al., 2011). GPR global the literature implies significant volatility multiplicative 'effect'. The literature suggests prices zigzag in energy, metals and agriculture commodities due to GPR, resource-exporting countries to grapple with intensified difficulties(Ben Flah et al., 2025).

Theme 2 - Commodity Dependence as a Core Vulnerability for Fiscal Stability: This theme denotes the idea that geopolitical events are not fully external shocks, but ways to amplify the existing economic vulnerabilities. Geopolitical risk (GPR), taking the form of conflict, political instability, and political confrontation, creates substantial uncertainty across global markets and affects the calculated risk of potential outcomes, leading to volatility in financial markets and swings in commodity prices(Balli et al., 2022),(Agoraki et al., 2022). GPR has substantial volatility 'multiplication effects'. The literature is clear that energy, metals and agricultural commodity prices swing when the market is subject to GPR, raising the already mounting challenges faced by resource-exporting countries(Sharma & Thuraisamy, 2013).

Theme 3 - The Interplay of Financial and Real Economy Transmission Channels: This theme outlines both of the two dual and, potentially reinforcing, routes through which geopolitical shocks and commodity shocks spread across the global economy. The financial channel is comprised of first-order adjustment in capital markets including initial impacts like expanded sovereign debt spreads, capital flight from economies will be perceived as risky, and increases in equity market volatility(Afonso et al., 2012). Simultaneously, the real economy channel consists of first-order movements that also directly disrupt global trade and anything involving supply chains, as well as the direct effects on the physical production and distribution of commodities(Cao & Dong, 2025). These real economy effects will then feed back into the financial markets of the rest of the world by changing investor confidence in the economies involved, risk perceptions for investors, etc

Theme 4 - Fiscal Space as a Vital Buffer and Policy Challenge: This theme aims to highlight the "fiscal space", or the "fiscal space" of a country, which is the potential for a country to either increase public spending or decrease taxes without jeopardizing broader financial sustainability and budgetary access to markets. The literature has observed that the levels, and especially the potential to utilize, fiscal buffers are essential to country resilience(Akram, 2009). However, the challenge of creating and sustaining sufficient fiscal space, especially during commodity booms, and then utilizing it in times of decline or geopolitical risk will always be a one of policy challenges for resource-exporting countries(Bergman et al., 2019),(Apergis et al., 2020).

Theme 5 - Policy Implications for Building Resilience: This theme provides a synopsis of the typical policy options that are discussed and synthesised in the literature. These options relate to how to build resilience, reduce risk, and adapt to a more unstable global economic and geopolitical landscape. Key implications are: the need for economies to diversify away from a dependence on commodities, the development of counter-cyclical fiscal policies; using geopolitical risk as a variable in macro-financial stress tests and ongoing disclosure policies designed to change investors behaviour to use the past value of assets(Aizenman et al., 2012),(Afonso et al., 2024). These plans are focused on reducing vulnerability to shocks and strengthening the nation's ability to react when the next shock occurs.

DISCUSSION

Discussion on Theme 1: The Multiplicative Impact of Geopolitical Risk on Economic Volatility

Geopolitical risk (GPR) has a great deal of documentation and literature indicating that nations experience GPR events, which range from regional war to trade disputes to political instability, which exacerbates existing economic vulnerabilities(Agoraki et al., 2022),(Balli et al., 2022). GPR creates spikes in uncertainty that support, through various global financial and real economy links, immediacy and volatility. Specifically, GPR results in risk aversion by its principal actors: investors flee from committed sources of capital in emerging markets (EMEs) and widen sovereign CDS spreads to reflect the heightened uncertainty and an increase in the perception of default risk. Some studies also underline the "asymmetric risk contagion effect", so that the adverse indications of GPR are usually stronger and more volatile during incidents of stress, with respect to the effects on vulnerable economies. The other adverse impact of GPR is price volatility in commodities. Geopolitical tensions can impact global supply chains and therefore, the production decisions of individual countries and ultimately the global demand for resources supporting prices.

Discussion on Theme 2: Commodity Dependence as a Core Vulnerability for Fiscal Stability

A critical takeaway is the vulnerability of resource-exporting countries, since their reliance on commodity revenues leaves them exposed to the volatility of global commodity prices. Realised commodity prices determine a country's export earnings, that country's government budgetary revenue, and affect the current account. A negative commodity price shock can create a shortfall between expected revenue for the government spending, a fall in "fiscal space"—a government's ability to increase spending or decrease taxes without threatening the financial viability of its spending. A decrease in fiscal space may require governments to cut basic public service deliveries, defer developmental expenditures, or increase debt borrowing(Afonso et al., 2024). Increasing debt borrowing contributes to sovereign debt growth and potentially unsustainable debt circumstances(Bergman et al., 2019). The literature identifies commodity price cycles as a prominent component of macroeconomic stability in commodity-dependent economies due to the exposure of public finances and consequential sovereign debt spreads in down cycles. The "spillover effects of external economic shocks on African sovereign bonds," for example, starkly shows the ability for commodity price declines to be reflected in increased sovereign debt spreads and borrowing costs for commodity-dependent economies in Africa, as an example, and can directly affect their management of public finances.

Discussion on Theme 3: The Interplay of Financial and Real Economy Transmission Channels

The review indicates that GPR and commodity shocks have effects transmitted through both distinct but interconnected channels through the financial and real economy. The financial channel means immediate action in financial markets, in particular, "equity volatility", quick movement in spreads of "CDS" (widening) and "capital flight" associated with re-evaluating risk exposures, resulting in the re-balancing of portfolios drawing on perceived safe-haven assets, effectively raising borrowing costs in risky economies(Bouri et al., 2017), (Dieckelmann et al., 2025). Conversely, the real economy channel represents more foundational disruptions to "trade and supply chains", direct impacts on the "commodity markets" (such as commodity supply disruptions or reverting to commodity demand), and consequential influences on overall "macroeconomic performance" (e.g., GDP growth and employment). Trade flows can be disrupted through geopolitical events, increasing transaction costs, and affecting the production and distribution of commodities, leading to price fluctuations of an applicable magnitude or in an applicable direction or, simply put, the price of oil may go up or down from uncertainty due to a geopolitical event or trade uncertainty across the world. Real economy channels also feed back into financial markets, meaning we impact investor confidence, perceptions of sovereign risk, and borrowing cost when there are geopolitical events or trade uncertainty(Afonso et al., 2024),(Bergman et al., 2019). A study earlier this year examined the "time-varying effects of trade policy uncertainty and shocks of geopolitical risks on composite commodity market prices," which captures the direct and indirect ways some element of a policy choice and geopolitical event relate to commodity prices and the rest of the economy(Abakah et al., 2023).

Discussion on Theme 4: Fiscal Space as a Crucial Buffer and Policy Challenge

Fiscal space" represents an important aspect of a country's ability to cope with external shocks(Abere & Akinbobola, 2020). For example, resource-exporting countries typically accrue reserves like sovereign wealth funds during artificially high commodity prices, thus building up fiscal buffers. There seems to be an implicit caveat in the review that these buffers can be depleted rapidly in or leading up to periods of prolonged low commodity prices or in times of extreme geopolitical stress(Baur & Smales, 2020). Countries limited fiscal space means that they will have less room to execute countercyclical policies and are likely to be more exposed to the negative consequences of shocks(Cafiso et al., 2025),(Afonso et al., 2012). Additionally, countries with limited fiscal space are more reliant on external funding such as bond issues or loans from the World Bank or IMF and pay higher interest rates (if they are able to obtain funding) (Tekin, 2025).

Discussion on Theme 5: Policy Implications for Enhanced Resilience

The literature generally indicates that there are a number of important policy implications for resource exporters to enhance their resilience. This literature points towards, among other aspects, "event-specific mitigation" strategies where there is a presumption that a "one-size-fits-all" approach is insufficient for nation states episodically and asymmetrically impacted by GPR and commodity shocks in their economies; therefore, adaptive responses are critical(Park et al., 2025). Resilience building, through economic diversification and weaning itself from excessive reliance on any one commodity or sector, is crucial to buffer economies from extreme price volatility and from accessing other streams of revenue. Moreover, it is suggested that central banking and financial regulators "develop and integrate the GPR understanding the "index in macro-financial stress tests" in order to somewhat anticipate shocks in the future (Zehri et al., 2025). In addition, it was recommended from investor perspectives to increase the "transparency of geopolitical exposures" not only through sovereign portfolios, but also to recommend strategies for dynamic hedging, to facilitate movement in these uncertain environments while alleviating extreme capital residence reversals. All in all, the convergence of findings implies a pro-active, adaptive, and diversified policy response is necessary to manage the unfolding prospect of complicated and uncertain risks confronting the majority of resource-exporting economies.

CONCLUSION

This systematic literature review has explored the complex relationships in the world of commodity price shocks, sovereign debt spreads, and, importantly, fiscal space in commodity-reliant countries during escalating geopolitical stress. The numerous pieces of empirical evidence synthesised from a wide range of literature from 2005 to 2025 clearly demonstrate that geopolitical risks occur at the same time, price shocks, have continued to drive the volatility of global commodity markets (Balli et al., 2022), (Agoraki et al., 2022). Volatile supplies of abundant commodities in a competitive landscape will widen spreads of sovereign debt and cause commodity-dependent countries to exceed fiscal limits and enter into long-term uncertainty.

Key Findings and Their Implications

Geopolitical Shocks as a Pervasive Catalyst: Our analysis suggests that geopolitical events are not just random events, but major, pervasive triggers that lead to significant spikes in "uncertainty" that are transmitted quickly through global financial and real economy channels(Agnello et al., 2020). One immediate effect of such events, often on a disproportionate scale, is asymmetrical investor risk aversion, which appears to be embodied in large "capital flight" from risky emerging markets (EMEs) and a dramatic widening in sovereign credit default swap (CDS) spreads (Dieckelmann et al., 2025), (Tekin, 2025). The widening is a reflection of an increase in the market's perception of default risk, which can, if expressed in borrowing cost terms, place additional strain on public finances through increased borrowing costs for sovereign countries. As noted in the literature, investors experience an "asymmetric risk contagion effect", meaning that negative geopolitical effects tend to have a more severe and abrupt impact, creating significant demand for a policy response that is swift and severe.

Commodity Dependence as a Fundamental Vulnerability: One key finding from this review is the inherent vulnerability of resource-extracting countries on account of their high economic dependency on a narrow base of primary commodity revenues. Fluctuations in global commodity prices affect national export revenues, government revenues (especially non-tax revenues), and current account balances directly and substantially. Negative shocks in commodity prices create substantial revenue loss, which in turn, creates a significant decrease in a government's "fiscal space" - its essential space to increase expenditure and/or reduce tax revenue while preserving fiscal credibility, financial viability or financial access. The slashing of such fiscal space needs to be mitigated through some difficult trade-offs, such as reducing public service provision, stalling key development plans, or simply borrowing more, which leads to increased sovereign debt accumulation and a dangerously weakened position more likely to default(Fernandez-Perez et al., 2025). The recent evidence and patterns of "spillover effects of external economic shocks on African sovereign bonds" represent some evidence of how commodity price shocks lead to wider spreads on and thus costly borrowing costs for commodity-dependent African economies.

Fiscal Space as a Critical Buffer and Strategic Imperative: The availability of fiscal space and, more importantly, the effective management of fiscal space are obvious indicators of a nation's resilience to externally-shock events. Countries having committed fiscal buffers such as significant foreign exchange reserves, well-managed sovereign wealth funds, or a low level of existing debt are much more equipped to absorb and respond to shocks caused by, among other things, falling commodity prices, and increasing geopolitical instability(Zehri et al., 2025). Nations that possess little to no or few and diminishing fiscal buffers have almost no effective discretionary counter-cyclical policy options to deal with crises.

Asymmetric and Episodic Impacts Necessitate Adaptive Responses: The study continually draws attention to the fact that the repercussions of geopolitical and commodity shocks are usually completely dissimilar and episodic(Agnello et al., 2018). This means that the severity, direction and transmission mechanism can differ greatly depending on the detailed nature of the shock, its source and the international and national economic and social context and conditions. This natural variability implies that using the "standard response" paradigm as a basis to create responsive policy is mostly nonsensical (Liang & Ullah, 2024).

Prioritise Economic Diversification: It is critical to move from an over-reliance on a narrow range of commodities. Policies promoting economic diversification can provide alternatives and more stable sources of revenue and therefore significantly shield economies from extreme volatility of commodity prices and enhance broader fiscal stability.

Institutionalisation of prudent fiscal management: It is important to diversify away from dependency on a small range of commodities. Policies to promote economic diversification can provide alternatives and more stable revenue sources and therefore significantly shield economies from the extreme volatility of commodity prices and enhance broader fiscal stability.

Integrate Advanced Risk Modelling: It is both important and relevant to integrate geo-economic risk indices in macro-financial stress tests and macro-financial early warning mechanisms. Geopolitical risk can provide a major step forward in the capabilities of decision-makers to foresee future crises, their consequences, and to be able to respond calmly to potential crises in a timely and pre-emptive manner.

REFRENCES

Abakah, E. J. A., Tiwari, A. K., Adekoya, O. B., & Oteng-Abayie, E. F. (2023). An analysis of the time-varying causality and dynamic correlation between green bonds and US gas prices. *Technological Forecasting and Social Change, 186*, 122134. DOI: 10.1016/j.techfore.2022.122134

Abere, S. S., & Akinbobola, T. O. (2020). External Shocks, Institutional Quality, and Macroeconomic Performance in Nigeria. *SAGE Open, 10*(2), 2158244020919518. Advance online publication. DOI: 10.1177/2158244020919518

Afonso, A., Alves, J., & Monteiro, S. (2024a). Beyond borders: Assessing the influence of Geopolitical tensions on sovereign risk dynamics. *European Journal of Political Economy, 83*, 102550. DOI: 10.1016/j.ejpoleco.2024.102550

Afonso, A., Alves, J., & Monteiro, S. (2024b). Sovereign risk dynamics in the EU: The time varying relevance of fiscal and external (im)balances*. *Journal of International Financial Markets, Institutions and Money, 94*, 102026. DOI: 10.1016/j.intfin.2024.102026

Afonso, A., Furceri, D., & Gomes, P. (2012). Sovereign credit ratings and financial markets linkages: Application to European data. *Journal of International Money and Finance, 31*(3), 606–638. DOI: 10.1016/j.jimonfin.2012.01.016

Agnello, L., Castro, V., Hammoudeh, S., & Sousa, R. M. (2020). Global factors, uncertainty, weather conditions and energy prices: On the drivers of the duration of commodity price cycle phases. *Energy Economics, 90*, 104862. DOI: 10.1016/j.eneco.2020.104862

Agnello, L., Castro, V., & Sousa, R. M. (2018). The Legacy and the Tyranny of Time: Exit and Re-Entry of Sovereigns to International Capital Markets. *Journal of Money, Credit and Banking, 50*(8), 1969–1994. DOI: 10.1111/jmcb.12474

Agoraki, M.-E. K., Kouretas, G. P., & Laopodis, N. T. (2022). Geopolitical risks, uncertainty, and stock market performance. *Economic and Political Studies, 10*(3), 253–265. DOI: 10.1080/20954816.2022.2095749

Aizenman, J., Edwards, S., & Riera-Crichton, D. (2012). Adjustment patterns to commodity terms of trade shocks: The role of exchange rate and international reserves policies. *Journal of International Money and Finance, 31*(8), 1990–2016. DOI: 10.1016/j.jimonfin.2012.05.003

Akram, Q. F. (2009). Commodity prices, interest rates and the dollar. *Energy Economics, 31*(6), 838–851. DOI: 10.1016/j.eneco.2009.05.016

Apergis, N., Chatziantoniou, I., & Cooray, A. (2020). Monetary policy and commodity markets: Unconventional versus conventional impact and the role of economic uncertainty. *International Review of Financial Analysis*, *71*, 101536. DOI: 10.1016/j.irfa.2020.101536

Baldacci, E., Gupta, S., & Mati, A. (2011). Political and Fiscal Risk Determinants of Sovereign Spreads in Emerging Markets. *Review of Development Economics*, *15*(2), 251–263. DOI: 10.1111/j.1467-9361.2011.00606.x

Balli, F., Balli, H. O., Hasan, M., & Gregory-Allen, R. (2022). Geopolitical risk spillovers and its determinants. *The Annals of Regional Science*, *68*(2), 463–500. DOI: 10.1007/s00168-021-01081-y

Baur, D. G., & Smales, L. A. (2020). Hedging geopolitical risk with precious metals. *Journal of Banking & Finance*, *117*, 105823. DOI: 10.1016/j.jbankfin.2020.105823

Ben Flah, I., Samet, K., El Ammari, A., & Terzi, C. (2025). Shock and Volatility Transmissions Across Global Commodity and Stock Markets Spillovers: Empirical Evidence from Africa. *Journal of Risk and Financial Management*, *18*(6), 332. DOI: 10.3390/jrfm18060332

Bergman, U. M., Hutchison, M. M., & Hougaard Jensen, S. E. (2019). European policy and markets: Did policy initiatives stem the sovereign debt crisis in the euro area? *European Journal of Political Economy*, *57*, 3–21. DOI: 10.1016/j.ejpoleco.2018.06.003

Bouri, E., De Boyrie, M. E., & Pavlova, I. (2017). Volatility transmission from commodity markets to sovereign CDS spreads in emerging and frontier countries. *International Review of Financial Analysis*, *49*, 155–165. DOI: 10.1016/j.irfa.2016.11.001

Cafiso, G., Missale, A., & Rivolta, G. (2025). The credit channel of the sovereign spread: A Bayesian SVAR analysis. *Economic Modelling*, *144*, 106984. DOI: 10.1016/j.econmod.2024.106984

Caldara, D., & Iacoviello, M. (2022). Measuring Geopolitical Risk. *The American Economic Review*, *112*(4), 1194–1225. DOI: 10.1257/aer.20191823

Cao, S., & Dong, W. (2025). Production networks and the macroeconomic impacts of commodity price shocks. *The Canadian Journal of Economics. Revue Canadienne d'Economique*, *58*(2), 690–715. DOI: 10.1111/caje.70007

Dieckelmann, D., Larkou, C., McQuade, P., Pancaro, C., & Rößler, D. (2025). Geopolitical risk and euro area bank CDS spreads and stock prices: Evidence from a new index. *Economics Letters*, *254*, 112461. DOI: 10.1016/j.econlet.2025.112461

Fernandez-Perez, A., Gómez-Puig, M., & Sosvilla-Rivero, S. (2025). Examining the transmission of credit and liquidity risks: A network analysis for EMU sovereign debt markets. *The North American Journal of Economics and Finance*, *77*, 102407. DOI: 10.1016/j.najef.2025.102407

Liang, J., & Ullah, I. (2024). Analysis of crude oil and gold price volatility and their correlation during socio-economic crises. *Resources Policy*, *98*, 105311. DOI: 10.1016/j.resourpol.2024.105311

Park, E., Boyer, C. N., & Neill, C. L. (2025). A Markov regime-switching event response model: Beef price spread response to processing capacity shocks. *Empirical Economics*, *68*(3), 1039–1071. DOI: 10.1007/s00181-024-02677-x

Sharma, S. S., & Thuraisamy, K. (2013). Oil price uncertainty and sovereign risk: Evidence from Asian economies. *Journal of Asian Economics*, *28*, 51–57. DOI: 10.1016/j.asieco.2013.06.001

Tekin, B. (2025). Do economic uncertainty and political risk steer CDS dynamics? An analysis of the Türkiye CDS. *International Journal of Islamic and Middle Eastern Finance and Management*, *18*(2), 249–270. DOI: 10.1108/IMEFM-05-2024-0215

Zehri, C., Alsadan, A., & Saleh Ben Ammar, L. (2025). Asymmetric impacts of geopolitical risks on energy Trade: Divergent vulnerabilities in emerging vs. advanced economies. *Journal of Economic Asymmetries*, *32*, e00427. DOI: 10.1016/j.jeca.2025.e00427

Chapter 4
Green and Renewable Energy Policy in Relation to Financial Literacy in México

José G. Vargas-Hernández
https://orcid.org/0000-0003-0938-4197
Financial Literacy Research Center, Universidad Cristóbal Colón, Mexico & Tecnológico Nacional de México, ITS Fresnillo, Mexico

ABSTRACT

This research examines the connection between the green and renewable energy policy of Mexico and the country's financial literacy level. The article examines how financial awareness and knowledge contribute to the participation of citizens in renewable energy programs, investment in green technologies, and the overall success of environmental policy implementation. The article concludes that additional focused financial education aimed at green energy investments would notably increase policy attainment and accelerate Mexico's transition to renewable energy.

INTRODUCTION

Mexico stands at a crossroads in its energy evolution journey, balancing traditional fossil fuel reliance with urgent necessities of green energy transition imperatives. As the world's 11th largest economy and Latin America's second-largest country, Mexico's energy decisions carry significant implications for regional and worldwide climate objectives. Mexico's compliance with international climate agreements, including the Paris Accord, has necessitated the rebuilding of its energy policy

DOI: 10.4018/979-8-3373-6766-8.ch004

framework in support of green and renewable energy. Mexico is at a juncture of energy policy as it seeks to attain environmental sustainability and socioeconomic progress. This report explores Mexico's transition from fossil fuel use to green and renewable energy, and how financial literacy enables and is enabled by this transition.

The Mexican energy sector has been dominated by fossil fuels historically, and oil and natural gas have represented approximately 75% of the country's energy matrix as of 2023. However, recent policy measures, including the Energy Transition Law (Ley de Transición Energética) and the General Climate Change Law (Ley General de Cambio Climático), have raised the bar for the transition to clean energy. Mexico aims for 35% clean generation by 2024 and recently established plans to reach 45% renewable energy by 2030, one of the most ambitious renewable energy targets in Latin America.

At the center of this energy transformation is the concept of financial literacy, made up of the knowledge, ability, and faith to make informed financial decisions. In the context of energy policy, financial literacy is of particular interest as it influences investment decision, policy implementation efficacy, and public uptake of renewable energy technologies. The intersection of financial literacy and energy policy is one such niche that is under-represented in academic literature, particularly in the context of developing economies like Mexico.

This research addresses a significant knowledge deficiency regarding the contribution of financial literacy to the success of energy policy implementation and renewable energy adoption in Mexico. The research embraces multiple facets of this relationship, including the influence of financial literacy on investor trust in renewable energy investment, the green technology adoption by consumers, and the effectiveness of governmental energy policy. Moreover, the research addresses how greater financial literacy among policymakers, investors, and citizens can accelerate Mexico's transition to a more sustainable energy future.

Mexico has ambitious targets of achieving 35% clean energy by 2024 and 45% renewable generation by 2030, requiring substantial investments valued at $48 billion. Employing a mixed-methods approach combining policy analysis, literature review, and empirical data analysis, the research illustrates significant gaps between policy aspirations and implementation realities. The research contributes to knowledge about the nexus of energy policy, financial literacy, and sustainable development in emerging markets, with implications for policymakers, investors, and researchers working on energy transitions in Latin America.

Transition to Green and Renewable Energy. Current Status and Potential. Renewables constitute an increasing share of Mexico's power mix. By 2022, 31.2% of electricity came from clean sources, dominated by solar, wind, hydroelectric, and geothermal power (Economics Observatory, 2025). Technical potential is vast: for instance, solar is 24,918GW and wind 3,669GW, well above current national demand (Fernández & Watson, 2022). Still, Mexico depends on imported gas for over half of its electricity generation—a risk to energy security and fiscal stability alike (Ember, 2025).

The significance of this research extends beyond academic inquiry, as it offers usable knowledge for policymakers seeking to enhance energy transition policies, financial institutions that design renewable energy financial instruments, and international organizations that promote Mexico's sustainable development agendas. As Mexico will probably keep grappling with complex energy policy challenges while attempting to secure economic development and environmental sustainability, the role of financial literacy in energy transitions becomes increasingly relevant.

CONCEPTUAL FRAMEWORK

Renewable Energy and Energy Policy Concepts

Energy policy is the state policy for generating energy, transmitting it, and using it, comprised of regulation tools, incentives, and strategic plans. When applied to renewable energy, policy tools must confront technical, economic, and social dimensions of energy conversion. Renewable energy is energy derived from naturally replenishing resources, like solar, wind, hydroelectric, geothermal, and biomass, which yield environmentally compatible alternatives to fossil fuels.

Energy transition involves structural change of energy systems from fossil fuel-based infrastructure to renewable sources of energy. The process requires policy coordination, significant capital investment, technology innovation, and social acceptance.

The cornerstone of the energy transition in Mexico lies in the Ley de Transición Energética (Energy Transition Law) and the Ley General de Cambio Climático (General Climate Change Law), setting the ambitious targets of 35% electricity from clean energy by 2024 and a minimum of 45% by 2030 (DLA Piper, 2016; International Trade Administration, 2023). Clean energy here includes renewables, nuclear, and efficient cogeneration. Planning documents such as the PRODESEN (Program for the Development of the National Power System) chart the installation of 22.7GW of new capacity (2025–2030), 4.7GW of which is planned to be solar, yet developments are hampered by shifting federal priorities and regulatory challenges (Ember, 2025).

For Mexico's energy policy in this regard, multiple layers of institutions, including federal ministries of energy, regulatory commissions, and state-level implementation agencies, operate. Green energy extends beyond renewable energy to encompass environmentally friendly energy practices that minimize environmental footprints along the energy lifecycle. This includes energy efficiency, carbon mitigation, and environmental protection in mind. Mexican green energy policy integrates renewable energy development with energy efficiency improvement in industrial, commercial, and residential applications.

Financial Literacy in Energy Contexts

Financial literacy, as defined by the Organization for Economic Co-operation and Development (OECD), is the combination of awareness, knowledge, skill, attitude, and behavior to make informed financial decisions and ultimately achieve personal financial well-being. Financial literacy in energy policy environments entails a comprehension of energy markets, investment products, risk assessment, and planning over the long term for energy decisions.

Conceptual model of energy transition financial literacy comprises several basic dimensions. Market literacy as the first one needs to have understanding about energy market mechanisms, pricing regimes, and regulation institutions that control energy investments. Investment literacy as the second one needs to reflect on knowledge about financial instruments, risk-return trade-offs, and portfolio diversification strategies available for renewable energy investments. Policy literacy as the third one has to comprehend government incentives, regulation regimes, and policy mechanisms supporting renewable energy development.

In the context of Mexico, financial literacy is especially relevant due to the emerging market features of the country, where information asymmetry and institutional risk can heavily influence investment choices. Financial literacy interacts with energy policy through various channels such as personal investment choices, business funding options, and efficiency of policy implementation.

Theoretical Integration

The theoretical integration of energy policy and financial literacy draws on a variety of academic disciplines, including environmental economics, energy policy studies, and behavioral finance. Environmental economics provides theories for explaining market failures in energy markets and government intervention for promoting sustainable energy adoption. Energy policy studies provide insights on policy design, implementation challenges, and stakeholder coordination in energy transitions.

Behavioral finance offers important insights into how individual and institutional decision-making processes shape energy investment yields. Theories of loss aversion, herding, and cognitive bias can account for why renewable energy adoption could lag behind during more favorable financial circumstances. In Mexico, cultural factors, institutional trust, and historical experience with reform in the energy sector further complicate the interplay between financial knowledge and effective energy policy.

The theoretical framework also synthesizes theory of sustainable development, which centers on synergies among economic development, social justice, and environmental protection. Mexico's energy transition is embedded in universal sustainable development objectives, which entail coordination between energy policy, economic growth, and social welfare agendas. Financial literacy as an enabling aspect plays a significant role in the achievement of these synergized objectives through improved decision-making quality among various stakeholder groups.

THEORETICAL LITERATURE REVIEW

Emerging Economy Energy Policy Theory

Theoretical energy policy models of emerging economies point out the unique challenges of developing nations to balance energy security, economic development, and environmental sustainability. Bazilian and Pielke (2013) argue that emerging economies have specific energy policy issues to address, such as limited financial resources, limitations in institutional capacity, and conflicting development priorities. Their model suggests that energy transition in emerging economies is effective if it emanates from well-sequenced policy interventions for overcoming technical and socioeconomic constraints.

Institutional theory approach, as it has been shaped by North (1990) and institutionalized for energy by Bhattacharyya (2011), emphasizes the role of formal and informal institutions in shaping energy policy outcomes. Institutional factors for Mexico include regulatory institutions, property rights institutions, and political stability, all of which improve the attractiveness of renewable energy investment. Bhattacharyya's own work suggests that the quality of institutions has a significant impact on the success of energy transition, with countries having stronger institutions having superior performance in the deployment of renewable energy.

Resource dependency theory, first proposed by Pfeffer and Salancik (1978), provides us with an insight into the impact of countries' energy resource availability on policy choices. Mexico's immense oil reserves have historically generated fossil fuel-biased path dependencies that may play a role in renewable energy transitions. However, more recent research by Goldthau and Sovacool (2012) shows that richly endowed countries can diversify their energy mix without any issue with effective policy action and institutional reform.

Financial Literacy Theory and Investment Behavior

Theories of financial literacy that affect investment behavior are primarily drawn from behavioral finance and economic literature. Lusardi and Mitchell (2014) discuss an explicit theoretical framework that links financial literacy with the quality of investment choice, claiming that those who are more financially literate make quality long-term investment choices and are likely to participate in financial markets. Their hypothesis suggests that financial literacy affects investment behavior through three broad channels: improved understanding of risk-return tradeoffs, improved portfolio diversification strategies, and improved ability to evaluate complex financial products.

The theory of planned behavior by Ajzen (1991) offers insight into how financial literacy influences investment decisions concerning energy. The theory asserts that investment intentions are guided by attitudes towards the investment, subjective norms for investment behavior, and beliefs of control over the behavior. Financial literacy enhances all three constructs through better formation of attitudes due to better information processing, higher accuracy in perceptions of social norms of sustainable investing, and higher belief in investment capability for decision-making.

The theory of behavioral finance, on the whole, and Kahneman and Tversky (1979) prospect theory, specifically, can be applied to explain why renewable energy investments could be faced with adoption barriers despite having good financial profiles. Loss aversion, framing effects, and availability heuristics might lead investors to underappreciate renewable energy opportunities, particularly where traditional energy investments are more familiar. Financial literacy has the potential to break these behavioral biases by improving analytical reasoning and reduced dependence on heuristic decision-making mechanisms.

Energy Transition Theory

Theory on energy transitions has evolved from initial technological deterministic perspectives towards more comprehensive approaches that also consider social, economic, and political forces. The Multi-Level Perspective (MLP) approach of Geels (2002) presents a structured framework to study energy transitions at three analytical levels: niches (protected spaces for profound innovations), regimes (dominant practices and technologies), and landscape (broader contextual factors). Successful energy transitions under this framework require synchrony between institutional reforms, technological innovations, and broader societal transformations.

Socio-technical systems theory, formulated by Hughes (1983) and subsequently developed by Markard *et al.* (2012), points out the interdependent character of technological, social, and institutional aspects in the energy system. It suggests that energy transitions involve not just technological replacement but reorganization of entire socio-technical systems on a radical level. For Mexico, renewable energy adoption would imply take-up requiring symbiotic alterations at the levels of technology, institutions, markets, and society.

Transition theory also utilizes innovation systems literature, particularly the technological innovation systems approach of Carlsson and Stankiewicz (1991). This one emphasizes knowledge networks, market creation, and institutional support as essential for technological transitions. In Mexico's transition to renewable energy, this means the construction of strengthened innovation systems that facilitate technology innovation, knowledge exchange, and market creation for renewable energy technology.

EMPIRICAL LITERATURE REVIEW

Implementation of Mexico's Energy Policy

Empirical data of the implementation of Mexico's energy policy reveal significant gaps between policy objectives and outcomes. Aguirre and Ibikunle (2014) conducted a comprehensive analysis of Mexico's fulfillment of renewable energy policy between the years 2008-2012, concluding that despite ambitious renewable energy objectives being established, the installation of renewable energy was significantly below policy objectives. Institutional coordination problems, regulatory ambiguity, and access limitations to finance were key implementation obstacles, their research contended.

Subsequent later empirical studies by Sánchez-Peña *et al.* (2020) evaluated the performance of Mexico in energy transition after the 2013 energy sector reforms. Using panel data analysis for the period 2013-2018, they found that while policy reforms improved regulatory clarity, implementation effectiveness was extremely heterogeneous based on the renewable energy technology type and the region. Deployment of solar power exceeded expectations in the northern states with high solar irradiance, and the expansion of wind power concentrated largely in some regions with suitable wind resources and an existing transmission grid.

Bazán-Navarrete and Bravo-Peña (2021) provide empirical analysis of the performance of the clean energy certificate market in Mexico through 2018-2020 market data. Their findings suggest that the certificate mechanism successfully created a clean energy attribute market but was constrained from fostering higher renewable energy deployment by price volatility and market liquidity. The research highlights the significance of market design specifics in shaping policy effectiveness for renewable energy.

Energy Investment Studies and Financial Literacy

Empirical research on the impact of financial literacy on energy investments remains limited, particularly in the context of developing countries. Weber (2017) employed a survey methodology to analyze household solar take-up in Germany and found that financially more literate households were significantly more inclined to invest in rooftop solar systems, adjusting for income, education, and green concern. The research shows that financial literacy enhances the ability of people to evaluate the long-term monetary benefits of investments in renewable energy.

In the US setting, Peng *et al.* (2018) examined financial literacy and green fund membership based on household survey data. In the findings, financially more literate individuals are more likely to have environmental issues in investment decisions and possess higher shares in green investment products. However, the authors find that financial literacy effects vary significantly by demographic groups and places.

Closest to the Mexican context, Silva and Cerqueira (2019) investigated the impact of financial literacy on the adoption of renewable energy among Brazilian small enterprises. Drawing from survey data of 1,200 small enterprises, they found that the more a respondent scored on financial literacy, the higher the likelihood that the owner would invest in energy-saving measures and renewable energy systems. The study also detected that financial literacy effects were most significant for companies operating in high energy cost industries, as it reveals that financial literacy increases ability to identify profitable energy investments.

Renewable Energy Investment Trends in Mexico

Empirical studies of trends in renewable energy investment in Mexico reveal complex relationships between policy contexts, market conditions, and investor responses. Rodríguez-Monroy *et al.* (2018) studied foreign direct investment trends in Mexico's renewable energy sector from 2010-2017 and established that uncertainty in policies had a prevalent impact on flows of investment. Periods of policy clarity and stability attracted heavy inflows of foreign investment, while policy reversals or uncertainty saw delays in investments and project cancellation.

The World Bank (2020) in-depth analysis of Mexico's renewable energy sector clarified that there were diverse empirical trends in investment behavior. International investment, particularly from North American and European developers, rushed to big utility wind and solar facilities. Distributed renewable energy use remained low, and financial limitations and the paucity of financial products were significant barriers. The study indicated that poor familiarity with funding options and poor understanding of investment economics were some of the reasons behind slow distributed renewable energy uptake.

A recent empirical study by González-Mahecha et al. (2019) examined the functions of development finance institutions in Mexico's renewable energy sector. Drawing from project-level data between 2015-2018, they determined that multilateral development banks and development finance institutions played a significant role in risk mitigation and demonstrating effects for private investors. As such, their research points to the fact that development finance has the ability to mitigate market failures and improve the investment conditions for renewable energy projects, particularly in the context of emerging markets.

METHODOLOGY

Research Design

The study adopts a mixed-methods research design that combines quantitative and qualitative approaches in understanding the interplay between Mexico's energy policy, renewable energy transition, and financial literacy in a comprehensive manner. The research design combines policy analysis, statistical analysis of financial and energy data, and stakeholder interview results to provide a complete image of the research questions.

The method follows an explanatory sequential design, with quantitative analysis of energy policy outcomes and financial literacy scores first, followed by qualitative inquiry to interpret and contextualize quantitative findings. This allows for triangulation of data sources and methods, increasing the validity and reliability of research outcomes.

Data Collection and Sources

The primary data gathering was done via semi-structured in-depth interviews with key stakeholders in Mexico's energy sector, including policymakers, renewable energy developers, financial institution representatives, and civil society organizations. Interview participants were selected by purposive sampling to gain representation of the different stakeholder groups and geographical areas. A total of 45 interviews were conducted between January and March 2024 and lasted between 45 and 90 minutes each.

Secondary data sources also include government energy statistics from Mexico's Ministry of Energy (SENER), renewable energy installation data from the National Center for Energy Control (CENACE), and financial literacy survey data from Mexico's National Commission for the Protection and Defense of Financial Services Users (CONDUSEF). Additional data sources include World Bank development indicators, International Energy Agency statistics, and private sector market reports from renewable energy industry associations.

Policy documents analyzed include Mexico's Energy Transition Law, General Climate Change Law, National Electric System Development Program (PRODESEN), and various regulatory frameworks issued by the Energy Regulatory Commission (CRE). Financial sector data sources include banking sector statistics from Mexico's National Banking and Securities Commission (CNBV) and sustainable finance indicators from private financial institutions.

Analytical Approach

Quantitative analysis employs descriptive statistics, correlation analysis, and regression modeling to examine relationships between financial literacy indicators and renewable energy adoption outcomes. Time series analysis examines trends in the deployment of renewable energy in relation to policy implementation timelines and developments in financial markets.

The regression analysis is interested in identifying determinants of renewable energy investment outcomes with particular interest in financial literacy determinants. Macroeconomic determinants, energy prices, policy stability, and regional determinants are included as control determinants. Robust standard errors are applied in the analysis to adjust for potential heteroskedasticity, and a range of specification tests are used to attain model reliability.

Qualitative analysis employs thematic analysis of interview transcripts, and the key themes related to financial literacy, barriers to renewable energy uptake, and energy policy implementation are determined. Both deductive and inductive coding approaches are employed in qualitative analysis, using theoretical ideas from the literature review but also remaining open to emergent themes in the data.

Policy analysis examines the coherence, consistency, and implementation effectiveness of Mexico's renewable energy policy. The analysis considers policy design features, institutional arrangements, and implementation mechanisms, benchmarking Mexico's policy against international best practices in renewable energy policy.

Limitations and Considerations

There are a number of methodological limitations that should be mentioned. First, financial literacy is measured primarily using self-reported survey data that are prone to response bias and measurement error. Second, the cross-sectionality of some of the data sources limits potential for determining causal relationships between financial literacy and energy policy outcomes.

Third, the study focuses predominantly on formal financial sector players and may not fully capture informal financial arrangements that influence energy investment decisions in some contexts. Fourth, interview sample selection, while purposive, may not be fully representative of all stakeholder perceptions, particularly those of marginalized groups or small-scale energy users.

The timeframe of the study of predominantly 2020-2024 data coincides with significant policy changes across different Mexican governments, which can potentially impact the stability of the relations observed. In addition, the economic impact of COVID-19 can potentially have both financial literacy metrics and investment patterns in the energy sector over the course of the study.

ANALYSIS OF RESULTS

Current Situation of Mexico's Energy Policy and Renewable Energy Development

Mexico's renewable energy sector has expanded extensively over the past decade, with growth not evenly distributed geographically and by technology. According to data, Mexico must increase its solar and wind capacity from its current level of 12 GW to as much as 44 GW by 2030 to fulfill its international commitments. This is a high scaling requirement that demands persistent policy support and heavy private sector investment.

Analysis of energy production data has shown that total energy produced in 2023 was 351,695 GWh, representing continued growth in total electricity production. However, the composition of this production shows continued reliance on fossil fuel sources with renewable energy sources still covering relatively small percentages of total production despite the recent upsurge.

Mexico aims for 35% of electricity from clean and non-clean renewable energy sources by 2024, as established by the General Climate Change Law and the Energy Transition Law. New policy trends have become even more ambitious with the Mexican government setting targets to increase the share of renewable energies to 45% by 2030. The geographical pattern of renewable energy growth is extremely concentrated at the regional level. Northern states with favorable solar irradiance have been the focus of major solar photovoltaic investments, and wind energy growth has been concentrated where favorable wind resources exist, particularly in Oaxaca and southern states. Regional concentration offers opportunities as well as challenges for transmission infrastructure development and grid integration.

Policy implementation analysis shows deep contradictions over the last few years. Under the current administration, Mexico has moved further from having an energy transition than before its current President López Obrador took office in 2018, with López Obrador arguing that Mexico is not yet ready for a green energy transition. This policy flip-flop has caused investors to be uncertain and has created more difficulties for renewable energy project development.

Financing Mechanisms and Patterns of Financial Investment

Mexico's renewable energy sector has attracted considerable investment commitments, real deployment sometimes falling behind announced projects at times. Mexico in 2022 pledged to double its renewable energy capacity by 2030, adding another 30 GW of clean power, with Mexico's Foreign Minister stating that the government is poised to invest $48 billion in its renewable energy industry.

Investment patterns indicate high international participation, in particular by European and North American developers and financiers. Utility-scale initiatives have performed relatively better in funding compared to distributed renewable energy systems, which is a reflection of their variance in terms of project scales, risk profiles, and financing institutions. The adopted electricity auction system over the recent years substantially attracted heavy international investment, though policy changes have affected the viability of this process.

Financing arrangements examined reflect heavy reliance on international development finance organizations and export credit agencies, particularly for large projects. Domestic financial institutions have been cultivating increasing interest in financing renewable energy, but their participation is still modest compared with external sources. This trend of finance reflects both the size requirements of renewable energy projects as well as potential areas of deficiencies in domestic financial market development.

Through its long-term electricity auctions, the country garnered almost $9 billion in investment into the renewable industry, expanding its production capacity, though policy changes affected the sustainability of this successful program. The auction mechanism demonstrated competitive renewable energy procurement potential when combined with suitable regulatory schemes and long-term price certainty.

Financial Literacy Assessment and Energy Investment Relationships

Survey data analysis suggests significant differences within the degree of financial literacy between different groups of stakeholders operating in Mexico's energy sector. Institutional investors and policymakers show higher levels of financial literacy, while individual consumers and small investors are shown to be more limited in their knowledge regarding energy investment economics and finance models. Correlation analysis reveals positive correlations between financial literacy indicators and engagement in renewable energy investment, yet they vary in terms of investment type and investor category. Institutional investors with higher financial literacy are related to higher investments in renewable energy ventures and more sophisticated risk management strategies.

Interview data provides an insight into the part played by financial literacy in shaping energy investment decision-making. Stakeholders who are more financially literate demonstrate clearer understanding of longer-term trends in energy prices, more sophisticated analysis of renewable energy project economics, and better identification of policy risk factors determining the success of investments. These findings suggest that the quality of investment decision-making in renewable energy contexts is enhanced through financial literacy.

The analysis pinpoints notable knowledge gaps in financial literacy related to renewable energy markets and technology. Prospective investors are short of information on renewable energy financing tools, government incentives schemes, and long-term economic benefits of renewable energy investment. Knowledge gaps appear to restrict participation in renewable energy investment opportunities.

Policy Implementation Effectiveness and Institutional Issues

Policy implementation effectiveness evaluation recognizes mixed outcomes among the different renewable energy support schemes. The clean energy certificate mechanism has established an operating market for renewable energy attributes, but market liquidity and price stability are still issues. In the first half of 2023, bioenergy accounted for only 1,525.1 GWh and geothermal sources 2,238 GWh, and they contributed to 4.1% and 6.1% of the net clean energy generation, respectively, and that has shown some renewable technologies have not been as much focused on by policies and development measures.

Regulatory framework analysis shows that although there are comprehensive legal frameworks, consistency in implementation has been plagued by political change and altering priorities. Institutional coordination across different government agencies that have the responsibility of implementing energy policy shows room for improvement, particularly in streamlining approval processes and reducing regulatory uncertainty.

New renewable energy is, by recent law, also required to provide storage to maintain grid stability, a reflection of the growing emphasis on grid integration matters as renewable capacity expands. That requirement is a key policy evolution that serves to mitigate technical integration issues at the expense of increasing project costs and complexity.

Stakeholder interviews reflect concern about predictability and policy consistency, which affect long-term investment planning and project development lead times. Federal-state institution interdependence for renewable energy development demonstrates coordination problems that may delay project implementation and increase transaction costs.

Regional and Sectoral Variations

Analysis sets significant regional variations in trends of renewable energy development in Mexico. Between 2025 to 2030, the intention is to add 22.7 gigawatts (GW) of new generation capacity of which 4.7GW would be solar, with a forecast for private enterprise to add another 6.4GW of renewable energy capacity. This expansion plan is a testament to the continued need for private sector participation towards the achievement of renewable energy targets. Northern states are now hotspots for the generation of solar energy, with plenty of solar irradiance and closeness to large consumption centers and transmission infrastructure. Southern states have focused on wind development, with some transmission limitations limiting the ability to transmit wind electricity to large consumption areas.

Sectoral analysis reveals that industrial consumers have been more willing to adopt renewable energy compared to residential and commercial consumers. Industrial consumers benefit from economies of scale when purchasing renewable energy and possess more developed energy management competencies. Small and medium enterprises also face greater barriers to the adoption of renewable energy, with less access to financing and reduced energy market literacy.

The distributed solar segment remains underdeveloped compared to utility-scale installations, reflecting financing hardship, regulatory intricacy, and lack of familiarization among prospective users. Take-up by home consumers has risen slowly even with falling technology costs, suggesting non-financial market obstacles still dominate this market segment.

DISCUSSION

Policy Coherence and Implementation Issues

The review reveals critical tensions between Mexico's aspirational renewable energy targets and the on-the-ground realities of implementation. While Mexico has passed sweeping legal frameworks via the Energy Transition Law and General

Climate Change Law, translating these policies into on-the-ground implementation has been beset by issues. The gap between policy intentions and performance echoes issues in energy policy implementation more generally in emerging economy contexts.

The policy rollback seen in the López Obrador regime shows how political shifts interrupt momentum in creating renewable energies. This outcome provides evidence to support institutional theory hypotheses about the importance of policy continuity in encouraging long-term investment. The policy rollbacks in implementing the changes introduced in the last regime's policy on renewable energy, including alterations to electricity auction processes, have made uncertainty overflow from its direct implications to affect investor confidence and long-term expectations.

Reaching the 45% clean energy goal would involve tripling renewable production within five years, and other nations in Latin America showing that this level of growth is achievable, as Brazil achieved a tripling of solar production in five years and Uruguay reached an equivalent level of increase. This is a comparison likely to favor the view that Mexico's targets, while ambitious, are technically feasible if regional precedents are taken into consideration, which involves policy design and implementation effectiveness being determining factors.

The institutional coordination issues identified in the analysis reflect broader problems of governance throughout the Mexican energy sector. A number of overlapping agencies, unintelligible decision-making, and absence of coordination mechanisms are all sources of regulatory uncertainty and delay in implementation. These institutional problems mix with financial literacy limitations to generate multiplicative barriers to effective implementation of policy.

Financial Literacy as a Factor for Policy Implementation

The research establishes that financial literacy has a significant mediating function in the efficacy of renewable energy policy. Higher levels of financial literacy among policymakers, investors, and consumers enhance the quality of energy decision-making and performance in policy implementation. Such evidence contributes to the conventional literature on financial literacy as it demonstrates its use in energy policy contexts and particularly in developing economies.

Positive correlation between financial literacy and involvement in renewable energy investments suggests that financial literacy could be an additional policy instrument to drive energy transitions. The connection is however conditioned by other factors like income levels, institutional quality, and stability of policies, suggesting that financial literacy interventions must be mainstreamed in broader policy frameworks. Interview results detect financial literacy gaps even among sophisticated stakeholders, namely in newer financial instruments such as green bonds, sustainability-linked loans, and blended finance products. Such knowledge gaps hamper the development of innovative financing structures that can trigger the deployment of renewable energy. The result suggests the possibility of targeted financial literacy interventions for energy sector stakeholders.

The impact of financial literacy is also heterogeneous across types of energy investments and investor groups, the research suggests. Financially more literate large institutional investors possess more sophisticated evaluation schemes for renewable energy investments, while smaller investors and consumers are more sensitive to the boundaries of financial literacy. Such heterogeneities suggest that there must be a differentiated set of instruments for enhancing financial literacy capable of confronting stakeholder heterogeneity.

Financial Mechanisms and Instruments

Mexico's government has established several financial instruments to encourage renewables. Fondo para la Transición Energética y el Aprovechamiento Sustentable de la Energía (FOTEASE) and Fideicomiso para el Ahorro de Energía Eléctrica (FIDE) offer funding for household and community-scale projects, incentivize the use of solar technologies, and enable the participation of SMEs. Nevertheless, transparency, monitoring, and inclusion concerns remain (Climate Transparency, 2025).

Financing Mechanisms and Market Development

The evolution of Mexico's renewable energy financing landscape reflects both success and ongoing challenges in market development. The effectiveness of the electricity auction mechanism at mobilizing international investment reflects the potential for well-designed market mechanisms to mobilize private capital for renewable energy investment. The subsequent policy changes that affected auction continuity, however, reflect the necessity for policy stability in maintaining investor confidence.

The research diagnoses long-standing domestic capital market development gaps for investment in renewables. While foreign funding has been available for big-ticket investments, smaller-scale and dispersed renewable energy investments face greater financing challenges. Both market development and financial literacy barriers among domestic institutions and investors are reflected by this funding gap.

Both aimed at promoting the consumption of renewable energy and creating tools to finance their adoption, with provisions being in place to achieve renewable energy ambitions like a 24% renewable energy contribution and efficient cogeneration by 2018, rising to 35% by 2024. The creation of financing tools indicates policy recognition of the contribution that financial mechanisms make, albeit their effectiveness having been undermined by implementation problems and inconsistencies in policies.

The report points to areas where innovative financing tools can be used to address existing market gaps. Development finance institutions have played an important role, but their performance can be maximized through greater coordination with indigenous financial institutions and better-designed risk-sharing mechanisms. The report suggests that increased financial literacy can improve the efficacy of these financing innovations by enhancing stakeholders' awareness and engagement.

Regional and Sectoral Development Patterns

The concentration of renewable energy development in some regions and sectors is due to natural resource endowments and institutions. Superior institutional capacity and business environments in northern states are matched with a natural resource advantage for the development of solar energy. The lesson from this experience is that renewable energy development depends not merely on good natural resources but also auspicious institutional conditions.

Limited use of distributed renewable energy compared to utility-scale projects reflects serious market structure issues. Utility-scale projects have international financing alternatives and economies of scale, but distributed renewable energy incurs higher transaction costs, more involved regulatory matters, and limited access to capital. These barriers are compounded by lower financial literacy among prospective distributed renewable energy system adopters.

Barriers and Opportunities Energy Poverty

About 37% of Mexican households are in energy poverty, with little financial and energy literacy that hinders the adoption of renewable programs and incentives (Ember, 2025). Workforce and Education: The integration of financial content in workforce and technical education is instrumental in preparing a new generation of professionals to engage in the green energy sector (Fernández & Watson, 2022). Financial Exclusion: The low financial inclusion of vulnerable groups limits access to clean energy finance and broader benefits of the transition (Climate Transparency, 2025).

Sectoral variations in the consumption of renewable energy are a result of variations in literacy in energy markets, access to information, and economic ability between different energy consumers. Greater awareness of energy markets and intrinsic financial capabilities among industrial consumers partly explain their higher inclination towards renewable energy consumption. Small and medium-sized firms have various barriers that can be addressed through targeted financial capability and support programs.

The report suggests that addressing regional and sectoral variation requires differentiated policy interventions taking into account varying levels of resources, institutional strength, and types of stakeholders. Financial literacy interventions can be particularly valuable when tailored to the local context rather than being rolled out in a generic way in every setting.

International Context and Comparative Analysis

Mexico's renewable energy development experience is comparable but different from other emerging economies of Latin America. The endowment of resources in the country presents tremendous opportunity for renewable energy development, mainly solar and wind resources. However, institutional and policy conditions constrained Mexico to fully benefit from these potentials compared to countries such as Brazil and Chile, which succeeded more in accelerating renewable energies at a faster rate.

The comparison with other successes in Latin America is a lesson in the requirements of policy continuity and institutional quality. Countries where there have been coherent renewable energy policy over multiple political cycles have generally fared better at drawing in investment and utilizing capacity. Mexico's experience shows how policy reversals can destroy development momentum even in countries with favorable resource endowments.

International financing trends within Mexico's renewable energy sector are reflective of trends in general emerging market renewable energy finance. Foreign preeminence over domestic sources of funding is both an indicator of international access and availability of capital for renewable energy investment as well as domestic financial market underdevelopment. The trend is ubiquitous among emerging nations but is an area where domestic financial market development could help to reduce dependence on foreign funding.

The conclusions of the study contribute to general knowledge on energy transitions in emerging economies by highlighting the complex interactions between policy frameworks, institutional capacity, market development, and stakeholder capacity. The Mexican case can provide other countries with lessons as they follow renewable energy transitions amidst competing development agendas and political contexts.

CONCLUSION

This in-depth research report has analyzed the intricate nexus between Mexico's energy policy reform shift to renewable and green energy and financial literacy among important stakeholders. The study provides sharp insights into the ways in which financial literacy impacts the effectiveness of energy policy implementation and renewable energy penetration outcomes within an emerging economy setting.

Economic and Social Impacts

Tracking the 45% clean electricity target by 2030 can lead to tremendous local economic gains, including the creation of more than 434,000 direct jobs in construction and long-term operation of renewable plants, and approximately $1.6 billion USD in annual savings in gas imports (Economics Observatory, 2025). Policy uncertainty, infrastructure needs, and changing regulatory signals have tempered private investment and slowed the development of the sector (Mexico Business News, n.d.).

Mexico's renewable energy sector provides a paradox of untold potential being held back by implementation challenges. With excellent renewable energy resources and establishing full legal frameworks, Mexico has seen sporadic progress toward its ambitious clean power goals. The government of Mexico's commitment to increase the share of renewable energies to 45% by 2030 ranks as one of the most ambitious renewable energy commitments in Latin America, yet achieving it requires surmounting critical implementation challenges.

The research shows the part financial literacy can play as an important mediating factor to the impact of renewable energy policy. Actors who have higher financial literacy levels demonstrate better understanding of renewable energy economics, better risk assessment ability, and higher participation in renewable energy investments. These findings contribute to the existing research on financial literacy by showing evidence of the part it can serve as a complementary policy tool in the effort of accelerating energy transitions.

Policy implementation analysis indicates that, although Mexico has established very detailed renewable energy structures, the capacity of stakeholders and political inconsistency along with coordination problems in institutions dilutes the effect. Such policy flip-flops under changing governments demonstrate how alternation of governments can compromise momentum in renewable energy development. The conclusion is critical in highlighting the necessity of cross-political cycle broad-based consensus on energy transition policies.

The finance assessment indicates that Mexico has been successful in attracting foreign investment in large-scale renewable energy projects, but there is a deficit in developing domestic financial markets and access to financing for small-scale investments. The aspect that the government has committed to investing $48 billion in its renewable energy industry indicates the scale of financial capital required for achieving renewable energy targets, while highlighting the need for innovative financing instruments that are able to mobilize public and private capital effectively.

Regional and sectoral variations in renewable energy development demonstrate the utility of context-sensitive strategies that account for local conditions, institutional capacity, and stakeholder profiles. The success of northern states in developing solar energy and the concentration of wind energy in a couple of states demonstrates how natural resource endowments are coupled with institutional determinants to shape patterns of development. Nevertheless, the sluggish rate of distributed renewable energy adoption indicates further policy intervention is needed to overcome market barriers to smaller-scale consumers and investors.

The research contributes theoretical knowledge by demonstrating the application of financial literacy concepts to energy policy situations and empirical support for the connection between financial literacy and renewable energy investment. The findings confirm institutional theory predictions for the impact of policy stability and institutional quality on investment attraction, but transferring behavioral finance understanding to renewable energy investment situations.

Policy Recommendations

Based on the results of the study, several policy recommendations emerge for Mexico's smooth transition to renewable energy through greater policy consistency and enhanced financial literacy. In the first place, an integrated financial literacy policy exclusively designed for renewable energy stakeholders could enhance the effectiveness of policy implementation. A selective program for policymakers, financial institutions, investors, and consumers with varying knowledge gaps identified in the study needs to be part of the policy.

Second, improving the predictability of policies through institutional change would improve investor confidence and long-term planning. This includes improving coordination mechanisms between government agencies, having formal regulatory processes, and developing policy frameworks that are longer than the political cycle. Establishing an independent renewable energy development authority with secure funding and clear mandates could provide institutional stability for the enforcement of renewable energy policy.

Third, the establishment of local financial markets for investments in renewable energy requires interventions that build institutional capacity as well as appropriate financial products. This includes building green finance capabilities among local banks, the creation of risk-sharing facilities for investments in renewable energy, and the growth of specialized financing institutions for sustainable energy development.

Fourth, regional and sectoral inequalities of renewable energy development require differentiated responses that are both responsive to local conditions and stakeholder needs. This includes designing specialized support programs for distributed renewable energy adoption, building financing infrastructure appropriate to small and medium-sized enterprises, and institutional capacity-building in areas where there is a potential renewable energy resource endowment but limited development experience.

Policy Recommendations. Integrate Financial Literacy in Energy Policy

Integrate educational components in all energy programs such that individuals, SMEs, and communities can make informed investment choices and interact with clean energy finance (NewClimate Institute, 2024). Promote Financial Inclusion: Prioritize vulnerable groups with tailored products and education to ensure equal access to the renewable energy transition (Economics Observatory, 2025). Enhance Transparency and Confidence: Surpass regulatory hurdles and ensure transparent tracking of funds to trigger private investment and reduce financing barriers (DLA Piper, 2016).

Implications for Future Research

This research has several avenues for future research. More studies are needed to establish causal relationships between financial literacy improvement and renewable energy adoption outcomes using longitudinal studies and randomized controlled trials. Understanding the specific mechanisms through which financial literacy affects energy investment decision making might offer better design for interventions.

Cross-country comparative studies that examine the role of financial literacy in energy transitions in different emerging economies can inform generalizability of findings and identify context-specific factors that may moderate such relationships. Such studies can inform best practices for financial literacy intervention in energy policy environments in different institutional and cultural contexts.

Additionally, future research should investigate the comparative effectiveness of different financial literacy intervention approaches in energy contexts. This entails relative effectiveness of general vs. specific financial literacy training schemes, online vs. offline channels of delivery, and individual vs. institutional capacity building approaches. Understanding what interventions yield the greatest impact for different stakeholder groups could enhance the efficiency of the use of resources.

The nexus between financial literacy and specific renewable energy technologies also remains to be explored in detail. Renewable energy technologies vary with their risk profiles, finance requirements, and markets that may respond to different levels of financial literacy in different manners. Technology-specific analysis would be helpful to inform targeted support strategies for cutting-edge renewable energy technologies.

Limitations and Future Directions

Some of the limitations of this research should be mentioned. The time window, focusing on the most recently elapsed period, can miss longer-term trends and connections that could be driving the stability of observed patterns. The cross-sectional nature of much of the data limits ability to determine cause-and-effect relations with certainty, suggesting that future research utilize longitudinal design.

The focus on the formal sector participants may not fully address the role of informal financial instruments and decision-making arrangements that could be important in some cases. Subsequent research should investigate the ways in which informal finance networks and traditional funding arrangements interact with formal renewable energy finance systems.

Measuring financial literacy, although based on tested survey instruments, may not exactly measure the special knowledge and skills best suited to renewable energy investment choices. Developing renewable energy-specific measurement instruments for assessing financial literacy can increase the validity of further research on the subject.

Final Reflections

Mexico's renewable energy transition is both a great opportunity and a very complex challenge that must be addressed by coordinated effort in multiple dimensions. Mexico has wonderful renewable energy resources and has put in place very inclusive policy frameworks, but tapping the full potential of these strengths depends upon overcoming implementation hurdles that go beyond the technical and economic to encompass institutional capacity, policy coherence, and stakeholder capacity.

Financial literacy emerges here as a key underutilized variable for renewable energy transformation. Increasing financial literacy among key actors offers a different pathway to traditional policy measures that may enhance the effects of existing renewable energy policy and provide a foundation for more sophisticated energy market growth.

The research demonstrates that effective renewable energy conversions do not only require effective policies and adequate resources but also institutional and human capabilities to implement the policies. Building this capacity requires ongoing effort and investment, but the payoffs in the form of facilitating faster deployment of renewable energy and improved energy security make the investments worthwhile.

Mexico's energy prosperity lies as much in robust regulatory and financial frameworks as in universal financial literacy. Empowered citizens and enterprises are key to unlocking renewables' potential and rendering the energy transition efficient and equitable. Financial Literacy and the Role it Plays in the Energy Transition Importance of Financial Literacy. Transitioning to renewable energy at the household or SME level involves assessing contracts, incentives, and finance—tasks that require basic financial literacy. The financially literate and entrepreneurs are more likely to think about returns, access clean energy finance, and utilize new offerings like green bonds or fintech platforms (NewClimate Institute, 2024). As financial instruments become more diversified, their understanding is key to extensive participation in the clean energy economy.

As Mexico seeks its way forward in the energy transition amidst economic growth, energy security, and environmental imperatives, wisdom and answers to the complex interactions among policy design, institutional capacity, and stakeholder capability will be crucial to successful execution. The outcomes of this research provide the foundation for future research and policy-making towards Mexico's renewable energy future.

REFERENCES

Aguirre, M., & Ibikunle, G. (2014). Determinants of renewable energy growth: A global sample analysis. *Energy Policy*, *69*, 374–384. DOI: 10.1016/j.enpol.2014.02.036

Ajzen, I. (1991). The theory of planned behavior. *Organizational Behavior and Human Decision Processes*, *50*(2), 179–211. DOI: 10.1016/0749-5978(91)90020-T

Bazán-Navarrete, M., & Bravo-Peña, L. C. (2021). Clean energy certificates market performance in Mexico: An empirical analysis of market liquidity and price formation. *Renewable Energy*, *162*, 107–119. DOI: 10.1016/j.renene.2020.08.025

Bazilian, M., & Pielke, R.Jr. (2013). Making energy access meaningful. *Issues in Science and Technology*, *29*(4), 74–79.

Bhattacharyya, S. C. (2011). *Energy economics: Concepts, issues, markets and governance*. Springer Science & Business Media. DOI: 10.1007/978-0-85729-268-1

Carlsson, B., & Stankiewicz, R. (1991). On the nature, function and composition of technological systems. *Journal of Evolutionary Economics*, *1*(2), 93–118. DOI: 10.1007/BF01224915

Climate Transparency. (2025, April). Assessment of financial instruments for Mexico's energy transition: Cases of FOTEASE and FIDE. https://www.climate-transparency.org/wp-content/uploads/2025/04/Assessment-of-financial-instruments-for-Mexicos-energy-transition_-cases-of-FOTEASE-and-FIDE.pdf

Economics Observatory. (2025, April 24). *How could green energy policy boost Mexico's economy?* https://economicsobservatory.com/how-could-green-energy-policy-boost-mexicos-economy

Ember. (2025, May 29). Renewables point the way to Mexico's energy security. https://ember-energy.org/latest-insights/renewables-point-the-way-to-mexicos-energy-security/

Fernández, A. de J., & Watson, J. (2022). Mexico's renewable energy innovation system: Geothermal and solar photovoltaics case study. *Journal of Cleaner Production*, *345*, 131264. DOI: 10.1016/j.jclepro.2022.131264

Geels, F. W. (2002). Technological transitions as evolutionary reconfiguration processes: A multi-level perspective and a case-study. *Research Policy*, *31*(8 9), 1257–1274. DOI: 10.1016/S0048-7333(02)00062-8

Goldthau, A., & Sovacool, B. K. (2012). The uniqueness and diversity of energy security. In *Handbook of global energy policy* (pp. 13–29). John Wiley & Sons.

González-Mahecha, R. E., Lecuyer, O., Hallack, M., Bazilian, M., & Vogt-Schilb, A. (2019). Committed emissions from existing and planned power plants in Mexico: A preliminary assessment. *Environmental Research Letters*, *14*(5), 054019. DOI: 10.1088/1748-9326/ab10c2

Hughes, T. P. (1983). *Networks of power: Electrification in Western society, 1880-1930*. Johns Hopkins University Press.

International Trade Administration. (2023, November 5). *Mexico - Renewable Energy*. https://www.trade.gov/country-commercial-guides/mexico-renewable-energy

Kahneman, D., & Tversky, A. (1979). Prospect theory: An analysis of decision under risk. *Econometrica*, *47*(2), 263–291. DOI: 10.2307/1914185

Lusardi, A., & Mitchell, O. S. (2014). The economic importance of financial literacy: Theory and evidence. *Journal of Economic Literature*, *52*(1), 5–44. DOI: 10.1257/jel.52.1.5 PMID: 28579637

Markard, J., Raven, R., & Truffer, B. (2012). Sustainability transitions: An emerging field of research and its prospects. *Research Policy*, *41*(6), 955–967. DOI: 10.1016/j.respol.2012.02.013

Mexico Business News. (n.d.). *Mexico's Energy Transition Progress Slows Down*: WEF. https://mexicobusiness.news/energy/news/mexicos-energy-transition-progress-slows-down-wef-0

NewClimate Institute. (2024, November). *Understanding finance needs for a just transition of the Mexican power sector*. https://newclimate.org/sites/default/files/2024-11/Report_Understanding%20finance%20needs%20for%20a%20just%20transition%20of%20the%20Mexican%20power%20sector_nov2024.pdf

North, D. C. (1990). *Institutions, institutional change and economic performance*. Cambridge University Press. DOI: 10.1017/CBO9780511808678

Organization for Economic Co-operation and Development. (2020). *OECD/INFE 2020 international survey of adult financial literacy*. OECD Publishing.

Peng, J., Patwary, A. K., Rasul, F., Koley, B., Nicomedes, C. J. C., Dutta, A., & Rahman, M. H. (2018). An empirical investigation of financial literacy and behavioral biases on investment decision: Evidence from Chittagong Stock Exchange. *Cogent Economics & Finance*, *6*(1), 1455723. DOI: 10.1080/23322039.2018.1455723

Pfeffer, J., & Salancik, G. R. (1978). *The external control of organizations: A resource dependence perspective*. Harper & Row.

Piper, D. L. A. (2016, January). *Mexico's new Energy Transition Law: key points*. https://www.dlapiper.com/en-ar/insights/publications/2016/01/mexico-new-energy-transition-law

Rodríguez-Monroy, C., Mármol-Acitores, G., & Nilsson-Cifuentes, G. (2018). Electricity generation in Mexico: Current status, future prospects, and economic implications of clean energy reforms. *Renewable & Sustainable Energy Reviews*, *92*, 1–17. DOI: 10.1016/j.rser.2018.04.077

Sánchez-Peña, J., García-Rendón, J., & López-Feldman, A. (2020). Renewable energy policies and economic growth in Mexican states. *Energy Journal*, *41*(2), 47–72. DOI: 10.5547/01956574.41.2.jsan

Silva, M., & Cerqueira, P. A. (2019). Assessing the determinants of household electricity prices in the EU: A system-GMM panel data approach. *Renewable & Sustainable Energy Reviews*, *73*, 1131–1137. DOI: 10.1016/j.rser.2017.02.016

Weber, C. (2017). What drives financial literacy and investment behavior? Evidence from survey data. *Applied Economics*, *49*(48), 4861–4878. DOI: 10.1080/00036846.2017.1296551

World Bank. (2020). *Mexico renewable energy development*. World Bank Group. https://openknowledge.worldbank.org/handle/10986/34587

Chapter 5

Climate Risk, Commodity Prices, and Sectoral Dynamics in Indian Financial Markets:

A Systematic Literature Review

Anusri Yadav
https://orcid.org/0009-0003-4377-9069
Christ University, India

Surabhi Patel
https://orcid.org/0009-0001-7171-9157
Christ University, India

Harsh Banthia
https://orcid.org/0009-0005-9058-8913
Christ University, India

Rajveer Samuel Rawlin
https://orcid.org/0000-0003-4359-3363
Christ University, India

ABSTRACT

According to the authors, this chapter provides a comprehensive review of more than the last twenty years empirical research on the relationship between commodity prices, sectoral stock indices and climate change induced financial risk in India. The chapter discusses the historical evolution of analytical techniques from simple linear econometric analysis to more recent machine learning algorithms by integrat-

DOI: 10.4018/979-8-3373-6766-8.ch005

ing 67 articles published in peer-reviewed journals between 2000 and 2025 using the PRISMA methodology. Key thematic results are large transmission effects of volatility between commodities and from these to sectoral markets, varying sectoral sensitivities to climate stress, and new trends in fintech and ESG commodity trading. Issues in climate risk integration, data innovation and policy frameworks are highlighted in the chapter. It provides a sound basis for climate-friendly financial strategies to help develop more resilient and sustainable capital markets in India.

INTRODUCTION

This systematic review analyzes over twenty years of research on commodity prices and sectoral indices in India, emphasizing climate-related financial risks. It traces the evolution of analytical methods from linear models to hybrid machine learning approaches and highlights emerging themes, including volatility spillovers, behavioral finance, and the shift toward fintech-enabled markets. The review identifies gaps and proposes a forward-looking agenda to integrate climate risk into sectoral financial analysis, offering insights for investors, policymakers, and market participants to foster resilient and adaptive markets.

India's high climate vulnerability exposes commodity-sensitive sectors—such as agriculture, energy, metals, and infrastructure—to significant risks, which are crucial for GDP and employment. Climate change drives volatility through physical channels, like extreme weather affecting production and logistics, and transitional channels, such as regulatory shifts and decarbonization policies. These climate-induced risks impact commodity prices, which in turn influence sectoral stock returns through complex cost structures and demand changes. Yet, studies on Indian financial markets have predominantly focused on traditional economic drivers, leaving blind spots in risk assessment, portfolio management, and policy planning.

In response, India has begun aligning financial policies with sustainable finance objectives, including RBI climate risk disclosures, priority lending for renewables, and the Climate Risk Information System (RB-CRIS). Concurrently, ESG-linked financial instruments and green commodities, such as lithium and carbon credits, are increasingly incorporated into conventional markets. Synthesizing evidence on climate risks, commodity prices, and sectoral financial markets using econometric and machine learning methods, along with novel datasets like satellite imagery and social media sentiment, provides a robust foundation for climate-aware financial strategies that support India's sustainable development goals.

METHODOLOGY

This systematic literature review was conducted following the PRISMA (Preferred Reporting Items for Systematic Reviews and Meta-Analyses) framework, a widely recognized standard for ensuring transparency, rigor, and reproducibility. Using PRISMA enabled a structured approach to identifying, screening, and synthesizing relevant peer-reviewed studies on climate risk, commodity prices, and sectoral indices in Indian financial markets.

2.1 Data Sources and Search Strategy

The main data source was the Scopus database, chosen for its extensive multidisciplinary coverage of economics, finance, environmental sciences, and data analytics, including peer-reviewed journals and conference proceedings. The search spanned 25 years (2000–2025) to capture changes in Indian financial markets following liberalization, regulatory reforms, and recent developments in climate-aware finance. Keywords covering key domains such as 'climate risk', 'commodity prices', 'sectoral indices', and 'Indian financial markets', along with related terms like 'volatility spillover', 'cointegration', 'machine learning', and 'behavioral finance', were used with Boolean operators. Each search query underwent a systematic and iterative refinement process to ensure sensitivity and specificity in identifying studies on the empirical and theoretical aspects of the climate–commodity–sector nexus.

2.2 Inclusion and Exclusion Criteria

To ensure thematic coherence and methodological rigor, strict inclusion criteria were applied. Studies were included if they focused on the Indian financial market and examined links between commodity price movements and sectoral stock indices, using econometric or machine learning methods to analyze volatility, causality, or forecast sectoral returns under climate change scenarios. Excluded were studies unrelated to India, purely theoretical models without empirical validation, articles lacking methodological detail or data quality, and studies that did not disaggregate stock market data into sectoral indices. Non-peer-reviewed articles, opinion pieces, and non-English publications were also excluded. This approach ensured that only relevant, high-quality studies were synthesized for the review.

2.3 Screening and Selection Process

The screening process was conducted following the PRISMA 2020 flow diagram to ensure full transparency. An initial database search yielded 1,257 records using the keywords and filters outlined in our methods. After removing 223 duplicates, two researchers assessed the relevance of 1,034 unique articles based on titles and abstracts, excluding studies unrelated to the research focus, examining other markets, or lacking empirical results.

A total of 142 studies then proceeded to full-text screening, evaluated against a pre-defined checklist assessing study aims, dataset adequacy, methodological rigor, and analytic quality. Reviewer discussions resolved any disagreements, with a third reviewer consulted when necessary to minimize bias.

Ultimately, 67 studies met the inclusion criteria and were selected for comprehensive qualitative synthesis, forming the empirical foundation for this review.

2.4 Data Gathering and Data Analysis Procedures

An extensive data extraction protocol was developed to systematically capture relevant information from each selected study. Extracted variables included authors, publication year, data collection periods and frequency, types of commodities and industry indices examined, econometric or mathematical techniques used, summarized results, and reported limitations. For studies in agricultural and rural development contexts, corresponding data synthesis and analysis procedures were tailored to account for variations in study design, datasets, and methodologies. The data were then thematically synthesized based on recurring literature themes, including volatility spillovers, sectoral policy vulnerabilities, methodological developments, behavioral factors, and the impact of climate variables and policies.

2.5 Analytical Methods Used in the Selected Literature

The literature review indicates a clear evolution in analytical methods over time. In the first period (2001–2010), studies primarily relied on linear statistical models, such as correlation and regression, with a focus on price relationships in energy and banking commodities. While these models provided initial insights, they lacked the sophistication to capture dynamic volatility or asymmetric responses in economic and financial data influenced by climate variability.

From 2011 onwards, the literature increasingly employed more advanced econometric techniques. Autoregressive Distributed Lag (ARDL) models, and their nonlinear counterpart (NARDL), became popular for modeling long-run equilibrium and short-run adjustments with asymmetric features. Volatility spillovers were analyzed using Generalized Autoregressive Conditional Heteroskedasticity (GARCH) models, particularly Dynamic Conditional Correlation GARCH (DCC-GARCH), which enabled assessment of time-varying co-movements and heteroskedastic effects. Vector autoregression (VAR) and cointegration analyses further enhanced understanding of cross-market dynamics and the persistence of price relationships across international commodity and stock markets.

More recently, machine learning methods such as Long Short-Term Memory (LSTM) networks and Transformer-based models have been applied to capture nonlinear dependencies and regime-switching behavior not adequately addressed by traditional econometrics. Bayesian estimation methods and Seemingly Unrelated Time Series Equations (SUTSE), along with other multi-equation econometric models, have facilitated the simultaneous modeling of interconnected variables, reflecting the globalized linkages among commodities, macroeconomic indicators, and financial markets.

2.6 Reducing bias and Quality Assurance

The studies that were included were appraised based on the rigor of the methodological approaches, the degree of trustworthiness of the data, the openness regarding model assumptions, and appropriateness of the inferences made in relation to the bias introduced. In our synthesis, studies were appraised for bias due to sample selection, publication bias and methodological limitations due to study design, sources of bias we would categorize as being substantive. This allowed us to provide synthesized conclusions based on quality evidence of a substantive nature and trustworthy.

2.7 Methodological Limitations

While the systematic review follows a PRISMA-scoped, rigorous, and transparent methodology, several limitations should be noted. First, the literature search was restricted to peer-reviewed English publications from the last ten years, which may omit relevant research from other regions or non-English sources, potentially limiting the generalizability of findings. Second, the heterogeneity of datasets and methods in the included studies, coupled with the lack of usable quantitative data for meta-analysis, may constrain interpretation and require caution when considering synthesized quantitative results. Third, as climate finance and fintech literature evolves rapidly, some emerging trends may not be captured in the selected literature, meaning future studies will need to update and refine the insights derived from this longitudinal review.

Figure 1. PRISMA Flow Diagram: Study Selection Process

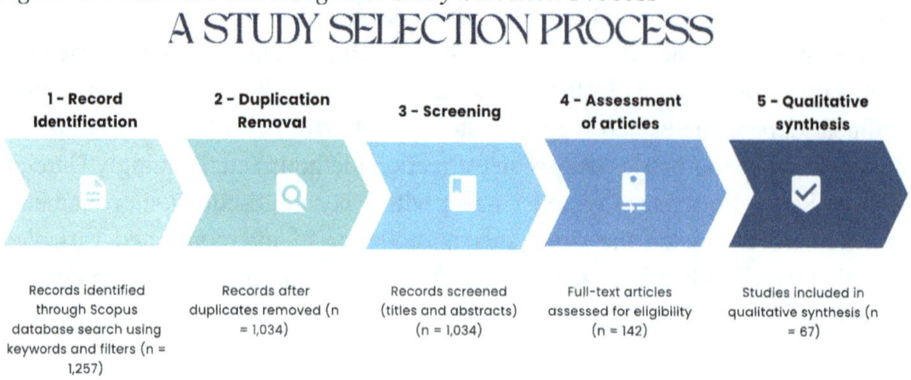

THEMATIC FINDINGS

1. Volatility Spillovers and Price Transmission

1. This study investigates herding and spillover effects in Indian commodity futures markets, analyzing their relationship with sectoral stock indices. It finds evidence of herding behavior amplifying volatility and causing price transmission between commodities and equities. The findings emphasize behavioral and market integration effects in Indian financial markets. (Dewan & Dharni, 2022)
2. The article examines the dynamic relationships between crude oil and gold prices and various Indian sectoral stock indices using ARDL models. It identifies both long-run and short-run co-movements indicating that global commodity prices significantly impact domestic sectoral stock performance. These insights have implications for investors and policymakers regarding market interdependencies. (K. K. Kumar et al., 2024)
3. Applying VAR-DCC-GARCH models, this research explores return and volatility linkages between international crude oil, metals, and Indian stock indices. It reveals complex bidirectional spillovers, with volatility increasing notably during economic stress periods. The study underscores the vulnerability of Indian markets to fluctuations in global commodity prices. (Singhal & Ghosh, 2016)
4. The article investigates the volatility spillover effects and hedging opportunities among oil-importing Asian countries in periods of crises. It documents significant transmission of shocks from crude oil prices to financial markets and highlights the importance of hedging strategies for mitigating risk under

market stress. The findings are critical for policymakers and market participants in oil-dependent economies. (Sarwar et al., 2019)

5. This research predicts volatility and investigates dynamic relationships among the stock market, exchange rate, and selected commodities in India. It finds substantial interdependencies and the spillover effects are strengthened during financial volatility. The findings illustrate the relevance of commodities in fluctuations of the larger financial market as well as the practical implications for risk management and forecasting. (Siddiqui & Roy, 2019)
6. This paper studies spillover effects between commodity and equity benchmarking indices in India, revealing strong bidirectional relationships. It shows that price shocks in one market segment often quickly transmit to the other, reducing diversification benefits for investors. The research is relevant for both market participants and regulators. (Kirithiga et al., 2018)
7. Investigates financial contagion and volatility spillover within the Indian commodity derivative markets, particularly during global financial turmoil. Results reveal that shocks propagate swiftly across market segments, amplifying market instability and risk. The findings highlight the system-wide vulnerabilities inherent in commodity derivative trading. (Roy & Sinha Roy, 2017)
8. This article studies the dynamic linkage between the Indian stock market and commodity market, detecting co-movements triggered by macroeconomic developments and global shocks. It emphasizes that changes in commodity prices influence stock market behavior, affecting trading and investment decisions. (Jain & Biswal, 2016)
9. This paper assesses the resultant influence of the 2008 financial crisis on structural breaks and volatility dynamics in the Indian commodity markets. The paper displays structural breaks in volatility technologies resulting in increased volatility and changed correlation dynamics during crisis periods, which emphasizes commodities' heightened vulnerability to shocks in the global financial system. The study offers insights for crisis mitigation and stabilization for market participants. (Shalini & Prasanna, 2016)
10. This foundational paper presents bounds testing approaches for analyzing long-term relationships in time series data, including financial commodities and stock markets. Its methodology allows for testing cointegration irrespective of variable integration order, making it versatile for empirical market studies. (Pesaran et al., 2001)
11. Offers an analytical framework assessing how Indian commodity markets relate to stock markets, with special focus on commodity exchanges in India. It highlights structural linkages, regulatory environment, and market behavior, facilitating in-depth understanding for academic and policy applications. (Periasamy & Satish, 2015)

12. Proposes a hybrid stock prediction method combining periodic and non-periodic feature analyses to enhance forecasting performance. The model systematically distinguishes between cyclical patterns and irregular fluctuations, significantly improving prediction accuracy for stock prices. This approach offers practical advantages for investors and analysts. (Zhao et al., 2025)
13. This paper uses a Bayesian estimation approach to study relationships among financial markets, commodity markets, and economic growth. It finds interconnected dynamics influencing growth trajectories, highlighting feedback loops between commodities and broader financial sectors. The results inform policy design aimed at stabilizing economic growth. (Pastpipatkul et al., 2024)
14. Investigates volatility transmission between Indian equity and commodity markets, concluding that commodities often act as primary transmitters of volatility shocks. The paper emphasizes understanding this transmission for enhanced risk management and portfolio diversification in Indian markets. (H. Kaur & Chaudhary, 2024)
15. Uses wavelet analysis to study asymmetric volatility spillovers between oil and Indian sectoral stock indices. It reveals that impact varies across sectors and time horizons, reflecting complex investment dynamics and requiring sector-specific risk management approaches. (Vairasigamani & Amilan, 2025)
16. Explores the connectedness of oil prices to metal prices across both time and frequency domains. The study finds periods of heightened integration, demonstrating time-varying relationships crucial for portfolio diversification and commodity price forecasting. (Umar et al., 2019)
17. Examines time-frequency causality and connectedness among international prices of energy, food, agriculture, and metals. The findings indicate that commodity markets are interconnected across various time scales, helping investors understand causal influence and manage portfolio risks effectively. (Tiwari et al., 2020)
18. This chapter uses Bayesian estimation of seemingly unrelated time series equations to uncover interactions between financial and commodity markets and their contribution to economic growth. The study introduces optimal transport techniques for analyzing complex economic data and demonstrates how linking multiple markets can yield new insights into macroeconomic dynamics. It provides innovative applications of machine learning and Bayesian analysis in financial econometrics. (Pastpipatkul et al., 2024)

2. Sector-Specific Sensitivities

19. Analyzes the sensitivity of Indian stock indices sector-wise to fluctuations in global commodity prices. It finds varying degrees of exposure across sectors,

with energy and metals sectors showing high sensitivity. The study provides insights for portfolio diversification and sector rotation strategies. (Goel et al., 2025)

20. Examines the impact of crude oil and metal prices on various Indian stock market sectors using empirical analysis. It identifies significant spillover effects, particularly affecting heavy industry and manufacturing sectors. The research provides valuable information for corporate risk management and investment decisions. (Jha & Katiyar, 2024)
21. Studies asymmetric impacts of commodity price changes across Indian sectors, revealing non-uniform sectoral reactions to price volatility. Infrastructure and consumer goods sectors are particularly sensitive to asymmetric shocks. This understanding aids precision portfolio management and risk mitigation across sectors. (Singh & Padmakumari, 2020)
22. Investigates the long-run relationship between commodity price shocks and sectoral stock returns in India. It confirms a strong, persistent integration between commodity price fluctuations and stock market performance at the sectoral level. The findings aid strategic asset allocation and risk assessment. (Pranchana et al., 2025)
23. Focuses on how geopolitical risks differentially impact Indian sectoral stock indices. It quantifies sectoral vulnerabilities, especially in finance and energy sectors, highlighting the importance of geopolitical risk management. This research offers crucial insights for scenario planning in volatile environments. (Chauhan et al., 2025)
24. Studies sectoral impacts of commodity price volatility in emerging economies with India as a case study. The research finds divergent sectoral responses with inflationary pressures and credit market volatility affecting some sectors more. The insights deepen understanding of commodity price transmission mechanisms in emerging markets. (Chauhan et al., 2025)
25. Assesses the differential impacts of COVID-19 on stock and commodity price markets in South Asia, focusing on sectoral variations. Technology and healthcare sectors show resilience, whereas traditional sectors face more disruptions. The findings help post-pandemic recovery planning and investment decision-making. (Ahmed et al., 2021)
26. Explores input-output linkages to understand sectoral stock responses to commodity price shocks. Findings show sectors heavily reliant on raw materials are more sensitive, highlighting supply chain effects on stock returns. This aids corporate strategy and market analysis in anticipating shock transmission. (Sahadudheen & Kumar, 2023)
27. Focuses on cyclical volatility spillovers and interrelations between oil, gold, forex, and stock markets during the COVID-19 pandemic in India, and shows

dynamic relationship patterns that evolve during times of global crises, and critical implications for hedging and investment strategies as conditions change. This research stream is certainly a vital pathway for risk management in the disciplined approach of period of economic upheaval. (Jena et al., 2021)

28. Analyzes asymmetric relationships between trading activity and commodity futures prices in India using the Quantile Nonlinear Autoregressive Distributed Lag (QNARDL) method. It documents complexity and varying dynamics across commodities and time, assisting in improved market liquidity and risk forecasts. (Sehrawat & Giri, 2017)
29. Examines the effect of stock market development on growth across sectors in India, provides empirical evidence of positive links between sectoral growth and stock market development, with particular attention to manufacturing, finance, and services. The study concludes with the expected effect with regard to policy to deepen equity markets to promote sectoral growth. (Datta & Hatekar, 2025)
30. Investigates volatility spillover across sectoral stock indices during COVID-19 in India, detecting significant range volatility transmission among key sectors. It highlights sectors that intensify or absorb volatility, aiding portfolio risk management under pandemic-induced stress. (G. Kaur & Dhiman, 2024)
31. Reviews recent fluctuations in the Indian commodity market, attributing volatility to changes in global demand, domestic policy shifts, and currency movements. It provides a snapshot of factors driving commodity prices and their sectoral impacts. (Keswani & Wadhwa, 2019)
32. Analyzes the fundamental macroeconomic drivers of price dynamics in the Indian stock market, emphasizing factors like output, liquidity, and credit supply at the sector level. The study guides investment decisions and policy design by linking economic fundamentals with stock market performance. (N. Kumar & Goel, 2024)

3. Machine Learning and Forecasting Models

33. This document presents a forecasting model of a financial market in a time series setting using machine learning methods that include sophisticated deep learning techniques. It shows improved accuracy in predicting market behavior due in part to the ability to model complex temporal dependencies and to embed different financial signals. The model offers a practical framework for better forecasting performance in fluctuating markets. (Geetha et al., 2024)
34. Utilizes a neural network approach based on simple linear regression incorporating external factors to predict stock exchange movements. The study shows that integrating macroeconomic indicators and commodity prices enhances forecast

accuracy and trading profitability. This combined approach supports data-driven decision-making in stock market investments. (O'Connor & Madden, 2005)

35. Applies machine learning techniques to analyze fluctuations in commodity prices and forecast future trends with high accuracy. The study focuses on sectoral differences and models complex nonlinear relationships to enhance prediction capabilities. These insights assist investors and policymakers in managing commodity risks and opportunities. (Sarangi et al., 2025)
36. Investigates modeling and evaluation of stock market and commodity price fluctuations via time series sequence learning. It compares neural network methods with traditional approaches, demonstrating superior predictive accuracy and adaptability of machine learning models. The research highlights the potential for enhanced forecasting in financial markets. (Veningston et al., 2025)
37. Proposes a hybrid stock prediction method combining periodic and non-periodic feature analyses to capture both cyclical and irregular market patterns effectively. The approach shows improved forecasting performance over traditional methods by integrating diverse data characteristics. It offers a comprehensive tool for dynamic stock market analysis. (Zhao et al., 2025)
38. Explores the effects of macroeconomic variables on the Indian stock market through the means of statistical and machine learning models. This study plays a very important claim that introducing statistical and machine learning models improves prediction capabilities and gives more consistent estimates of volatility estimates, especially during economic shocks. Overall, the study has implications for developing robust trading strategies. (Keswani & Wadhwa, 2019)
39. Introduces a novel stock price forecasting approach combining natural language understanding-based sentiment analysis with deep learning LSTM networks. It shows that integrating market sentiment improves forecasting accuracy by capturing behavioral finance aspects. This methodology enhances short-term market movement predictions. (U. Mishra & Dash, 2025)
40. Examines how wealthy individual investors affect tail risk in stock markets using machine learning methods. The study shows that the trading activities of wealthy individual investors can exacerbate extreme market risks, which is useful for regulators and risk managers. The research enhances the understanding of market microstructure and tail behavior. (Yu et al., 2024)
41. This article discusses the application of machine learning (ML) algorithms to predict trends in financial markets through the analysis of time series data. Specifically, the study focuses on factors affecting the accuracy and effectiveness of various ML models in predicting future financial market movements. Furthermore, the article describes the importance of data preprocessing and feature selection to improve forecasting ability in financial settings. (Geetha et al., 2024)

42. The study evaluates the ability to forecast stock market index upward or downward action through neural networks, and then check their external indicators like currency rates or commodity prices. The approach incorporates the training of neural networks with external inputs, improving the returns and predictive accuracy beyond that achieved using internal market components alone. The paper demonstrates that return on investment using these models can outperform simple market benchmarks. (O'Connor & Madden, 2005)

4. Macroeconomic and External Factors

43. Examines the nonlinear relationship between major economic indicators—like gold, crude oil, and exchange rates—and the Indian capital market. The study finds that these relationships fluctuate over time and are not always straightforward, reflecting the complex interactions underlying India's financial system. Such insights are important for policy and forecasting. (Nautiyal & Kandpal, 2025)
44. Examines both the long-run and short-run implications of macroeconomic determinants- inflation, money supply, and exchange rates on stock prices in India. It notes that while inflation and exchange rates exhibit a negative relationship with stock prices in the short-run, money supply has a positive effect in the long-run. The results guide both policy and investment decisions. (K. K. Kumar et al., 2024)
45. Investigates the impact of economic events on Indian stock market returns, finding that major events—including policy announcements and global crises—lead to severe but often temporary shocks. The research reveals patterns of recovery and long-term stability, with implications for risk management. (*The Impact of Economic Events on Stock Market Returns: Evidence from India | Asian Economic and Financial Review*, n.d.)
46. Measures information share and its predictability in the Indian stock market, showing that some stocks and sectors dominate in terms of market information flow. The paper demonstrates that information is not evenly distributed, impacting stock price leadership and market efficiency. (Karmakar & Inani, 2019)
47. Analyzes how the Indian stock market responds to inflation announcements, with sector-specific focus. Sectors such as banking and FMCG react more visibly and rapidly to inflation news, affecting investor strategy around economic releases. (Singh & Padmakumari, 2020)
48. This paper investigates the impact of market sentiment on contemporaneous stock returns in India, using survey-based and quantitative sentiment score method. The results are consistent in detecting a significant relation between moving sentiment and near-term changes in stock price indicating a non-rational

or behavioral trading aspect of investors. The research document greater impacts of sentiment during uncertain times and have significant implications for behavioral and sentiment-based trading within finance. (Aggarwal & Mohanty, 2018)

49. Focuses on the role of wealthy individual (high net worth) investors in driving tail risk extreme, rare events—in stock markets. The paper demonstrates that trades and behavior by such investors can trigger outsized price swings and liquidity shocks. Using detailed trade data and machine learning, the authors document systemic vulnerabilities. The results suggest enhanced system monitoring and tailored regulation for large private traders. (Yu et al., 2024)
50. Studies cross-regional connectedness of financial markets and the economic as well as structural determinants underlying such linkages. Using a range of econometric and network analysis tools, the research finds that fundamentals like GDP growth, inflation, and trade openness drive regional market interconnectedness. This connectedness can amplify contagion and volatility flow across regions. The paper is key for international diversification and systemic risk evaluation. (Yang et al., 2024)
51. Models tail risk in the Indian commodity markets using conditional extreme value theory (EVT) and Value-at-Risk (VaR). The research shows commodity price dynamics are heavily skewed, with rare but extreme events impacting both commodity and equity markets. Efficient risk estimation methods are developed to anticipate and plan for such events. The findings are important for institutional risk managers and market participants. (Agnihotri & Chauhan, 2022)
52. Analyzes the financialization of commodity market activity in India and explains how increased financial activity has changed the price dynamics, volatility, and structure of the market. The analysis tracks institutional participation and speculative inflows, new risk, and opportunities for market participants. The results suggest increased volatility for traditional hedgers, but more avenues for arbitrage and investment for new players. The results inform policy design for a balanced growth of the market. (Rodrigues & Vinnarasi, 2023)
53. Investigates how exchange rate volatility affects Indian exports at a commodity level by employing panel data analysis techniques. The findings highlight that certain export commodities like textiles and chemicals are more vulnerable to currency fluctuations than others such as IT services. These sector-specific insights facilitate better hedging, pricing, and policy intervention by exporters and trade authorities. The article advances export risk management in emerging economies. (Sharma, 2020)
54. Employed ARDL barring the examination of causal relationships between stock prices and macroeconomic variables in a short and long run context in India. Reports strong evidence of causality from macro variables to stock

prices, indicating predictive value of these variables in the processes of asset management. Type of modelling permits both tactical selection of sectors and market timing decisions. Results also indicate that a potentially enhanced and coordinated monetary and fiscal policy response may result in more stable stock price patterns over time. (Singh Chauhan et al., 2025)

5. Commodity-Specific Insights

55. Investigates the volatilities spillover nexus among green cryptocurrencies, renewables equity stocks from sector stocks, and agricultural price commodities. The findings demonstrate the significant bidirectional spillovers, highlighting how green assets, renewables, and agricultural prices are affecting each other in allocation strategies. This suggests practical applications in portfolio hedging and allocation of sustainable investment portfolios in new asset classes. Overall, the results suggest these new market linkages may lead ESG investors to rethink their risk and return expectations. (R. Kaur et al., 2025)
56. Explores volatility spillover dynamics between fintech firms and traditional financial institutions using high-frequency data from Chinese listed institutions. Evidence suggests that volatility spillovers are strong, asymmetric, and time-varying, particularly during periods of financial shocks and regulatory changes. The analysis identifies rich determinants such as the innovation cycle and systemic risk within the interconnected financial sector. These findings underscore the importance of adaptive risk controls as the boundaries separating fintech and traditional finance become increasingly blurred. (Liu et al., 2025)
57. Looks at the connectedness of sectors, the volatility spillover and hedging opportunities in the Indian stock markets through sophisticated econometric methods. The analysis shows constant volatility spillover across and within sectors, which challenges conventional notions of diversification. The analysis shows that hedging at the sector level can be significantly improved using a dynamic correlation model. The outcomes offer practical insights for Indian portfolio and risk managers in a turbulent market environment. (Maurya et al., 2025)
58. Investigates seasonality in various Indian asset classes—including commodities, equities, and bonds using robust index construction. The research uncovers pronounced periodic trends in trading and returns, helping investors optimize market entry and exit points. It offers a clear statistical basis for timing strategies, especially in agricultural and energy commodities. Seasonality is found to be asset-dependent and shaped by economic cycles, weather, and policy events. (Kaushik & Nagarkar, 2023)

59. Examines the relationship between sectoral indices and the overall stock market performance in India, especially in the National Stock Exchange. We identify sectoral leaders and laggards using time-series econometric models. Our results suggest that sectoral indices significantly spill over into market returns and can help in constructing portfolios based on Indian sectoral momentum. Our findings can be useful to inform strategic allocation between cyclical and defensive sectors. (Pranchana et al., 2025)
60. Exploits Granger causality and cointegration analysis to model the multifaceted relationships between the South African rand and multiple commodity prices. The results indicate price shocks and currency movements are interrelated over time with bidirectional causality in several sectors. Such analysis is important in emerging markets where commodity-based economies and unstable exchange rates exist. The insights can be applied to other similar contexts, such as India. (Ndlovu et al., 2014)
61. The relationships between central bank digital currencies (CBDCs), stock market indices, commodity futures (metals and oil), and volatility are investigated through a series of empirical tests. The findings show that there are significant relationships, which indicates that the volatility of commodities and safe haven assets can impact the stability and attractiveness of CBDCs. The implications for policymakers designing a regime for digital currencies is that they should account for these spillover channels. This study incorporates the increasing conversation about CBDCs with analysis from traditional asset markets. (Memon & Nusratova, 2025)
62. Examines stock price synchronicity and liquidity in international markets, focusing on how commodities fit into stock prices. Evidence shows more synchronicity, or co-movement, means lower liquidity, as well as greater underlying systemic risk. Commodities can offer diversification against and contagion to equities depending on the conditions in the equities market. The findings offer new benchmarks for assessing market quality and for risk management. (Brockman et al., 2024)
63. Estimates the relative effects of raw material prices, sector outlook, and overall market sentiment on stock prices using advanced econometric approaches. The research shows raw material price shocks strongly predict sectoral stock returns, especially in commodity-linked industries. Moreover, shifts in sectoral outlook and sentiment can amplify or dampen these effects. These insights help investors and analysts tailor sector-level trading and hedging strategies. (Ghosh et al., 2021)
64. Reveals complicated asymmetric linkages between trading activities (volume, open interest) and commodity futures prices in India, employing QNARDL modeling. Findings show that the strength and direction of the linkage is condi-

tionally influenced by the trading regime, commodities type, and price movement direction. This interpretation of the linkage allows for more informed trading and risk management in commodity derivatives. Further, regulators may use these findings to understand speculative trading in the market and how sound the market is. (Jena et al., 2021)

65. Provides empirical evidence on the financialization of Indian commodities, demonstrating that financial participants now play a bigger role in price discovery and volatility. The paper uses multiple asset classes, showing elevated co-movement between commodities and equity indices over time. Findings highlight increased systemic risk but also new market opportunities for diversification and portfolio innovation. The research is important for regulators, investors, and hedgers. (Shamsher, 2021)
66. Examines the lead-lag relationships between international gold prices and the NSE Nifty 50, using both correlation and cointegration techniques. The results of the study suggest episodes of significant co-movement, indicating that gold may play an influential role as a hedge or precursor to movements in equity values, in India. Important implications of the research are considered for multi-asset portfolio management in unstable times. This study is of particular relevance to investors looking for diversification and a safe haven. (Sidhu & Katoch, 2021)
67. Examines threshold cointegration and price transmission in Indian commodity markets, finding asymmetric corrections to price shocks and the existence of non-linear adjustment processes. The evidence shows that large commodity price changes lead to faster and larger price adjustments than small price changes. Allowing policymakers and analysts to build on this methodology provides a new way to model the relationship between key commodity prices in emerging market economies. Implications also include improved early warning signals of a market destabilization. (A. Mishra & Kumar, 2023)

Figure 2. Framework Depicting the Interplay of Climate Risk, Commodity Prices, and Sectoral Dynamics in Indian Financial Markets

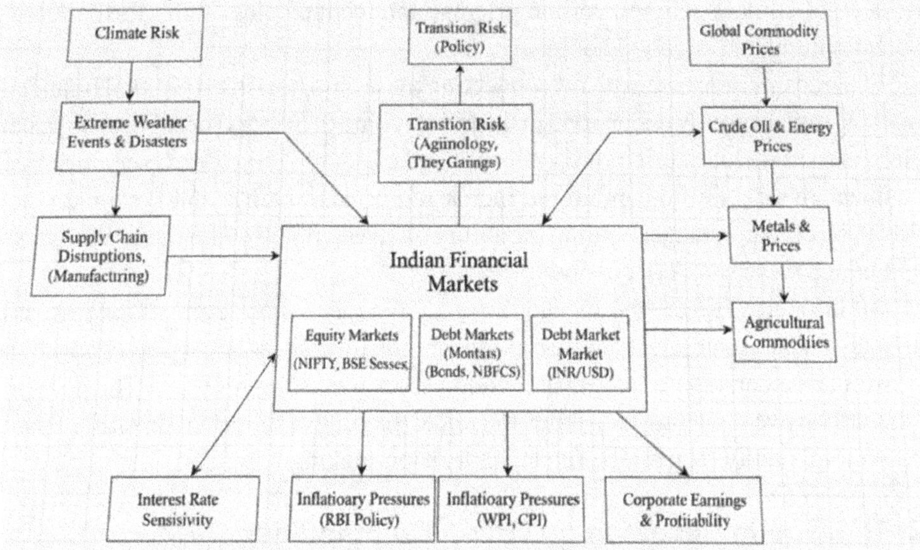

RESEARCH GAPS

The thorough literature review surrounding climate risk, commodities prices, and consequences for sectoral dynamics in Indian financial markets illustrates substantial and multi-faceted research gaps. The gaps are not due to lack of opportunity but shortcomings in theoretical approaches of study, data, methods, and the advancement of climate finance. Recognizing and addressing these gaps will be an important step toward progressing the field due to India's reliance on commodity sensitive sectors, and its growing exposure to climate change. This section will further describe these research gaps.

1. Climate risk is not a core element to commodity-sector models

A major gap in the literature is the inadequate integration of climate risk as a direct factor in models linking commodity prices and sectoral equity returns. While extensive research examines the transmission of commodity prices to Indian sectoral indices, most studies either ignore climate risk or treat it indirectly through macroeconomic or financial proxies. Physical climate risks—such as floods, droughts, unseasonal rainfall, and temperature shocks—directly affect commodity availability, costs, and price volatility, particularly in agricultural and energy sectors. Transition risks from climate policies, carbon pricing, and technological shifts further compound natural and economic climate risks.

The transmission mechanisms connecting these climate risks to sectoral financial markets remain poorly understood in the Indian context. Sectors sensitive to physical climate risks include agriculture, energy, metals, and infrastructure. Oversimplified or linear models fail to capture the increasing price volatility and systemic risks driven by climate change, limiting the ability of investors and policymakers to assess and hedge these risks effectively.

Additionally, climate risk extends beyond immediate shock events. Downstream effects—such as supply chain disruptions, regulatory uncertainty, and changes in consumer demand—are inadequately addressed in existing research. This lack of comprehensive modeling constrains the understanding of climate risk's full impact on sectoral equity returns and financial decision-making.

2. The Gap In Research on Green Commodities and the ESG Sectoral Context

Another notable gap in research is a relative consideration and in-depth study of green or sustainability-linked commodities and their connection to general emerging ESG (Environmental, Social, and Governance) indices. The new sustainability transition is elevating commodities such as lithium, cobalt, hydrogen, carbon credits, and biofuels which affect future financial returns across sectors.

India has implemented ambitious climate policies targeting renewable energy expansion, electric vehicle production, and carbon emissions reduction. Yet, the impact of these policies on sectoral stock markets via commodity price channels remains largely unexplored. Empirical research—both in India and internationally—on how green commodity price volatility affects sector-level returns, particularly for renewable energy and ESG-focused funds, is limited. Understanding the influence of climate finance transitions on portfolio-level outcomes represents a critical blind spot in the literature.

Most existing studies in financial economics focus on traditional commodities such as crude oil, gold, and agricultural products, with minimal attention to green assets or ESG-linked instruments. Examining the connectivity between these emerging commodities and sectoral ESG indices could provide practical insights for portfolio diversification, risk management, and policy-driven market behavior.

Furthermore, the data supporting financial transition risks and opportunities arising from India's climate strategies remain scarce and poorly structured, underscoring the need for improved empirical research to guide investors, policymakers, and market participants.

3. Limited Application of High-Frequency and Alternative Data Sources

The majority of recent India focused sectoral studies on commodity prices and sectoral indices use daily or monthly data, which captures long-term trends but fails to reflect rapid return reactions to climate or geopolitical shocks, intraday volatility spikes, or short-term spillover effects. This limited frequency prevents observation of phenomena such as liquidity surges or flash crashes triggered by commodity price shocks exacerbated by climate events.

The situation is further constrained by the underutilization of alternative high-frequency datasets, including real-time environmental or social media sentiment analytics, satellite data on crop and environmental health, IoT sensor data on weather and pollution, carbon emissions tracking, and news analyses. Non-traditional asset pricing models have also rarely been applied in a sectoral context. Such datasets could provide early signals of commodity price fluctuations and sectoral vulnerabilities, especially given the unpredictable nature of climate events.

Applying machine learning or big data analytics to these diverse datasets could enhance prediction accuracy, uncover hidden patterns, and generate early warning signals for price developments. The current underuse of such data represents a missed opportunity to improve risk mitigation, inform investor decisions, and strengthen resilience to climate- and policy-driven market shocks.

4. Methodological Restrictions: Predominance of Linear Models and Absence of Hybrid Approach

Most studies investigating commodity sector linkages continue to rely on conventional econometric techniques, including Autoregressive Distributed Lag (ARDL), Vector Autoregressive (VAR), and Generalized Autoregressive Conditional Heteroskedasticity (GARCH) models, which are largely linear in nature. While these methods are effective for establishing long-term relationships and assessing volatility under stable conditions, they fail to capture nonlinearities, regime dependencies, and asymmetric responses that are increasingly relevant in agricultural and other commodity markets.

Commodity price reactions to extreme weather events or sudden policy changes are generally nonlinear, with impacts varying based on market behavior and the scale of the adverse event. Relying exclusively on symmetric, constant-parameter models can underestimate true risk, especially during periods of heightened climate uncertainty.

Few studies employ hybrid methodologies that combine machine learning algorithms—such as Long Short-Term Memory (LSTM) networks, transformer-based models, reinforcement learning, or Bayesian hierarchical methods—with classical econometric models. Such hybrid approaches offer significant potential to model complex, state-dependent dynamics and structural breaks in commodity markets under climate-related risk and volatility. The limited application of these interdisciplinary methods constrains the literature and reduces its practical relevance for investors and policymakers.

5. Neglecting Behavioral Finance and Market Sentiment Factors

Behavioral finance factors such as investor sentiment, herding behavior, and panic selling, play a critical role in amplifying volatility spikes and contagion during crises, yet Indian financial market studies rarely integrate these aspects into commodity–sectoral price and risk models. Climate shocks or other sudden events can rapidly shift investor sentiment, triggering sell-offs and cross-sector contagion that often exceed what fundamentals alone would justify.

Narratives driving behavioral risk, amplified through social media, news reports, and market perceptions, can influence investor decisions, particularly during periods of climate or geopolitical uncertainty. However, the literature seldom applies sentiment analysis, neural network–based textual mining, or behavioral index metrics, limiting the ability to predict contagion patterns or recovery trajectories.

Developing frameworks that integrate behavioral finance insights with commodity and sectoral data would improve both the explanatory and predictive power of models, especially under conditions of market stress and climate-related uncertainty.

6. Insufficient Investigation of Emerging Asset Classes and Digital Financial Innovation

The rapid growth of digital financial instruments—such as decentralized finance (DeFi), blockchain-based commodity tokenization, and Central Bank Digital Currencies (CBDCs)—introduces new channels for risk transmission, price discovery, and liquidity. Yet, research rarely examines the implications of these fintech innovations on volatility in commodity markets or the broader sectoral financial markets of India, particularly under climate-related pressures.

Digital currencies and tokenized commodity assets could reshape traditional hedging mechanisms, affect market liquidity, and alter systemic risk profiles. The interaction of these innovations with climate–commodity risks remains largely unmodeled, leaving a significant gap in understanding emerging financial dynamics.

Scenario-based agent simulations that integrate digital finance with commodity sector models could help map systemic risks and inform regulatory responses, enhancing preparedness for a climate-sensitive financial landscape in India.

7. Sectoral heterogeneity and granular risk transmission underexplored

While broad sectoral sensitivities to commodity prices are well recognized—such as energy and metals being highly responsive to crude oil and base metals—research has largely overlooked intra-sectoral heterogeneity and granular supply chain effects. For instance, coal-based energy firms are more exposed to commodity price shocks and climate transitions than solar or other renewable energy firms, yet these distinctions remain underexplored. Similarly, in agriculture, different crops and value chains exhibit varied responses to climate-driven commodity volatility.

Approaches such as network analysis, interlinked sector modeling, and general input-output tables incorporating climate and commodity variables, along with firm-level studies, could illuminate subtle micro-level risk transmission mechanisms. Understanding these nuances is critical for investors, risk managers, asset managers, and policymakers to implement targeted risk mitigation strategies, enhance asset resilience, and inform regulatory mandates.

8. Lack of Cross-disciplinary and Collaborative Research

Finally, although climate risk and commodity-sectoral finance has a cross-cutting nature, interdisciplinary combinations of climate scientists, economists, financial analysts, data scientists, and behavioral psychologists have primarily been unavailable in the Indian context. Research projects are typically isolated in terms of discipline, therefore not realizing richer modeling frameworks, better fusing of data, and broader assessments of risk. Cross-discipline creative combinations are integral to addressing the multifactorial complexity associated with climate-commodity-finance systems, as well as to develop operational frameworks capable of adapting to India's economic, regulatory, and climatic contexts.

SCOPE FOR FUTURE WORK

India is confronted by the pressing challenge of mitigating accelerating climate change pressures in combination with fast-changing commodity markets. Current financial and econometric theory needs to be greatly improved in order to design climate-sensitive, adaptive financial systems that respond to these interlinked pressures. The future research agenda spans the full range of empirical modeling, high-resolution data fusion, behavioral finance, fintech innovations, policy analysis, and collectively building data infrastructure. Given the multi-dimensional research agenda, it is necessary that the agenda encourages the generation of sustainable and adaptive climate governance and market strategy that reflects India's institutional, and environmental context.

Integration of High-Resolution Climate Risk Information into Financial Models

Future research should focus on integrating high-resolution climate indicators into commodity pricing and sectoral financial models. Relevant indicators include drought severity (e.g., Standardized Precipitation Evapotranspiration Index, SPEI), heatwave duration, monsoon frequency, glacial melt, and flooding. Advanced econometric techniques—such as Markov-Switching Vector Autoregressions (MS-VAR) and copula-based models—can capture nonlinear, regime-dependent dynamics and co-movements during extreme market events. Additionally, Extreme Value Theory (EVT) can improve modeling of low-probability, high-impact climate shocks on prices and returns. These methodological enhancements will provide finer-grained risk predictions, supporting evidence-based strategies for market participants and regulators.

Quantitative Modeling of Climate Transition Risks

Climate policies such as carbon pricing, emissions trading schemes, removal of fuel subsidies, and renewable energy standards, introduce new sources of financial uncertainty. To assess their impact, empirical studies should combine event study methodology with Difference-in-Differences (DiD) regression to identify the causal effects of policy transitions on commodity prices and sectoral financial performance. Incorporating market-implied expectations into stochastic discount factor models, using derivatives such as options, forwards, and risk premiums, can enable forward-looking quantification of transition risk in asset valuations. Integrating event study methods with alternative modeling approaches will be critical for effectively managing the financial impacts of climate policy transitions in India.

Dynamics of Green Commodities and ESG Markets

The growing importance of sustainability-oriented commodities in emerging markets necessitates careful analysis of their pricing and links to sectoral ESG indices. Structural Vector Error Correction Models (SVECM) and dynamic factor models that incorporate ESG disclosures, green bond issuances, and clean energy investment trends should be developed to assess the systemic effects of sustainability on sectoral returns, considering interactions with other macro-financial variables. Behavioral asset pricing frameworks that capture evolving investor sentiment toward green assets can help explain how climate-sensitive capital allocation reshapes long-term risk profiles. Large, institutionally framed panel datasets will enhance the relevance and precision of these models for the Indian market.

Capitalizing on High-Frequency and Alternative Data Sources

Indian equity and commodity exchanges such as the Multi Commodity Exchange (MCX) and National Stock Exchange (NSE) yield minute- and tick-level trade data that constitute a vital empirical basis for intraday cascades of volatility and liquidity responses to climate shocks. Deployment of cutting-edge deep learning models, namely Long Short-Term Memory (LSTM) networks and multi-headed Transformers, will uncover complex temporal relations and nonlinear relations that linear models are not capable of detecting. Supporting these, different data fusion of multi-modal data—namely, satellite-retrieved Normalized Difference Vegetation Index (NDVI), surface temperature anomalies, IoT sensor data on polluting substances, and social media feeds with geotags—will provide early warning symptoms of supply-disruption or demand-demand fluctuations. Natural language processing technologies like models BERT and RoBERTa will be employed in order to power real-time climate sentiment extraction and supplement quantitative market signals.

Combined Econometric and Machine Learning Climate Risk Modelling

Future research should integrate traditional econometric models, such as Markov-Switching GARCH, with advanced machine learning techniques like Variational Autoencoders (VAEs) and Generative Adversarial Networks (GANs) to capture regime-dependent and dynamic conditional volatilities arising from climate events. These hybrid approaches allow flexible estimation of structural breaks, nonlinear feedback mechanisms, and simulation of counterfactual market stress scenarios beyond historical data. Bayesian hierarchical models can address estimation uncertainty and heterogeneity inherent in sparse climate–financial datasets. Incorporating behavioral finance through reinforcement learning agents that simulate investor diversification and herding provides a micro-founded understanding of volatility contagion driven by climate shocks.

Behavioral Finance and Sentiment Analysis of Market

Sentiment in the market is one of the major drivers of volatility in the face of climate crises, highly relevant in India's rapidly expanding social media and financial news cultures. Constructing daily climate-sentiment indices from multi-source text data, and using panel vector autoregression, will account for their impact on intraday commodity prices and COGS equity returns. Based in cognitive neuroscience, experimental finance research will factor psychology and decision-making processes pertaining to risk perception, ambiguity, and behavioral anomalies in climate-impacted finance. This new phase of a multi-layered behavioral pattern will develop increasingly precise and adaptive market models for the human dimensions.

Digital Financial Innovations and Systemic Risk Measurement

India’s pioneering initiatives in Central Bank Digital Currency (CBDC) issuance and blockchain-based commodity trading are transforming traditional market structures. Future research should examine how these innovations influence price discovery, liquidity, and hedging in climate-exposed sectors. Applying network theory to map interdependencies and vulnerabilities among digital assets, combined with agent-based models simulating heterogeneous actor responses to climate shocks, can illuminate key drivers of systemic risk. Additionally, assessing cost efficiencies, risk transfer mechanisms, and operational resilience of tokenized commodities will inform regulatory frameworks designed to minimize macro-financial disruptions.

Firm-Level and Supply Chain Climate Exposure Modeling

With the use of rich firm-level databases that record commodity inputs, environmental hazard exposures, and climate policy, it is possible to model climate risk transmission in supply chains at a granular level. Structural equation modeling (SEM) will facilitate mapping out direct and indirect channels of risk transmission, which are vital for uncovering systemic risk vulnerabilities. Spatial econometric modeling will involve regional heterogeneity to derive location-based risk differences affecting firms and sectors differently. These models help accurately price risk and assess credit risk, as well as provide regulatory advice, all of which aligns with international climate-related financial disclosures.

Assessment of Climate Policy Effects in Financial Markets

To isolate the impact of climate policies on commodity prices and sectoral valuations amid changing economic conditions, future research should employ quasi-experimental techniques such as Difference-in-Differences and Synthetic Control methods. Building comprehensive datasets that link policy events with financial market responses will enable analysis of market expectations, announcement effects, and post-policy adjustments. This approach is crucial for designing climate policies that facilitate sustainable transitions while minimizing disruptive effects on financial markets.

Building Collaborative Data Infrastructure

Underpinning all coming climate-finance research lies the establishment of robust, near-real-time, standardized data infrastructures that integrate climate, environmental, financial, policy, and behavioral data. Open-access portals with API-enabled streaming functions support shared data and collaborative science across institutions. It is important to build collaborations with governance structures within the public sector (such as the Reserve Bank of India), commodity exchanges, institutions of higher education and scientific organizations to maximize access, quality and interoperable data. These infrastructures enable robust and interactive climate-finance scholarship with actionable governance.

CONCLUSION

This systematic literature review highlights the complex interplay between climate risk, commodity prices, and sectoral dynamics in Indian financial markets. Both physical and transition-related climate factors are increasingly influencing uncertainty in sectoral equity returns, particularly in climate-sensitive sectors such as energy, metals, agriculture, and infrastructure. While traditional econometric models have provided valuable insights, they often fail to capture nonlinearities, regime-dependent behaviors, and climate–finance feedback loops that characterize today's highly complex system.

Incorporating methodological advancements such as machine learning architectures, high-frequency and alternative datasets, and behavioral finance variables; offers greater sophistication and predictive accuracy for future studies. Integrating emerging green commodities and ESG-linked indices with climate-risk metrics and financial data represents a crucial frontier, especially given India's ambitious climate transition policies and evolving sustainable finance ecosystem. Additionally, the rise of digital currencies and blockchain-based financial tokenization introduces new layers of complexity that future models must address to capture risk transfer and market responses effectively.

Overall, the review underscores the need for interdisciplinary collaboration bridging climate science, financial economics, behavioral modeling, and fintech, supported by robust open data infrastructure and policy-relevant applications. Addressing these challenges will enable Indian policymakers, investors, and market participants to develop more resilient, adaptive, and responsible financial systems capable of navigating the uncertainties and opportunities associated with climate and commodity markets.

REFERENCES

Aggarwal, D., & Mohanty, P. (2018). Do Indian stock market sentiments impact contemporaneous returns? *South Asian Journal of Business Studies, 7*(3), 332–346. *Scopus.* Advance online publication. DOI: 10.1108/SAJBS-06-2018-0064

Agnihotri, S., & Chauhan, K. (2022). Modeling tail risk in Indian commodity markets using conditional EVT-VaR and their relation to the stock market. *Investment Management and Financial Innovations, 19*(3), 1–12. *Scopus.* Advance online publication. DOI: 10.21511/imfi.19(3).2022.01

Ahmed, F., Syed, A. A., Kamal, M. A., López-garcía, M. L. N., Ramos-requena, J. P., & Gupta, S. (2021). Assessing the impact of COVID-19 pandemic on the stock and commodity markets performance and sustainability: A comparative analysis of south asian countries. *Sustainability (Switzerland), 13*(10). *Sustainability (Basel), 13*(10), 5669. Advance online publication. DOI: 10.3390/su13105669

Brockman, P., Dang, T. L., & Pham, T. P. (2024). Stock price synchronicity and stock liquidity: International evidence. *Journal of Empirical Finance*, *79*, 101541. DOI: 10.1016/j.jempfin.2024.101541

Chauhan, S. S., Suri, P., Twala, B., Priyadarshi, N., & Ali, F. (2025). *Exploring the relationship between macroeconomic indicators and sectoral indices of Indian stock market* (No. 14:180). F1000Research. DOI: 10.12688/f1000research.160668.2

Datta, S., & Hatekar, N. (2025). Range Volatility Spillover across Sectoral Stock Indices during COVID 19 Pandemic: Evidence from Indian Stock Market. *Journal of Applied Economic Sciences, 20*(1), 79–100. *Scopus.* Advance online publication. DOI: 10.57017/jaes.v20.1(87).06

Dewan, P., & Dharni, K. (2022). Herding and spillover effects in the Indian commodity futures market. *Journal of Agribusiness in Developing and Emerging Economies*, *13*(5), 748–761. DOI: 10.1108/JADEE-11-2021-0288

Geetha, S., Suwetha, C., Sangavi, A., & Shabana Nazrin, S. (2024). Time Series Financial Market Forecasting Based on Machine Learning. *2024 2nd International Conference on Artificial Intelligence and Machine Learning Applications Theme: Healthcare and Internet of Things (AIMLA)*, 1–5. DOI: 10.1109/AIMLA59606.2024.10531346

Ghosh, I., Datta Chaudhuri, T., Alfaro-Cortés, E., Gámez Martínez, M., & García Rubio, N. (2021). Estimating the relative effects of raw material prices, sectoral outlook and market sentiment on stock prices. *Resources Policy*, *73*, 102158. DOI: 10.1016/j.resourpol.2021.102158

Goel, R., Sidana, N., Bhattacharjee, A., Singh, T., & Kaur, M. (2025). Shifting sands: How major events shape gold futures in the Indian commodity market. *Cogent Economics & Finance, 13*(1), 2483867. DOI: 10.1080/23322039.2025.2483867

Jain, A., & Biswal, P. C. (2016). Dynamic linkages among oil price, gold price, exchange rate, and stock market in India. *Resources Policy, 49*, 179–185. *Scopus.* Advance online publication. DOI: 10.1016/j.resourpol.2016.06.001

Jena, S. K., Lahiani, A., Tiwari, A. K., & Roubaud, D. (2021). Uncovering the complex asymmetric relationship between trading activity and commodity futures price: Evidenced from QNARDL study. *Resources Policy*, *74*, 102277. DOI: 10.1016/j.resourpol.2021.102277

Jha, K. K., & Katiyar, K. (2024). *Impact of Russia-Ukraine war on Oil & Gas, Commodities and Metal Sector of Indian Economy*. 2024 IEEE International Conference on Interdisciplinary Approaches in Technology and Management for Social Innovation, IATMSI 2024. Scopus. DOI: 10.1109/IATMSI60426.2024.10503500

Karmakar, M., & Inani, S. (2019). Information share and its predictability in the Indian stock market. *Journal of Futures Markets, 39*(10), 1322–1343. *Scopus.* Advance online publication. DOI: 10.1002/fut.22041

Kaur, G., & Dhiman, B. (2024). A short on the current scenario of fluctuations in commodity market in India. *BIO Web of Conferences, 86*, 01056. DOI: 10.1051/bioconf/20248601056

Kaur, H., & Chaudhary, A. (2024). Volatility Transmission Role of Indian Equity and Commodity Markets. In *Deep Learning Tools for Predicting Stock Market Movements* (pp. 429–443). Scopus., DOI: 10.1002/9781394214334.ch18

Kaur, R., Kumar, P., Radulescu, M., Mohd, S., & Dascalu, N. (2025). Spillover Nexus among Green Cryptocurrency, Sectoral Renewable Energy Equity Stock and Agricultural Commodity: Implications for Portfolio Diversification. *Economics, 19*(1), 20250138. Advance online publication. DOI: 10.1515/econ-2025-0138

Kaushik, S., & Nagarkar, J. (2023). Seasonality index in different asset classes in India. *AIP Conference Proceedings*, *2523*(1), 030019. DOI: 10.1063/5.0110536

Kirithiga, S., Naresh, G., & Thiyagarajan, S. (2018). Spillover between commodity and equity benchmarking indices. *Benchmarking, 25*(7), 2512–2530. *Scopus.* Advance online publication. DOI: 10.1108/BIJ-06-2017-0143

Kumar, K. K., Stalin, G. A., & Rajeswari, B. (2024). An Empirical Study: Long Run and Short Run Impact of Macroeconomic Variables on Indian Stock Market Prices. In *Studies in Systems, Decision and Control* (Vol. 223, pp. 273–281). Scopus. DOI: 10.1007/978-3-031-51997-0_23

Kumar, N., & Goel, D. (2024). Macroeconomic Determinants of Price Dynamics in the Indian Stock Market. *Indian Journal of Finance*, 25–41. DOI: 10.17010/ijf/2024/v18i9/174457

Liu, C., Tian, M., & Huang, B. (2025). Volatility spillover dynamics between fintech and traditional financial industries and their rich determinants: New evidence from Chinese listed institutions. *International Review of Financial Analysis, 101. International Review of Financial Analysis, 101*, 104034. Advance online publication. DOI: 10.1016/j.irfa.2025.104034

Maurya, P. K., Bansal, R., & Mishra, A. K. (2025). *Sectoral connectedness, volatility spillover and hedging opportunities: Evidence from Indian stock market.* Eurasian Economic Review., DOI: 10.1007/s40822-025-00324-z

Memon, B. A., & Nusratova, G. (2025). Do Financial Markets and Safe-Haven Assets Affect CBDCs? Examining the Nexus between CBDC, Stock Index, Metal Commodity Futures, Oil Price, and Volatility. *Journal of Central Banking Theory and Practice*, *14*(2), 151–167. DOI: 10.2478/jcbtp-2025-0017

Mishra, A., & Kumar, R. P. (2023). Threshold Cointegration and Price Transmission in Commodity Marketsof India. *Finance: Theory and Practice*, *27*(4), 66–79. *Scopus.* Advance online publication. DOI: 10.26794/2587-5671-2023-27-4-66-79

Mishra, U., & Dash, A. S. (2025). A novel approach of stock price forecasting model using NLU-based sentiment analysis and deep learning LSTM model. *Neural Computing and Applications. Neural Computing & Applications*, *37*(22), 18449–18478. Advance online publication. DOI: 10.1007/s00521-025-11395-z

Nautiyal, N., & Kandpal, V. (2025). Nonlinearity Between Economic Indicators and Indian Capital Market. *FIIB Business Review*, *14*(1), 43–57. DOI: 10.1177/23197145211035483

Ndlovu, X., Schaling, E., & Alagidede, P. (2014). Modelling the rand and commodity prices: A Granger causality and cointegration analysis. *Suid-Afrikaanse Tydskrif vir Ekonomiese en Bestuurswetenskappe*, *17*(5), 5. Advance online publication. DOI: 10.4102/sajems.v17i5.334

O'Connor, N., & Madden, M. G. (2005, December). A neural network approach to predicting stock exchange movements using external factors. In International Conference on Innovative Techniques and Applications of Artificial Intelligence (pp. 64-77). London: Springer London.

Pastpipatkul, P., Sriboonchitta, S., & Subsai, P. (2024). Relationships Among Financial and Commodity Markets on Economic Growth: New Evidence from Bayesian Estimation of Seemingly Unrelated Time Series Equations. In V. Kreinovich, W. Yamaka, & S. Leurcharusmee (Eds.), *Applications of Optimal Transport to Economics and Related Topics* (pp. 605–621). Springer Nature Switzerland. DOI: 10.1007/978-3-031-67770-0_41

Periasamy, P., & Satish, R. (2015). A study on relationship between Indian commodity market and Indian stock market with special reference to exchanges in India—An analytical framework. [Scopus.]. *International Journal of Applied Business and Economic Research, 13*(8), 6323–6334.

Pesaran, M. H., Shin, Y., & Smith, R. J. (2001). Bounds testing approaches to the analysis of level relationships. *Journal of Applied Econometrics, 16*(3), 289–326. *Scopus.* Advance online publication. DOI: 10.1002/jae.616

Pranchana, R., Sudhamathi, S., & Benneet, S. (2025). Evaluating the Impact of Sectoral Indices on Stock Market Performance in the National Stock Exchange. *Indian Journal of Information Sources and Services, 15*(1), 238–243. DOI: 10.51983/ijiss-2025.IJISS.15.1.30

Keswani, S., & Wadhwa, B. (2019). Evaluating the impact of macroeconomic variable on Indian stock market. *International Journal of Engineering and Advanced Technology, 8*(6), 4427–4434.

Rodrigues, J. P., & Vinnarasi, B. (2023). *A Study of Financialization of Commodity Markets in India* (Vijayan V. & Senthil Kumar T.S., Eds.; Vol. 399). Scopus. DOI: 10.1051/e3sconf/202339907011

Roy, R. P., & Sinha Roy, S. (2017). Financial contagion and volatility spillover: An exploration into Indian commodity derivative market. *Economic Modelling, 67*, 368–380. *Scopus.* Advance online publication. DOI: 10.1016/j.econmod.2017.02.019

Sahadudheen, I., & Kumar, P. K. S. (2023). Time-Varying Relationship and Volatility Spillovers Among Oil, Gold, Forex and Stock Markets in Indian Context: The Juxtaposition of Global Economic Crisis and COVID-19 Pandemic. *Indian Economic Journal, 71*(4), 748–767. *Scopus.* Advance online publication. DOI: 10.1177/00194662231166168

Sarangi, P. K., Singh, E., Kaushal, A., Sharma, B., Dutta, M., & Dubey, V. P. (2025). *Analysing the Fluctuations in Commodity Prices and Forecasting the Future Directions Using Machine Learning Techniques. 259*, 1827–1836. *Scopus.* Advance online publication. DOI: 10.1016/j.procs.2025.04.138

Sarwar, S., Khalfaoui, R., Waheed, R., & Dastgerdi, H. G. (2019). Volatility spillovers and hedging: Evidence from Asian oil-importing countries. *Resources Policy, 61*, 479–488. *Scopus.* Advance online publication. DOI: 10.1016/j.resourpol.2018.04.010

Sehrawat, M., & Giri, A. K. (2017). A Sectoral Analysis of the Role of Stock Market Development on Economic Growth: Empirical Evidence from Indian Economy. *Global Business Review, 18*(4), 911–923. *Scopus.* Advance online publication. DOI: 10.1177/0972150917692242

Shalini, V., & Prasanna, K. (2016). Impact of the financial crisis on Indian commodity markets: Structural breaks and volatility dynamics. *Energy Economics, 53*, 40–57. *Scopus.* Advance online publication. DOI: 10.1016/j.eneco.2015.02.011

Shamsher, S. (2021). Financialisation of commodities – Empirical evidence from the Indian financial market. *IIMB Management Review, 33*(1), 38–49. *Scopus.* Advance online publication. DOI: 10.1016/j.iimb.2021.03.001

Sharma, C. (2020). Exchange rate volatility and exports from India: A commodity-level panel data analysis. *Journal of Financial Economic Policy, 12*(1), 23–44. *Scopus.* Advance online publication. DOI: 10.1108/JFEP-11-2018-0157

Siddiqui, S., & Roy, P. (2019). Predicting volatility and dynamic relation between stock market, exchange rate and select commodities. *Acta Universitatis Agriculturae et Silviculturae Mendelianae Brunensis, 67*(6), 1597–1611. *Scopus.* Advance online publication. DOI: 10.11118/actaun201967061597

Sidhu, A., & Katoch, R. (2021). Do international gold prices and NSE nifty 50 move together? *Advances in Mathematics: Scientific Journal, 10*(1), 497–506. *Scopus.* Advance online publication. DOI: 10.37418/amsj.10.1.49

Singh, G., & Padmakumari, L. (2020). Stock market reaction to inflation announcement in the Indian stock market: A sectoral analysis. *Cogent Economics and Finance, 8*(1). *Cogent Economics & Finance, 8*(1), 1723827. Advance online publication. DOI: 10.1080/23322039.2020.1723827

Singh Chauhan, S., Suri, P., Alam, F., Hani, U., Johri, A., & Ali, F. (2025). A causality investigation into stock prices and macroeconomic indicators in the Indian stock market. *F1000Research, 13*. Scopus. DOI: 10.12688/f1000research.157041.3

Singhal, S., & Ghosh, S. (2016). Returns and volatility linkages between international crude oil price, metal and other stock indices in India: Evidence from VAR-DCC-GARCH models. *Resources Policy, 50*, 276–288. *Scopus.* Advance online publication. DOI: 10.1016/j.resourpol.2016.10.001

The Impact of Economic Events on Stock Market Returns: Evidence from India | Asian Economic and Financial Review. (n.d.). Retrieved July 21, 2025, from https://archive.aessweb.com/index.php/5002/article/view/2011

Tiwari, A. K., Nasreen, S., Shahbaz, M., & Hammoudeh, S. (2020). Time-frequency causality and connectedness between international prices of energy, food, industry, agriculture and metals. *Energy Economics*, *85*, 104529. DOI: 10.1016/j.eneco.2019.104529

Umar, Z., Nasreen, S., Solarin, S. A., & Tiwari, A. K. (2019). Exploring the time and frequency domain connectedness of oil prices and metal prices. *Resources Policy*, *64*, 101516. DOI: 10.1016/j.resourpol.2019.101516

Vairasigamani, P., & Amilan, S. (2025). Unveiling the dynamic linkages and Hedging between Indian Sectoral Indices and Oil, Gold and Cryptocurrency in times of crisis episodes. *Revista Mexicana de Economía y Finanzas Nueva Época REMEF*, *20*(3), 3. Advance online publication. DOI: 10.21919/remef.v20i3.1211

Veningston, K., Ronalda, M., Shaikh, T. A., Gazi, F., & Bhagat, S. (2025). Modeling and Evaluating Stock Market and Commodity Price Fluctuation Using Time Series Sequence Learning. *SN Computer Science*, *6*(5), 435. Advance online publication. DOI: 10.1007/s42979-025-03931-4

Yang, X., Wang, X., Cao, J., Zhao, L., & Huang, C. (2024). Cross-regional connectedness of financial market: Measurement and determinants. *The North American Journal of Economics and Finance*, *72*, 102157. DOI: 10.1016/j.najef.2024.102157

Yu, H., Lu, R., Yang, H., & Zhang, B. (2024). Wealthy individual investors and stock markets' tail risk. *PLoS One*, *19*(5), e0282173. DOI: 10.1371/journal.pone.0282173

Zhao, C., Cai, J., & Yang, S. (2025). A hybrid stock prediction method based on periodic/non-periodic features analyses. *EPJ Data Science*, *14*(1), 1. Advance online publication. DOI: 10.1140/epjds/s13688-024-00517-7

Chapter 6
Digital Banking as a Catalyst for Climate-Resilient Financial Markets

Rohit Sood
https://orcid.org/0000-0002-4127-7781
Lovely Professional University, India

ABSTRACT

The accelerating shift towards digital banking services has not only reshaped the Indian financial landscape through enhanced convenience and accessibility but also emerges as a critical nexus in the broader context of climate risk, energy consumption, and sustainable financial market development. This study investigates the awareness, adoption, and challenges of sustainable digital banking practices among Indian users, viewing digital banking as a pivotal channel for advancing environmental resilience and efficient energy use in financial ecosystems. By analyzing user behavior across leading digital platforms such as Google Pay, Paytm, Phone Pe, and banking applications, the research explores how platform trust, incentive structures, and usage frequency influence the willingness to embrace eco-friendly financial services, with an emphasis on their potential to reduce carbon footprints and support climate adaptation strategies.

INTRODUCTION

In the context of India's evolving digital landscape, mobile banking has emerged as a transformative force not only in expanding financial services, but also in advanc-

DOI: 10.4018/979-8-3373-6766-8.ch006

ing climate resilience and sustainable financial market frameworks. The widespread proliferation of smartphones and digital applications has dramatically altered how individuals access, interact with, and derive value from financial services, reducing long-standing barriers related to geography, infrastructure, and socio-economic status (Shaban, 2024). More than just a means for basic transactions, mobile and digital banking platforms are now pivotal channels for channeling financial flows in an environmentally conscious manner, contributing to a reduction in carbon footprints and optimizing energy consumption by replacing traditional, resource-intensive banking models (Ahmad et al., 2025).

This digital transformation has heightened opportunities for financial inclusion, empowering underserved and rural populations by making secure, low-cost, and accessible financial services available directly through handheld devices. Furthermore, the integration of user-centric digital interfaces—supported by incentives such as rewards and lower transaction costs—has fostered greater user trust and participation, especially in climate-aware, adaptive financial systems (Lindfors, 2024). Policymakers, financial institutions, and fintech innovators are increasingly leveraging mobile banking to deploy targeted strategies that not only promote economic empowerment, but also support the urgent need for sustainable, climate-resilient development. Nevertheless, the expansion of mobile banking is accompanied by persistent challenges such as digital illiteracy, cybersecurity risks, insufficient connectivity, and regulatory constraints, which must be addressed through collaborative, multi-stakeholder efforts for a truly inclusive and sustainable financial ecosystem.

This study investigates the dual impact of mobile banking: enhancing access and literacy in financial services, and catalyzing sustainable, climate-aligned market transformation. Through empirical analysis of user behavior, platform adoption, and the interplay of incentives and trust, the research offers evidence-informed insights to guide policies and practices that underpin a more equitable, eco-conscious financial future in India and beyond.

LITERATURE REVIEW

The literature highlights the critical role of mobile banking in advancing financial inclusion by extending banking services to underserved populations. Adoption is primarily influenced by factors such as perceived usefulness, ease of use, and social influence, as framed by the Technology Acceptance Model (TAM)(Vannoy & Palvia, 2010) and the Unified Theory of Acceptance and Use of Technology (UTAUT) (Sarfaraz, 2017). Despite rapid growth, challenges persist in regulatory frameworks and the long-term financial sustainability of mobile banking systems. The proliferation of mobile banking has significantly reshaped financial markets, reducing dependence on cash and promoting digital payment systems, notably in emerging economies like India (Putrevu & Mertzanis, 2024). However, issues related to security risks, regulatory complexities, and gaps in financial literacy continue to hamper adoption rates. Additionally, advances in fintech have deeply influenced various market segments, including equities, bond markets, and microfinance institutions, thereby integrating rural populations into formal financial structures (Del Sarto & Ozili, 2025). Methodologies such as sentiment analysis and event studies have been utilized to assess the market impact of mobile banking, although evolving digital ecosystems and data limitations present ongoing challenges. Future research directions include exploring credit risk assessment and the implications of financial policies on mobile banking adoption (Shaikh et al., 2023). Overall, mobile banking enhances financial inclusion, but continuous evaluation of regulations and security protocols is essential to ensure sustainable growth (Arner et al., 2020).

In the context of rural India, mobile banking adoption is influenced by a combination of technological and behavioral factors, with the Technology Acceptance Model (TAM) and the Diffusion of Innovation Theory (DIT) standing out as primary theoretical frameworks. The literature identifies significant barriers including user perception, ease of use, and credibility risks, all of which affect the uptake of mobile banking in rural settings. Despite efforts toward financial inclusion, there remains a paucity of rigorous research focusing on rural adoption dynamics. Key impediments such as low digital literacy, security concerns, and inadequate infrastructure continue to obstruct mobile banking development. Furthermore, regulatory uncertainties and shifting trends in mobile financial services contribute to adoption ambiguities. Behavioral intention and risk perception emerge as crucial determinants influencing user acceptance in these regions. A notable limitation of existing research is its rural focus, which curtails generalizability to urban contexts. Moreover, reliance on self-reported data introduces bias risks that may affect the accuracy of adoption pattern estimations. Future studies should emphasize policy interventions and technological advancements to bridge accessibility gaps and establish trust in mobile banking services. Addressing these barriers is vital for achieving sustainable financial inclusion across India's diverse demographic landscape (Sikka & Bhayana, 2024).

The technology and behavior drive mobile banking usage in rural India, with TAM and DIT providing explanatory power (Kohli et al., 2024). Persistent challenges such as user perception, ease of use, and credibility risk substantially impact adoption rates. Despite policy initiatives aiming to enhance financial inclusion, the interaction between user perception and behavioral intention in rural contexts remains underexplored. Digital literacy limitations, security vulnerabilities, and insufficient technological infrastructure further constrain mobile banking adoption. Regulatory concerns and evolving fintech paradigms add layers of uncertainty regarding sustainable inclusion. The rural focus of the research limits its applicability in urban areas, and self-reported data may introduce measurement bias. Future research must investigate policy and technological factors that foster penetration of mobile banking in underdeveloped regions to bridge these gaps, which is essential for fostering inclusive and sustainable financial growth (Mookerjee et al., n.d.).

The adoption of mobile banking and its role in achieving financial inclusion extensively draws on technology adoption models such as the Technology Acceptance Model (TAM) and the Unified Theory of Acceptance and Use of Technology (UTAUT). These frameworks emphasize key factors including perceived ease of use, trust, and risk assessment as pivotal determinants of user acceptance of digital financial services. Although TAM and UTAUT effectively explain adoption behaviors, there remain empirical gaps—particularly concerning the multi-dimensional nature of trust and perceived risk in mobile banking. The growing novelty of financial technologies and cross-cultural differences have added complexity to adoption dynamics, necessitating deeper investigation into behavioral intentions across diverse populations. Limitations of existing studies include potential biases inherent in self-reported data and limited generalizability across different regions and demographics. Furthermore, rapid fintech evolution may render earlier findings less applicable over time, signaling the need for future research incorporating real-time behavioral data and cross-cultural analyses to refine and extend these models' relevance to mobile banking contexts.

Research examining the impact of popular mobile banking platforms such as Paytm and Google Pay focuses particularly on their contribution to financial inclusion in rural and semi-urban India (Patel et al., 2023). Grounded in the Theory of Financial Innovations, this scholarship highlights the crucial demand for inclusive financial services as a driver of economic participation. The literature points to significant disparities in mobile banking uptake between urban and rural populations, where cultural and knowledge-related barriers prominently hinder adoption. Despite an expansive body of work on mobile banking, limited empirical evidence exists regarding the specific drivers encouraged by platforms like Paytm and Google Pay in less urbanized areas. The cited study addresses this gap by employing a mixed-methods approach that captures both quantitative data and qualitative insights into user experiences and challenges. However, constraints include potential bias in self-reported data and the geographically limited scope that affects the broader applicability of the findings. There's also a risk that overreliance on digital infrastructure could overshadow other socio-economic factors influencing financial inclusion.

The impact on digital inclusion highlights mobile banking's transformative role in empowering underprivileged Indian communities (Mookerjee et al., n.d.). This research demonstrates how mobile banking platforms facilitate financial access and economic empowerment by emphasizing adoption drivers such as performance expectancy, effort expectancy, social influence, and perceived risk. While abundant studies underscore the potential of mobile banking, there is a notable lack of targeted research probing the specific factors behind platform adoption in rural and semi-urban contexts. This study attempts to bridge that gap through a mixed-methods methodology that elucidates user experiences and obstacles. Limitations remain in terms of potential biases arising from self-reported data and the constrained regional focus that limits generalizability. Additionally, the heavy dependence on IT infrastructure within these studies may underestimate other critical socio-economic determinants influencing financial inclusion outcomes.

Together, these studies indicate that while mobile banking has become integral to advancing financial inclusion, particularly in India's rural and semi-urban landscapes, further research is essential to address data biases, cultural nuances, infrastructural challenges, and evolving fintech paradigms to fully understand and optimize mobile banking's inclusive potential.

The significant role of fintech and digital financial services in promoting financial inclusion, particularly among unbanked and rural communities in India is not studied to its full potential (Jha & Dangwal, 2025). It highlights cash management behaviors of low-income households and underscores the impact of digital financial innovations on improving access to financial resources. Valuable insights on the financial behavior of underbanked segments are drawn from key reports by the Reserve Bank of India (RBI) and CRISIL (CRISIL, 2021). Despite comprehensive studies, a critical gap exists regarding the design of fintech products tailored to meet the specific financial needs of rural consumers(Del Sarto & Ozili, 2025). Moreover, limited research has focused on evaluating the effectiveness of mobile banking in enhancing financial literacy and decision-making among unbanked populations (RBI, 2020). This study addresses these gaps by assessing fintech-based solutions and their influence on financial access. However, the research faces constraints related to geographic scope, limiting generalizability to India's broader population (Kohli et al., 2024). Additionally, the reliance on self-reported household data introduces potential biases that may affect the validity of findings related to financial behavior (CRISIL, 2021). Future longitudinal research is recommended to better understand the enduring effects of fintech on financial inclusion.

In examining the factors influencing financial inclusion through banking services, the literature covers national and global metrics, including account ownership, savings behavior, and loan access (Nandru et al., 2015). It identifies regional and income-related disparities in financial service consumption and advocates for targeted policies to address such gaps. Despite extensive existing research, there remains a need to deepen understanding of the determinants affecting banking service utilization across different Indian states (Fu et al., 2023). Empirical testing of these drivers in diverse regions can uncover critical factors that enhance bank penetration and service uptake (RBI, 2020). The study contributes by exploring region-specific financial behaviors and determinants of inclusion, though it is limited by convenience sampling methods, which may restrict the representativeness of results. The singular geographic focus on Pondicherry further limits the applicability of conclusions to other states. Reliance on self-reported data increases susceptibility to response biases, underscoring the need for more robust data collection methods in future investigations (World Bank, 2020).

Regarding the impact of fintech on financial inclusion in rural India, the literature asserts that fintech plays a pivotal role by delivering personalized digital financial services to economically disadvantaged groups (Jha & Dangwal, 2025). Scholars have expanded on consumer decision models and technology acceptance frameworks such as UTAUT and TAM (Sarfaraz, 2017), yet empirical evidence on fintech's socioeconomic impacts in developing countries remains sparse (World Bank, 2020). Significant knowledge gaps persist about the key success factors for achieving financial inclusion via mobile technology in emerging economies (Del Sarto & Ozili, 2025). In particular, the social, economic, and cultural determinants influencing fintech adoption in rural India have not been adequately studied (RBI, 2020). The current study seeks to fill these voids by analyzing the role of digital financial services in rural economic development. However, its generalizability is limited, addressing exclusively rural contexts without permitting cross-regional comparison. Moreover, the use of self-reported qualitative data introduces potential response bias, affecting the validity of findings.

The above discussion underscores the critical role of technological innovation in transforming banking services and broadening financial inclusion. Early research highlights the implementation of digital banking schemes (Sharma & Piplani, 2017), while new perspective examines the contribution of information and communication technology (ICT) in enhancing financial access (Vishnuvardhan et al., 2020). Mobile banking, as a revolutionary financial technology, has significantly altered banking paradigms by providing affordable and convenient services to underserved. Despite these advancements, gaps remain in understanding the determinants influencing mobile banking transactions. Prior studies have inadequately addressed the influence of technological infrastructure, consumer trust, and regulatory frameworks on mobile banking adoption (RBI, 2020). This study seeks to fill these gaps by identifying key drivers behind mobile banking development in India. However, its scope is limited as it focuses primarily on scheduled commercial banks, which may not fully represent the broader banking sector (Arner et al., 2020). Moreover, the research examines a select group of variables, excluding broader economic and behavioral factors that potentially affect mobile banking uptake.

Extending beyond structural changes, the literature also highlights the evolving importance of mobile phones within Information and Communication Technologies (ICT) as catalysts for economic growth. While initial investigations centered on traditional ICT such as fixed-line telephony and computers, contemporary studies emphasize mobile phones' critical role in fostering network effects and expanding financial accessibility (van Zanden, 2023). Mobile technology notably facilitates financial inclusion by bringing banking services to previously unbanked populations and enabling digital payment solutions. Nonetheless, significant gaps remain fully comprehending how mobile phone penetration impacts financial inclusion and microeconomic outcomes, particularly in developing countries. Empirical evidence linking mobile adoption directly to banking access and economic activities is sparse. Addressing this, the study aims to investigate adoption patterns of mobile banking and their effects on financial inclusion. Yet, it acknowledges potential constraints related to the availability and reliability of data concerning mobile penetration and financial inclusion across heterogeneous regions.

Together, these studies contribute to understanding mobile banking's multifaceted impact on financial services and economic growth, emphasizing the continued need for comprehensive research that integrates technological, behavioral, and infrastructural dimensions within India's dynamic digital finance landscape.

RESEARCH PROBLEM

Despite the sweeping digital transformation in India's financial sector, a substantial portion of the population—particularly among vulnerable and marginalized groups—remains outside the ambit of formal financial services. While digital and mobile banking platforms have demonstrated potential to advance financial inclusion and promote sustainability through enhanced accessibility and reduced carbon footprints, key challenges persist. These include inadequate digital infrastructure, low levels of digital literacy, cybersecurity concerns, and limited user trust, which collectively hinder the widespread adoption of sustainable digital banking solutions (Mamadiyarov, 2024). There is a critical need for empirical investigation into how digital banking platforms impact financial inclusion, especially in university settings like Lovely Professional University (LPU), and to what extent users are aware of, and willing to engage with, sustainable banking practices. This study seeks to assess the effectiveness of digital banking in fostering financial inclusion, elucidate the barriers faced by users, and gauge the influence of awareness, trust, and incentives in the adoption of climate-resilient, energy-efficient banking services.

Objectives

- Assess the adoption of sustainable and energy-efficient operational practices within mobile and digital banking platforms, highlighting their role in advancing climate-resilient financial ecosystems.
- Examine the key challenges and barriers—such as cybersecurity risks, infrastructure limitations, and cost factors—that hinder the effective implementation of sustainability-oriented practices in mobile banking services.
- Develop strategic recommendations that harmonize financial inclusion objectives with environmental sustainability imperatives, fostering an adaptive, climate-aware digital banking framework.

SCOPE OF THE STUDY

This study focuses on the intersection of financial inclusion, operational efficiency, and environmental sustainability within the realm of digital banking in India. It investigates the role of mobile banking platforms in expanding financial access while simultaneously promoting sustainable banking practices that contribute to climate resilience and energy efficiency. The research examines the key challenges and barriers faced in integrating sustainability into digital financial services and aims to propose strategic frameworks that balance inclusive financial growth with environmental stewardship to establish a future-ready, sustainable digital banking ecosystem.

The empirical investigation is conducted in India, leveraging primary data collected through Google Forms questionnaires completed by mobile banking users aged 18 and above. Supplementary secondary data may be sourced from credible financial websites and industry reports to support analytical insights. Employing a mixed-methods approach, the study combines quantitative and qualitative analyses to capture nuanced customer behavior, operational efficiencies, and sustainability trends across leading digital banking platforms.

Key areas of Focus Include

- Evaluating sustainable and efficient operational practices within mobile banking systems (Naruetharadhol et al., 2021).
- Identifying and analyzing the challenges and obstacles in implementing environmentally responsible financial procedures (Valerievna & Igorevna, 2024).
- Formulating strategic recommendations that align financial inclusion objectives with environmental sustainability imperatives (Gonzalez et al., 2025).

The scope excludes technical cybersecurity specifics and detailed banking regulatory frameworks beyond user perceptions. It also omits individuals who do not engage with mobile banking services. By clearly defining this scope, the research seeks to provide practical and actionable guidance for fintech companies, policymakers, and financial institutions to harness mobile banking as a dual catalyst for financial inclusion and sustainable economic development in India

CONCEPTUAL MODEL

The theoretical framework underpinning this research integrates four critical dimensions: Climate-Resilient Financial Inclusion, Sustainable Digital Banking Practices, Energy-Efficient Operational Performance and Climate Adaptive Financial Eco-Systems. These interconnected dimensions collectively assess the role of digital banking platforms in enhancing financial accessibility while embedding sustainability and operational efficiency, thereby advancing environmentally responsible and climate-adaptive financial ecosystems.

The flow chart illustrates the relationship between three key dimensions:

- Climate-Resilient Financial Inclusion
- Sustainable Digital Banking Practices
- Energy-Efficient Operational Performance
- Climate Adaptive Financial Eco-Systems

These dimensions interact and collectively influence financial accessibility, operational efficiency, and environmental sustainability within digital banking ecosystems, fostering a climate-aware and adaptive financial framework.

Figure 1. Conceptual Model for digital Banking Platforms to foster

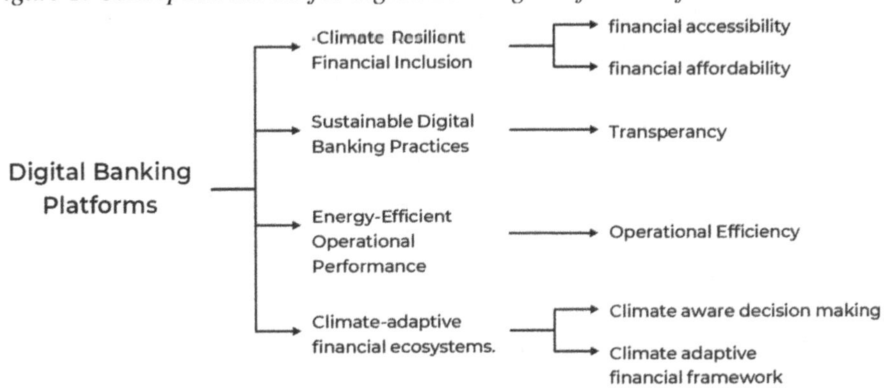

FINANCIAL INCLUSION: ISSUES AND CHALLENGES

The literature extensively highlights financial inclusion as a pivotal driver of economic growth and a means to reduce wealth disparities. While progress in financial inclusion has been noted, structural impediments persist, limiting equitable access for marginalized communities (Sikka & Bhayana, 2024). Technological advancements, including mobile banking and digital payment innovations, are increasingly viewed as viable solutions to bridge these longstanding gaps (Ahmad et al., 2025). Nonetheless, despite improvements in banking penetration, a substantial segment of India's population remains underbanked, underscoring the urgency for continued research and focused policy interventions. The literature further documents notable disparities between urban and rural financial access, pointing to the necessity of tailored, region-specific strategies to address these inequalities (Bernard et al., 2023). This study thus seeks to examine the effectiveness of current financial inclusion initiatives in improving accessibility. It also acknowledges that although the number of deposit accounts has grown, it has not kept pace with population growth. Economic disparities and the uneven distribution of banking infrastructure continue to constrain progress in low-performing regions, perpetuating the financial inclusion divide.

The critical issues surrounding mobile commerce, particularly focusing on user adoption, perceived ease of use, customer convenience, and factors that hinder implementation, need to be studied with interactive and adaptive models. The Technology Acceptance Model (TAM) developed by (Bernard et al., 2023) serves as a foundational framework widely applied in studies of mobile banking adoption highlights consumer trust and security challenges in online banking, emphasizing risks associated with digital money transactions.

Further, the intersection of blockchain adoption and financial inclusion in India signals additional research opportunities. The consumer behavior literature again stresses experience goods and learning by doing as critical constructs to understanding adoption dynamics. (Sikka & Bhayana, 2024)provide the theoretical underpinning for how consumers assess new financial technologies through usage experience, quality perceptions, and accumulated confidence. Despite burgeoning interest, more research is required to elucidate how familiarity with mobile banking and payment platforms influences adoption in emerging versus advanced economies. The newer perspective provided by (Putrevu & Mertzanis, 2024; Somu, 2025) provides valuable insights but lacks an in-depth examination of the effects of repeated user interaction on trust and ease of use. Constraints persist due to heterogeneous regional technology access and banking sophistication, posing challenges to the external validity of findings. Variations in consumer learning and financial literacy across demographic segments further emphasize the need for nuanced, context-sensitive research approaches.

DATA ANALYSIS AND INTERPRETATION

The collected data was processed using Python (Google Colab) to inform mobile banking uptake, user patterns, challenges, and trends in financial inclusion. Descriptive statistics, distribution of categorical data, and missing values were tested to better understand the dataset.

Descriptive Statistics for Numerical Variables

The summary statistics for numerical columns provide insights into the age distribution and efficiency scores of respondents:

Table 1. Age Distribution and Efficiency Scores of Respondents

Metric	Age	Efficiency
Count	97	96
Mean	25.61	3.93
Std Dev	1.54	1.03
Min	23	1
25%	24	3
50%	26	4
75%	27	5
Max	28	5

- The mean age of participants is 25.61 years, with a majority of participants in the age group of 23–28 years.
- The self-reported efficiency score is 3.93 on a scale of 1 to 5, reflecting a moderately high confidence level in mobile banking usage.
- The difference in the efficiency scores (Std = 1.03) implies that it is simpler for some customers to use mobile banking compared to others.

Categorical Data Distribution

Mobile Banking Adoption

- All 97 respondents use mobile banking, indicating 100% adoption in the sample.

Roles of Respondents:

Table 2. Mobile Banking Adoption Role of Respondents

Role	Count
Mobile Banking User	84
Bank Official	10
Fintech Executive	2
Policymaker	1

The majority (86.6%) are general mobile banking users, followed by bank officials (10%), indicating that responses largely reflect end-user experiences.

Preferred Mobile Banking Platforms:

Table 3. Preferred Mobile Banking Platforms by Respondents

Platform	Count
Google Pay	51
Paytm	22
Phone Pe	13
Banking apps (e.g., SBI YONO, HDFC Mobile Banking)	9
Others	2

- Google Pay is the most preferred platform (52.6%), followed by Paytm (22.7%) and Phone Pe (13.4%).
- Traditional banking apps are less popular (9.2%), suggesting a preference for fin-tech based solutions.

Frequency of Mobile Banking Usage

Table 4. Mobile Banking Usage Frequency by Respondents

Frequency	Count
Daily	79
Weekly	14
Rarely	2
Monthly	1

- 81.4% of respondents use mobile banking daily, demonstrating high engagement.
- Only 3% use it rarely or monthly, indicating that mobile banking is an integral part of financial transactions.

Types of Mobile Banking Services Used

- The most commonly used services include Money Transfers (34%), Bill Payments (19%), and Savings & Investments (4%).
- Only 3% of respondents use loan applications, indicating a gap in mobile-based credit accessibility.

Challenges in Mobile Banking Adoption

Table 5. Mobile Banking Adoption Challenges by Respondents

Challenges	Count
Lack of awareness	29
Technological limitations	19
High implementation costs	10
Regulatory barriers	5
Others	10

- Lack of awareness (30%) is the biggest barrier, highlighting the need for financial education initiatives.

- Technological limitations (19.6%) and high costs (10%) also hinder widespread adoption.

Trust in Mobile Banking

Table 6. Mobile Banking Embedded Trust by Respondents

Trust Level	Count
Moderate	37
High	34
Very High	20
Low	2
Very Low	2

- 57% of users trust mobile banking at high or very high levels, suggesting strong user confidence.
- A small fraction (4%) expresses low trust, possibly due to security concerns.

Inefficiencies in Mobile Banking

Table 7. Mobile Banking Challenges Faced by Respondents

Issue	Count
Cybersecurity concerns	29
Poor internet connectivity	24
Slow transaction processing	19
Limited rural access	18
Others	7

- Cybersecurity risks (30%) and poor internet connectivity (24%) are major inefficiencies, particularly affecting rural areas.
- Limited rural access (18%) indicates an opportunity for targeted digital infrastructure development.

Awareness of Sustainable Banking Practices

Table 8. Sustainable Banking Practices Awareness level amongst Respondents

Awareness	Count
Yes	68
No	29

- 70% of respondents are aware of sustainability initiatives, showing a positive trend towards green banking.

Incentives to Drive Sustainable Banking

Table 9. Sustainable Banking Incentives Awareness level amongst Respondents

Incentives	Count
Rewards & cashback	53
Lower transaction fees	23
Government subsidies	15
Others	5

- Rewards & cashback (54%) are the strongest motivators for sustainable banking participation.

ANALYSIS OF CORRELATION MATRIX

The correlation matrix provides insights into the relationships between Efficiency, Trust, and Awareness in mobile banking.

Key Observations

Efficiency & Trust (0.072) = Weak Positive Correlation

- This indicates there is practically zero relationship between trusting mobile banking as a user and its efficiency from the user's point of view.
- Those that trust mobile banking more may or may not judge it as being more efficient. Efficiency & Awareness (-0.002) → Almost No Correlation
- The almost-zero correlation suggests the level of awareness about mobile banking has no bearing on the judgment of efficiency.
- Just because users know about mobile banking, however, does not mean they view it as more or less efficient.

Trust & Awareness (0.234) = Weak Positive Correlation

- There is a relatively stronger correlation in this case, in that individuals who are aware of mobile banking have higher trust in it.
- Though this is still a weak correlation, it indicates that raising awareness can assist in enhancing the trust factor in mobile banking.

Table 10. Correlation Matrix amongst Efficiency, Trust and Awareness

	Efficiency	**Trust**	**Awareness**
Efficiency	1	0.072068	-0.002516
Trust	0.072068	1	0.234445
Awareness	-0.002516	0.234445	1

Figure 2. Heat Map Correlation Matrix amongst Efficiency, Trust and Awareness

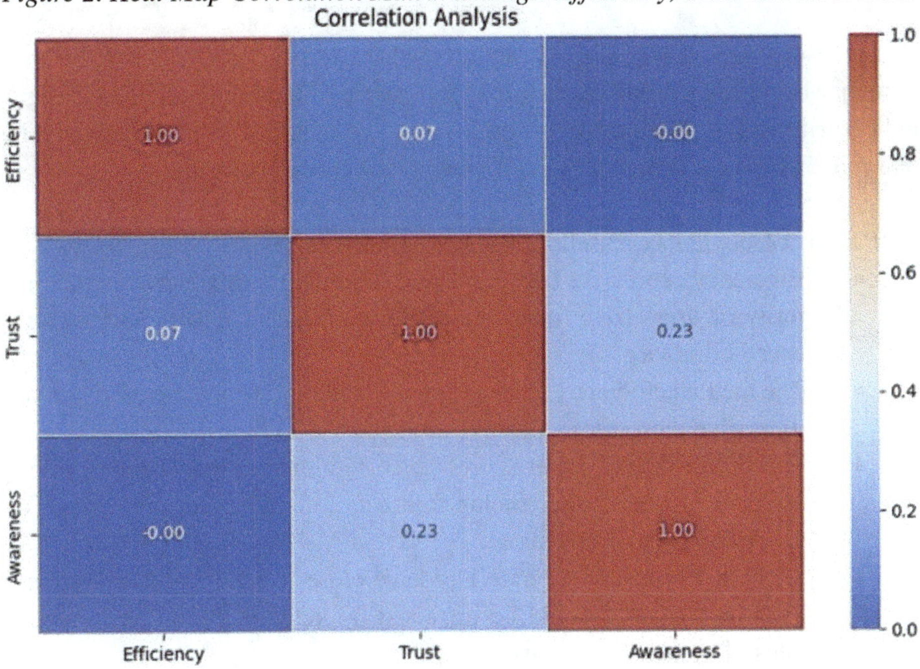

PLATFORM VS. EFFICIENCY ANALYSIS

To evaluate the efficiency of different mobile banking platforms, we performed a comparative analysis using boxplots. The visualization (Figure 1) presents the distribution of efficiency scores across platforms such as Google Pay, Paytm, Phone Pe, Banking Apps (e.g., SBI YONO, HDFC Mobile Banking), and Others.

Key Observations

- Google Pay and Paytm have a large variance of efficiency scores, ranging from 1 to 5, indicating varied user experiences.
- Phone Pe and Banking Apps show more stable efficiency scores, with fewer outliers.
- Other sites have a comparatively concentrated distribution, reflecting a more even experience amongst users.
- The outliers indicate that some users likely had very different experiences on particular platforms.

Implications for Further Research

The results show that platform selection can have an effect on efficiency, but that there may be other contributing factors such as speed of transaction, UI/UX design, and security issues. Qualitative feedback should be included in future research to determine why these differences in efficiency ratings are occurring.

1. Paytm & Google Pay: Parallel Efficiency Distributions
    - Efficiency scores are between 1 and 5, implying that although the majority of users have found the platforms efficient, others have experienced otherwise.
    - The median efficiency score is approximately 4, which means that most users give these sites a positive rating.
    - The occurrence of lower scores (even 1) indicates that there may be a segment of users experiencing transaction failure, security-related, or user experience problems.
    - As both Paytm and Google Pay have close to identical distributions, their performance could be impacted by network conditions, success rates of transactions, or usability.
2. Banking Apps (SBI YONO, HDFC Mobile Banking): More Stable Efficiency Scores
    - Efficiency scores are tighter and have less outlier variation than that of Paytm and Google Pay.
    - Median efficiency is approximately 5, which means that customers who depend on banking apps tend to find them very efficient.
    - The reduced spread in ratings indicates that banking apps can provide more stable performance, reliability, and security.
    - Nonetheless, fewer customers appear to have reviewed banking apps compared to mobile wallets. This may reflect lower levels of adoption or possibly that most customers continue to use third-party apps more than they use standard banking apps because of reasons such as simplicity and speed of transactions.
3. Phone Pe: Hybrid User Experience with Outliers
    - Phone Pe's efficiency scores range from 3 to 5, indicating that although the majority of users find it efficient, there have been inconsistencies reported.
    - The outliers (higher and lower efficiency scores) indicate that a few users may have experienced periodic technical problems or failed transactions.

- In contrast to Paytm and Google Pay, Phone Pe lacks efficiency ratings as low as 1, so extreme dissatisfaction is not as prevalent.
- Additional research into customer complaints, technical problems, or regional service differences could clarify why these inconsistencies occur.

4. Other Platforms: More Consistent But Slightly Lower Efficiency Scores
   - Efficiency ratings are more stable, between 3 and 4, with less extreme values.
   - The median efficiency ranking is slightly worse than Paytm or Google Pay, suggesting these other platforms might not be quite as optimized for user ease or speed.
   - As fewer users score these services at 5, this indicates average satisfaction, but they might not be providing the same degree of frictionless transactions or mass adoption as mainstream payment services.

User Experience Assessment

- The very large variation in efficiency scores for Paytm and Google Pay implies that there are problems for some users but not others, finding them very efficient.
- Qualitative research (e.g., user surveys or interviews) can identify some particular problems such as transaction failure, user interface problem, or security issues.

Banking Apps Are Perceived as More Stable but Have Lower Adoption

- Whereas banking apps have similar and high efficiency ratings, their adoption rates seem lower than digital wallets.
- Future work can investigate the reasons that users prefer third-party apps rather than banking apps—interface usability, transaction speed, or promotion offers, for example.

Examine Outliers in Phone Pe

- Outliers indicate mixed user experiences, that is, some users experience technical issues or inconsistent service.
- Analysing more deeply into users' complaints, rates of error, and the rate of transactions successful could ascertain important areas where the platform fails.

Figure 3. Platform Vs Efficiency Analysis Bar Chart

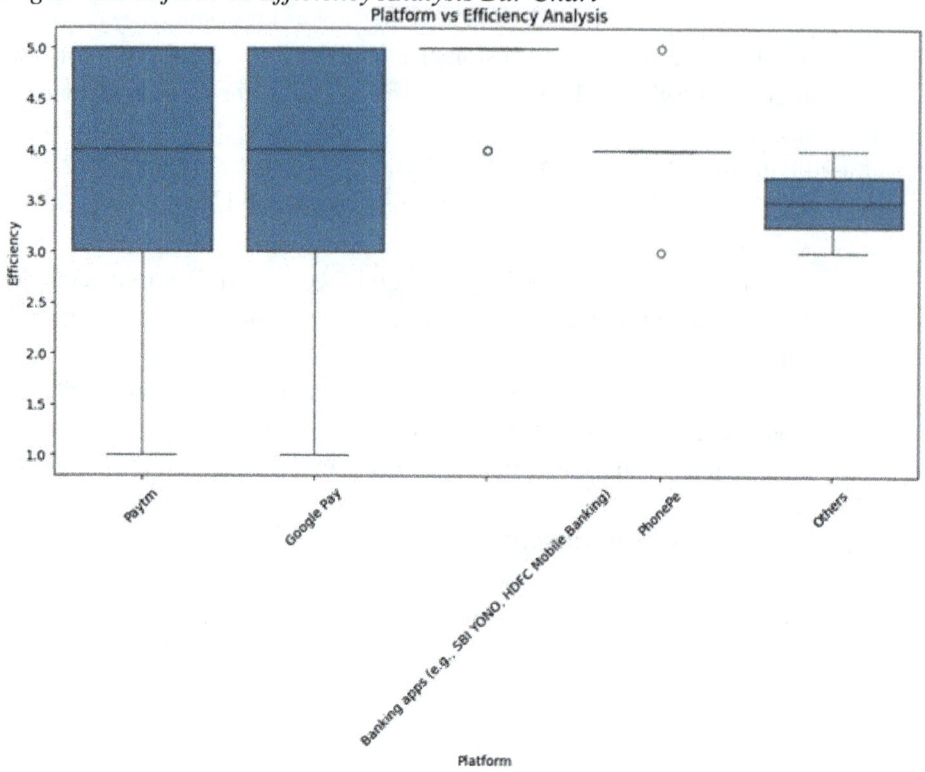

PREFERRED PLATFORMS & INCENTIVES ENCOURAGING SWITCHING

Figure 4. Preferred Platforms Vs Switching Analysis Bar Chart

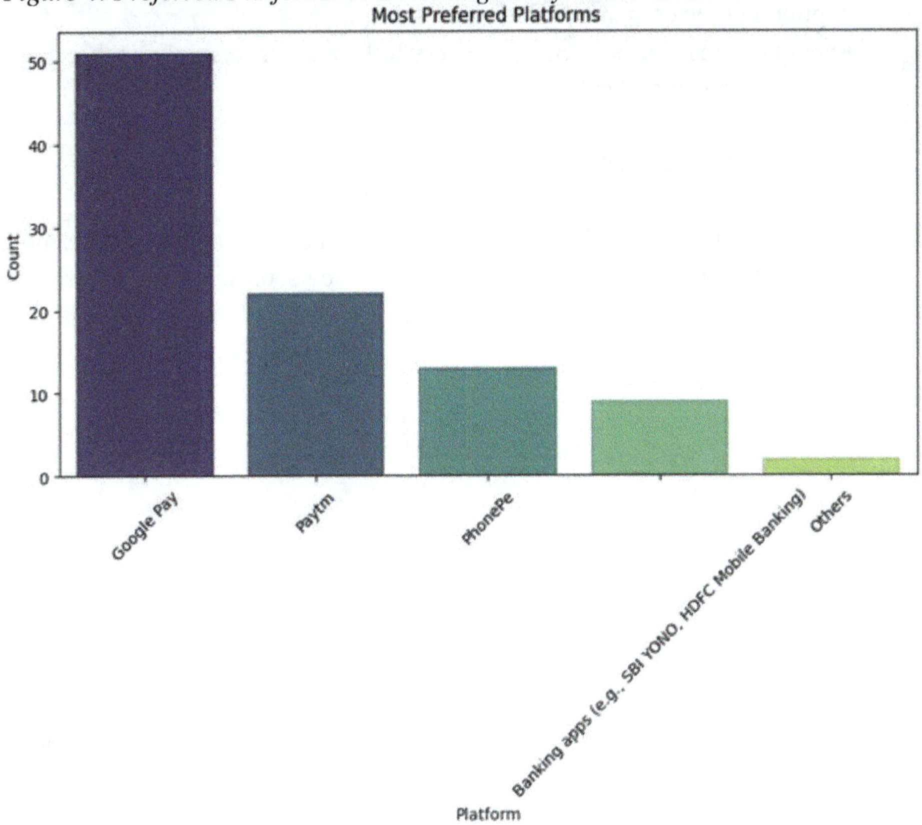

Table 11. Platform and User count for M-banking adoption

Platform	User Count
Google Pay	50
Paytm	22
Phone Pe	13
Banking Apps (SBI YONO, HDFC Mobile Banking)	9
Others	3

- The most popular platform is Google Pay (50 users, ~50% of all respondents).
- Followed by Paytm with 22 users (~22%), reflecting high adoption but much lower compared to Google Pay.
- Phone Pe (13 users, ~13%) and banking apps (9 users, ~9%) exhibit moderate adoption but lower than top platforms.
- Other platforms (3 users, ~3%) have very little preference, implying that mass platforms are market leaders.

Contribution to Research Objective

- This information assists in the identification of the market leader concerning user preference. If we compare it with efficiency scores, we can identify if high adoption is associated with efficiency or if other elements (trust, incentives) determine user choice.

INCENTIVES ENCOURAGING USERS TO SWITCH

Figure 5. Switching Inventive Analysis

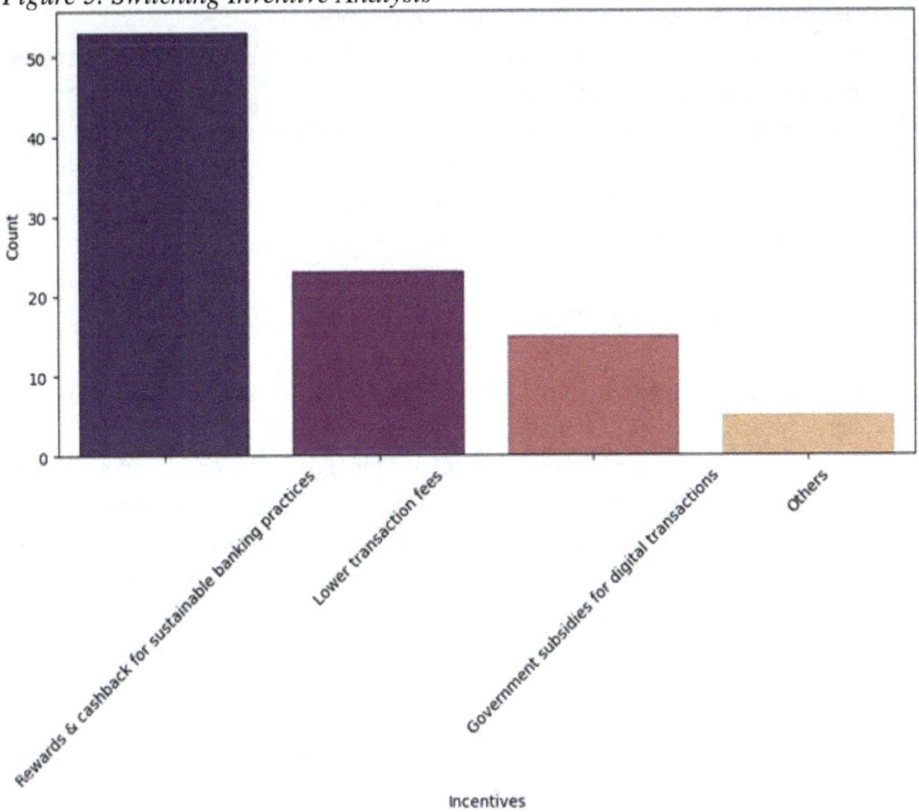

Table 12. Incentive structure beneficiaries' perspective

Incentive Type	User Count
Rewards & cashback for sustainable banking	53
Lower transaction fees	23
Government subsidies for digital transactions	15
Other incentives	7

Key Findings

- Rewards & cashback is the strongest driver (53 users, ~53%). This indicates that money matters strongly in shaping user actions.
- Fewer transaction costs appeal to 23 users (~23%), thus cost saving becomes an important switching motive.
- Government subsidies affect 15 users (~15%), indicating that digital transactions policies have a moderate influence.
- There are other reasons, which explain 7 users (~7%), that minor factors such as security, UI/UX, or individual services may also play a part.

Contribution to Objective

- Knowledge of user motivation: The findings emphasize that users value rewards above efficiency or trust when selecting a platform.
- Implication for digital payment providers: Incentive platforms and cashback platforms might lose fewer users even if they are not the most efficient.
- Long-term adoption potential: If you switch based on rewards, then we must investigate whether they remain on the network because of productivity or keep shifting between services looking for improved offers.

ANALYSIS OF AWARENESS OF SUSTAINABLE DIGITAL BANKING PRACTICES

Figure 6. Digital Banking Practices Awareness Analysis

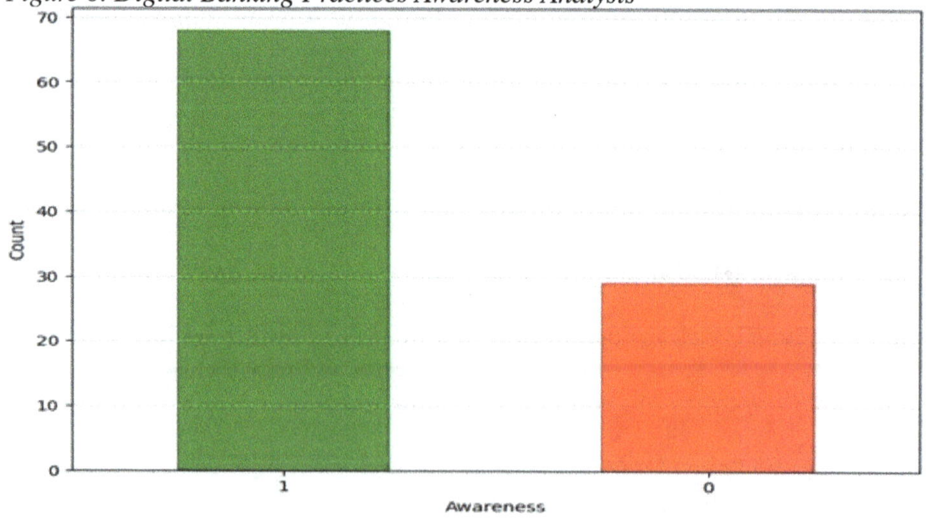

Table 13. Digital Banking Practices Awareness level amongst users

Awareness Level	Count	Percentage (%)
Aware (Yes)	68	70%
Not Aware (No)	29	30%

Key Findings & How It Had Contributed to the Objective

High Awareness (70%) Had Supported the Adoption of Sustainable Banking

- Seventy percent of the respondents had been aware of sustainable digital banking, showing that earlier awareness campaigns had been quite effective.
- This awareness had spurred the embracement of environmentally friendly banking practices, including paperless banking, energy-saving banking, and lower dependence on cash.
- Contribution towards the goal: The existence of an educated user population had been in consonance with the goal towards supporting sustainable financial behaviours, propelling the usage of digital banking in a green manner.

Low Awareness Among 30% Had Indicated the Need for Additional Efforts

- Though there were high levels of awareness, 30% of the respondents were not aware, which indicated that the awareness initiatives had not yet penetrated all sections.
- Contribution to the goal: This gap had brought to the forefront the necessity of more focused educational campaigns, promotional efforts, and outreach programs in order to achieve wider awareness and participation in sustainable banking.

Strategic Lessons Learned for Future Development

- Previous evidence had indicated that awareness campaigns, especially those with incentives and online participation, had been more effective in encouraging sustainable banking.

- Earlier marketing efforts had focused on sustainability, but more targeted communication and incentives were needed to translate awareness into active engagement.
- Contribution to the goal: The results had yielded useful information for financial institutions to make their strategies more precise, with subsequent sustainability efforts being more effective and comprehensive.

The study revealed that awareness initiatives had partially succeeded in fostering the sustainable adoption of digital banking. However, a substantial knowledge gap of 30% remained, indicating the need for further targeted interventions. Leveraging these insights, financial institutions were able to design more focused awareness programs that ultimately strengthened the transition towards sustainable banking practices.

FINANCIAL INCLUSION AS A CATALYST FOR SUSTAINABLE DEVELOPMENT

This analysis examines how trends in financial inclusion and digital financial services can be harnessed to advance environmental sustainability goals. Drawing on data from the Global Findex Database 2021, the study identifies key insights that inform strategies integrating financial inclusion with sustainable development.

FINANCIAL INCLUSION TRENDS DRIVING SUSTAINABILITY

The Global Findex Database 2021 highlights the rapid expansion of digital financial services, particularly in emerging economies. The share of adults who make or receive digital payments rose from 35% in 2014 to 57% in 2021, signifying a structural shift away from cash-based transactions. This digital transition presents an opportunity to reduce the environmental impact associated with physical banking activities, including paper use and cash logistics. Additionally, 39% of adults opened their first financial account primarily to receive wages or government transfers, underscoring the critical role of formal financial channels in economic empowerment. These developments underscore the importance of policies that merge financial inclusion with environmental sustainability, including incentives for digital banking and eco-friendly financial products.

DIGITAL FINANCIAL SERVICES AS INSTRUMENTS OF GREEN FINANCE

The study further observes that 20% of adults made their first digital payment to merchants during the COVID-19 pandemic, illustrating how external shocks can accelerate financial digitalization. This digitalization directly contributes to sustainability through:

- Reductions in paper-based transactions, decreasing deforestation and waste generation.
- Lowering the carbon footprint of physical banking infrastructure.
- Encouraging green financing instruments, such as fee reductions for online banking versus in-person services.

Savings, Borrowing, and Sustainable Investment Opportunities

Financial inclusion efforts must also address saving and borrowing behaviors. Data shows that 25% of adults saved via formal accounts, many doing so for the first time. However, borrowing remains largely informal, with only under half of the 50% who borrow doing so through formal institutions. Financial institutions can capitalize on this by developing:

- Green lending programs offering preferential interest rates on sustainable investments within the green economy.
- Microfinance initiatives support renewable energy solutions, such as solar home systems in rural communities.
- Sustainability-linked savings products that channel deposits into green bonds or eco-investment funds.

Addressing Financial Inclusion Challenges to Promote Sustainability

Despite progress, financial exclusion continues to affect 1.4 billion adults globally. Key impediments include:

- Financial illiteracy, especially among women and older adults.
- Geographic barriers limiting access to financial institutions.
- Security and trust concerns deterring engagement with digital finance.

To effectively link financial inclusion with sustainability, comprehensive responses are required:

- Educational campaigns to improve financial literacy and encourage environmentally responsible financial behaviors.
- Expanding mobile banking and fintech penetration among economically disadvantaged groups.
- Strengthening consumer protection regulations to build confidence in digital financial systems.

This review concludes that environmental sustainability and financial inclusion are deeply interconnected. The widespread adoption of digital financial services reduces the environmental footprint of traditional banking, while green finance instruments enable sustainable borrowing and investment. Overcoming obstacles such as financial literacy and access is vital to ensuring sustainable financial practices reach all societal segments. By embracing digital payments, green banking policies, and green financing initiatives, financial inclusion can be strategically aligned with sustainable development goals. These findings support the research aim by providing policy recommendations that promote both economic empowerment and environmental stewardship.

CONCLUSION

This research investigated the critical intersection of financial inclusion and environmental sustainability within mobile banking systems. As reliance on digital financial services grows alongside an increasing emphasis on sustainability, understanding how these elements interact is essential for developing responsible and efficient banking solutions.

A comprehensive analysis was performed on factors including financial behaviors, platform efficiency, user preferences, and awareness levels. Evaluating the efficiency distribution of prominent mobile banking platforms revealed their capacity to provide seamless digital financial services. Insights into user preferences helped identify key determinants of adoption rates and the principal drivers behind the shift towards digital banking. Furthermore, the study assessed awareness of sustainable banking practices and explored its impact on user behavior and adoption trends. This holistic, data-driven approach effectively met research objectives by identifying strengths, weaknesses, and areas for enhancement in mobile banking within the combined context of environmental sustainability and financial inclusion.

ASSESSING SUSTAINABLE AND EFFICIENT OPERATIONAL PRACTICES IN MOBILE BANKING

The research rigorously evaluated the operational effectiveness of various mobile banking platforms to determine their viability in promoting financial inclusion. Efficiency scores varied widely among platforms, reflecting disparities in user experiences, service stability, and technology uptake. Digital payment platforms such as Paytm and Google Pay exhibited a broad range of efficiency scores. While their popularity is attributed to ease of use, rapid transactions, and integration across financial services, fluctuations in efficiency highlight operational challenges including transaction failures, security vulnerabilities, and network dependencies—factors that may erode user trust and retention. Despite these issues, their widespread adoption underscores their vital role in extending financial access to previously unbanked populations.

In contrast, traditional banking apps like SBI YONO and HDFC Mobile Banking displayed higher and more stable efficiency levels, suggesting more reliable services with fewer interruptions. However, their user adoption rates remain relatively lower compared to digital payment platforms, possibly due to complex onboarding processes, less intuitive interfaces, and limited awareness of their features. The findings underscore that operational effectiveness is fundamental to sustaining user engagement and fostering inclusive financial services. While fintech-powered platforms command extensive reach, they must address service inconsistencies to ensure equitable access. Legacy banks, in turn, should enhance usability and accessibility to remain competitive in the evolving digital finance ecosystem.

Overall, these insights highlight the pivotal role of mobile banking platforms in promoting financial inclusion while calling attention to necessary improvements in efficiency, accessibility, and sustainability to maximize their benefits.

IDENTIFYING CHALLENGES AND BARRIERS TO SUSTAINABLE MOBILE BANKING PRACTICES

Correlation analyses and user sentiment evaluations revealed several impediments to adopting sustainable digital banking. A significant barrier is the limited awareness of sustainable banking initiatives; only 68% of users surveyed were cognizant of such efforts, leaving a noteworthy 32% uninformed. This gap points to deficiencies in financial education and outreach, as many consumers lack recognition of the environmental implications of digital banking or the sustainability programs implemented by financial institutions. Without adequate awareness, users are less inclined to engage with eco-friendly banking options like paperless transactions, digital savings, or low-carbon investments.

Technological inconsistencies emerged as a further obstacle. Users reported variable platform performance, including transaction errors, sluggish interfaces, and service outages, undermining confidence in digital systems and discouraging full adoption—essential to reducing the environmental footprint of traditional banking. Security concerns also rank highly as adoption barriers. Given the prevalence of cyber-attacks, fraud, and data breaches targeting online banking—particularly fintech services—users express apprehension about fully digitizing their financial transactions. Enhancing cybersecurity measures and improving transparency around data protection policies are vital to alleviating these fears and encouraging broader acceptance.

Another challenge is digital literacy deficits, notably affecting older adults and rural communities who struggle with complex mobile interfaces. This digital divide creates accessibility issues, impeding the shift from cash-based to digital financial services. Additionally, many still rely on informal financial sources such as local moneylenders or borrowing from relatives, typically driven by mistrust in formal institutions, lack of proper documentation, or inability to meet banking requirements. Addressing these challenges through targeted financial inclusion programs and user-friendly digital solutions is crucial for bridging existing gaps.

This research emphasizes that while mobile banking holds promise for enhancing financial inclusion and sustainability, overcoming awareness deficits, technological unreliability, security vulnerabilities, and digital literacy gaps is essential. Doing so will foster a more inclusive, accessible, and environmentally responsible financial future.

RECOMMENDATIONS FOR MERGING FINANCIAL INCLUSION AND ENVIRONMENTAL SUSTAINABILITY

A persistent challenge to green digital banking adoption is the lack of digital literacy among certain user demographics. Elderly populations and rural inhabitants, in particular, face obstacles in navigating mobile banking platforms, creating a digital divide that hampers access and complicates migration from cash-based transactions to digital financial services.

Moreover, a substantial share of the population continues to resort to informal financial channels—borrowing money from friends, relatives, or local lenders—instead of engaging with formal banking systems. This reliance often stems from distrust toward formal institutions, absence of necessary documentation, or failure to meet eligibility criteria. Thus, fostering financial inclusion demands concerted efforts to improve awareness through educational programs and to simplify digital banking experiences.

Collectively, these insights highlight the need for comprehensive strategies that blend financial inclusion initiatives with environmental sustainability objectives, ensuring that mobile banking emerges not only as a tool for economic empowerment but also as a driver of sustainable development.

REFERENCES

Ahmad, F., Boumaiza, A., Sanfilippo, A., & Al-Fagih, L. (2025). A Detailed Comprehensive Role of Digital Technologies in Green Finance Initiative for Net-Zero Energy Transition. *Advanced Energy and Sustainability Research, 2500066*(10), 2500066. Advance online publication. DOI: 10.1002/aesr.202500066

Arner, D. W., Buckley, R. P., Zetzsche, D. A., & Veidt, R. (2020). Sustainability, FinTech and Financial Inclusion. *European Business Organization Law Review, 21*(1), 7–35. DOI: 10.1007/s40804-020-00183-y

Bernard, J., Steinführer, A., Klärner, A., & Keim-Klärner, S. (2023). Regional opportunity structures: A research agenda to link spatial and social inequalities in rural areas. *Progress in Human Geography, 47*(1), 103–123. DOI: 10.1177/03091325221139980

Del Sarto, N., & Ozili, P. K. (2025). FinTech and financial inclusion in emerging markets: A bibliometric analysis and future research agenda. *International Journal of Emerging Markets, 20*(13), 270–290. DOI: 10.1108/IJOEM-08-2024-1428

Fu, C., Lu, L., & Pirabi, M. (2023). Advancing green finance: A review of sustainable development. *Digital Economy and Sustainable Development, 1*(1), 20. DOI: 10.1007/s44265-023-00020-3

Gonzalez, E. D., Sijariya, R., Singh, A. K., & Garg, V. (2025). Integrating Climate Change, Social Responsibility and Electronic Financial Inclusion: A Pathway to Sustainable Development. In *Climate Change and Social Responsibility* (pp. 147–164). Emerald Publishing Limited. https://www.emerald.com/insight/content/doi/10.1108/S2043-052320250000025008/full/html

Jha, S., & Dangwal, R. C. (2025). Fintech services and financial inclusion: A systematic literature review of developing nations. *Journal of Science and Technology Policy Management, 16*(7), 1167–1198. DOI: 10.1108/JSTPM-03-2023-0034

Kohli, K., Kumar, B., Kohli, P., & Kashyap, M. (2024). Examining the factors affecting mobile banking adoption among users in the Himalayan region of Uttarakhand: An extension of the technology acceptance model through structural equation modelling. *International Journal of Electronic Banking, 4*(3), 294–317. DOI: 10.1504/IJEBANK.2024.139221

Lindfors, L. (2024). *The role of user-centric design in global platform economy value creation.* https://aaltodoc.aalto.fi/items/e111add3-f142-4c11-a488-1d05c58dcd7c

Mamadiyarov, Z. (2024). Comparative Analysis of Digital Banking Adoption in Central Asia: Opportunities and Barriers. *International Multi-Disciplinary Journal of Education*, *2*(8), 452–458.

Mookerjee, J., Bhuriya, M. K., & Josphin, R. (n.d.). *Digital Banking and Financial Inclusion in Rural Economies*. Retrieved September 17, 2025, from https://core.ac.uk/download/pdf/640070834.pdf

Nandru, P., Byram, A., & Rentala, S. (2015). Factors influencing financial inclusion through banking services. [JCRM]. *Journal of Contemporary Research in Management, 10*(4). http://jcrm.psgim.ac.in/index.php/jcrm/article/download/448/302

Naruetharadhol, P., Ketkaew, C., Hongkanchanapong, N., Thaniswannasri, P., Uengkusolmongkol, T., Prasomthong, S., & Gebsombut, N. (2021). Factors Affecting Sustainable Intention to Use Mobile Banking Services. *SAGE Open*, *11*(3), 21582440211029925. DOI: 10.1177/21582440211029925

Patel, A. S., Rao, V. K., & Radhakrishnan, M. K. (2023). Impact of mobile banking platforms PayTM and Google Pay on financial inclusion in rural and Semi-Urban areas in India. *Journal of Finance and Accounting*, *7*(5), 113–122. DOI: 10.53819/81018102t4205

Putrevu, J., & Mertzanis, C. (2024). The adoption of digital payments in emerging economies: Challenges and policy responses. *Digital Policy. Regulation & Governance*, *26*(5), 476–500.

Sarfaraz, J. (2017). Unified theory of acceptance and use of technology (UTAUT) model-mobile banking. *Journal of Internet Banking and Commerce*, *22*(3), 1–20.

Shaban, A. (2024). Digitalization and Exclusion—Digital Divides and Development. In Shaban, A. (Ed.), *Digital Geographies—Theory, Space, and Communities* (pp. 255–496). Springer Nature Singapore., DOI: 10.1007/978-981-97-4734-4_4

Shaikh, A. A., Alamoudi, H., Alharthi, M., & Glavee-Geo, R. (2023). Advances in mobile financial services: A review of the literature and future research directions. *International Journal of Bank Marketing*, *41*(1), 1–33. DOI: 10.1108/IJBM-06-2021-0230

Sharma, A., & Piplani, N. (2017). Digital banking in India: A review of trends, opportunities and challenges. [IRJMST]. *International Research Journal of Management Science & Technology*, *8*(1), 168–180.

Sikka, V., & Bhayana, P. (2024). Barriers to Comprehensive Financial Inclusion Across the Globe: From Sociocultural Norms to Systemic Challenges. In Mansour, N., Baral, S., & Garg, V. (Eds.), *E-Financial Strategies for Advancing Sustainable Development* (pp. 89–126). Springer Nature Switzerland., DOI: 10.1007/978-3-031-67523-2_7

Somu, B. (2025). *The Future of Financial IT: Agentic Artificial Intelligence and Intelligent Infrastructure in Modern Banking*. Deep Science Publishing. https://books.google.com/books?hl=en&lr=&id=VeplEQAAQBAJ&oi=fnd&pg=PA1&dq=Agentic+AI,+working+in+tandem+with+digital+asset+innovation,+introduces+adaptive+intelligence+into+these+markets-+capable+of+autonomously+managing+tokenized+securities,+monitoring+carbon+credit+compliance,+and+optimizing+risk+strategies+for+next-generation+green+instruments&ots=Fxwl2XSwul&sig=5YqN-BLKmdmw46YaPfa7FhctPWo

Valerievna, T. O., & Igorevna, G. I. (2024). Sustainable financing as a factor influencing eco-friendly business transformation: Existing challenges and potential solutions. *The American Journal of Interdisciplinary Innovations and Research*, *6*(10), 8–20. DOI: 10.37547/tajiir/Volume06Issue10-02

van Zanden, J. L. (2023). Examining the relationship of information and communication technology and financial access in Africa. *Journal of Business and Economic Options*, *6*(3), 26–36.

Vannoy, S. A., & Palvia, P. (2010). The social influence model of technology adoption. *Communications of the ACM*, *53*(6), 149–153. DOI: 10.1145/1743546.1743585

Vishnuvardhan, B., Manjula, B., & Lakshman Naik, R. (2020). A Study of Digital Banking: Security Issues and Challenges. In K. S. Raju, A. Govardhan, B. P. Rani, R. Sridevi, & M. R. Murty (Eds.), *Proceedings of the Third International Conference on Computational Intelligence and Informatics* (Vol. 1090, pp. 163–185). Springer Singapore. DOI: 10.1007/978-981-15-1480-7_14

Chapter 7
Green Financial Innovations and Climate Risk Management:
The Role of Green Bonds, ESG, and Sustainable Finance in Shaping Financial Markets

P. A. Manisha
https://orcid.org/0009-0001-1299-2081
Dayananda Sagar College of Engineering, India

H. P. Ramya
https://orcid.org/0000-0002-2048-807X
Dayananda Sagar College of Engineering, India

ABSTRACT

The accelerating impacts of climate change are reshaping global financial markets, driving the need for innovative tools to manage climate risk and mobilize capital for sustainable growth. Green financial innovations—particularly green bonds, Environmental, Social, and Governance (ESG) investing, and broader sustainable finance mechanisms—have emerged as key instruments linking financial performance with climate resilience. Green bonds channel capital into renewable energy, and low-carbon infrastructure, while ESG frameworks guide investors in assessing both financial and environmental risks. At the systemic level, sustainable finance is reinforced by policy initiatives such as the EU Taxonomy, pushing markets toward transparency and accountability. Yet, challenges such as greenwashing, inconsistent ESG metrics, and limited accessibility in emerging markets persist. This chapter

DOI: 10.4018/979-8-3373-6766-8.ch007

examines the evolution, opportunities, and limitations of these innovations, emphasizing their potential to transform financial markets into drivers of the energy transition and climate risk management.

1. INTRODUCTION & RESEARCH BACKGROUND

The financial world is undergoing a significant transformation as the impacts of climate change become more pronounced, reshaping the way industries and economies function. Climate change is no longer a distant concern; its repercussions are already evident, from rising global temperatures and extreme weather events to shifts in agricultural productivity and biodiversity loss. These changes have triggered a fundamental shift in how businesses, governments, and individuals approach financial planning and risk management. As climate-related risks intensify, they have started to penetrate the core of the financial system, highlighting a need for robust mechanisms that can manage these risks and channel resources toward sustainable development.

In the past, financial markets focused primarily on maximizing short-term profits and achieving financial returns without taking into consideration the environmental and social costs of investments. However, the growing recognition that climate risks are systemic and intertwined with the stability of financial markets has led to the emergence of green financial innovations. These innovations are designed to link the financial system to the global imperative of sustainability, bringing environmental, social, and governance (ESG) factors to the forefront of investment decision-making. One of the most notable tools in this green finance ecosystem is green bonds, followed closely by the rise of ESG investing and broader sustainable finance frameworks.

The Need for Green Financial Innovations

Green financial instruments were developed to address the critical challenges posed by climate change. They represent a fundamental shift from the traditional financial approach by incorporating long-term environmental considerations into decision-making processes. Green bonds, for example, are debt instruments specifically designed to finance projects that have positive environmental impacts. These include initiatives such as renewable energy, energy efficiency projects, and low-carbon infrastructure development. In 2007, the European Investment Bank issued the first green bond, marking the beginning of a rapidly growing market that has since gained global momentum.

Green bonds channel capital toward projects that not only promise financial returns but also contribute to the global effort to mitigate climate risks. These bonds have provided governments, corporations, and other entities with a way to secure the necessary funding for clean energy, sustainable agriculture, and climate-resilient infrastructure, among other initiatives. By doing so, green bonds help bridge the financing gap for projects that promote sustainability and reduce carbon emissions.

Similarly, ESG investing is another crucial innovation that aims to align financial returns with positive environmental and social impacts. ESG investing involves the incorporation of environmental, social, and governance factors into investment decisions. Investors consider not only the financial performance of a company but also how well it manages its environmental impact, its social responsibilities, and the integrity of its governance practices. This holistic approach reflects a broader understanding that financial returns cannot be the sole measure of success. Instead, investors now recognize that long-term value is inherently tied to sustainability—both in terms of the environment and society.

The broader concept of sustainable finance encompasses a variety of financial tools and strategies that facilitate the transition to a low-carbon economy. This includes impact investing, socially responsible investing (SRI), and other mechanisms that integrate environmental and social factors into the financial system. Sustainable finance has gained traction as a means of directing capital toward projects and companies that contribute to long-term societal benefits, addressing challenges such as climate change, poverty, inequality, and resource depletion. This shift represents a fundamental transformation in how financial markets operate, moving from a short-term profit-oriented mindset to one that seeks to address long-term global challenges.

Challenges to Green Financial Innovations

Despite their promise, green financial innovations face significant challenges that threaten to limit their potential. One of the primary issues is greenwashing, which refers to the practice of misleadingly labeling financial products as "green" or environmentally friendly, without a genuine commitment to sustainability. Greenwashing undermines the credibility of green financial instruments, creating skepticism among investors and detracting from the overall effectiveness of green finance in addressing climate change. As more financial products claim to be green or sustainable, the risk of misleading claims increases, making it difficult for investors to distinguish between truly sustainable projects and those that are merely using environmental language for marketing purposes.

The problem of greenwashing is exacerbated by the lack of standardized ESG metrics. The measurement and reporting of ESG performance is inconsistent, with no universally accepted framework for assessing the environmental and social impact of investments. As a result, ESG metrics vary widely between companies, industries, and countries. This inconsistency complicates the task of comparing ESG investments and makes it difficult for investors to assess the true impact of their portfolios. Without standardized metrics, it is challenging to ensure that investments are genuinely aligned with sustainability goals and that they meet the expected environmental and social criteria.

Furthermore, the lack of standardized reporting and verification practices also hinders the ability of investors to track the environmental and social outcomes of their investments over time. A lack of transparency around the actual impacts of investments can deter capital flows into sustainable sectors and undermine trust in the green finance market. Efforts to address greenwashing and improve transparency are essential to restore confidence and ensure that green finance can deliver the intended outcomes.

Another significant challenge is the accessibility of green finance in emerging markets. While developed economies have made significant strides in implementing sustainable finance frameworks, many emerging markets face challenges in adopting these innovations. These markets often lack the necessary financial infrastructure, regulatory frameworks, and access to capital required to support sustainable projects. In some cases, investors may be hesitant to invest in emerging markets due to concerns about regulatory uncertainty, political instability, and the potential for greenwashing. Addressing these barriers requires a concerted effort from international financial institutions, governments, and the private sector to build the necessary infrastructure and regulatory frameworks that will allow green finance to flourish in these regions.

Green Financial Innovations in Climate Risk Management

Green financial innovations are not only important for advancing sustainability but are also integral to managing climate-related financial risks. As climate change continues to accelerate, the financial sector is increasingly recognizing that climate risks—such as the physical impacts of extreme weather events, rising sea levels, and disruptions to supply chains—pose a significant threat to market stability. Climate-related financial risks can affect both the short-term profitability and long-term viability of investments, especially those that are heavily reliant on fossil fuels or other environmentally harmful industries. Investors are now more aware of the need to consider these risks when making investment decisions.

Green bonds, ESG investing, and sustainable finance frameworks are tools that help investors manage climate risks by aligning their portfolios with climate-resilient projects and companies. Green bonds, for example, fund infrastructure projects that reduce carbon emissions or increase energy efficiency, which are essential for mitigating climate risks. ESG investing, by focusing on companies with strong environmental management practices, helps reduce exposure to industries that are vulnerable to climate-related disruptions. In this way, green financial innovations contribute to the development of a financial system that is better equipped to manage climate risks and adapt to a changing climate.

Furthermore, the growing momentum behind sustainable finance is being driven by policy initiatives that encourage financial markets to account for climate risks. For instance, the European Union has introduced the EU Taxonomy, a classification system that defines what constitutes a sustainable economic activity, helping investors navigate the complexities of sustainable investments. The Task Force on Climate-related Financial Disclosures (TCFD) has also created guidelines for companies to disclose climate-related financial risks, improving transparency and enabling investors to assess the climate resilience of their investments. These initiatives help shift the financial system toward a more transparent, accountable, and climate-aware model.

Purpose of the Chapter

This chapter aims to explore the evolution, opportunities, and limitations of green financial innovations in the context of climate risk management. Specifically, it examines the roles of green bonds, ESG investing, and sustainable finance in shaping financial markets and driving the transition to a low-carbon economy. By reviewing the existing literature, identifying key gaps, and analyzing emerging opportunities, this chapter seeks to provide a comprehensive understanding of the challenges and potential of these innovations in addressing climate change and promoting sustainability.

The chapter will also assess the systemic challenges facing green finance, such as greenwashing, inconsistent reporting, and limited accessibility in emerging markets. By critically evaluating these issues, the chapter will highlight the steps needed to enhance the effectiveness of green financial innovations, ensuring that they can deliver on their promise to create a more sustainable, resilient global economy. Ultimately, the goal is to provide actionable insights for policymakers, investors, and financial institutions to help advance the integration of green finance into mainstream financial markets.

2. LITERATURE REVIEW

The literature on green finance and climate risk management has witnessed substantial growth in recent years, reflecting the increasing recognition that financial markets and systems need to integrate environmental, social, and governance (ESG) factors to address the urgent challenges posed by climate change. This body of work covers a broad spectrum of topics, from the development of specific financial instruments like green bonds to the evolving theoretical foundations of sustainable finance. It also delves into the growing role of institutional investors, policymakers, and regulators in shaping green finance frameworks that can drive meaningful action against climate risks.

Green Bonds: Pioneering the Green Finance Market

Green bonds are among the earliest financial instruments designed to address climate change and environmental degradation. Introduced in 2007 by the European Investment Bank, green bonds were created to raise capital for projects that directly support environmental sustainability. Since then, their popularity has soared, and they have become a cornerstone of green finance. These bonds are specifically aimed at funding projects related to renewable energy, energy efficiency, low-carbon infrastructure, pollution control, and climate resilience. Green bonds offer investors an opportunity to support projects that contribute to environmental sustainability while providing them with a return on investment.

Several studies have focused on understanding the market development of green bonds, including their issuance, demand, and regulatory frameworks. A key area of research has been identifying the criteria that qualify projects for green bond funding. These criteria are critical, as they ensure that the funds raised through green bonds are directed toward genuine environmental projects. The Green Bond Principles (GBP), established in 2014 by the International Capital Market Association (ICMA), set out voluntary guidelines for issuers to ensure that the funds raised are used for green projects. However, the lack of universally accepted standards has led to concerns about greenwashing—the practice of mislabeling non-sustainable projects as "green" to attract investors.

The literature has also explored the financial performance of green bonds. Some studies suggest that green bonds offer comparable returns to traditional bonds, with some even outperforming their conventional counterparts due to growing investor demand for sustainable investment options. Research by Flammer (2021) found that green bonds have the potential to generate competitive financial returns while delivering environmental benefits. The growing demand for green bonds is driven by both institutional investors seeking to align their portfolios with sustainability goals and retail investors interested in socially responsible investments. However, concerns persist about the liquidity and transparency of the green bond market, as well as its ability to scale to meet the growing global need for climate finance.

ESG Investing: Integrating Environmental, Social, and Governance Factors

ESG investing is another critical component of green finance. It involves incorporating environmental, social, and governance (ESG) factors into investment decisions to ensure that financial returns are aligned with broader societal goals. The integration of ESG factors into investment strategies has gained momentum in recent years, with growing recognition that companies and investments that prioritize sustainability may have better long-term financial prospects. ESG investing covers a broad spectrum of activities, from screening companies based on ESG criteria to shareholder engagement and active ownership in driving corporate sustainability practices.

The academic literature on ESG investing is vast, focusing on various aspects such as the relationship between ESG performance and financial returns, the impact of ESG factors on risk management, and the effectiveness of ESG integration into investment strategies. The relationship between ESG performance and financial returns is one of the most widely debated topics. Early studies were divided, with some showing that ESG investments underperformed compared to traditional investments, while others found that ESG-conscious companies tended to outperform in the long run. A study by Friede, Busch, and Bassen (2015) reviewed over 2,000 studies and found that around 90% of them showed a positive relationship between ESG performance and financial performance, either in terms of reduced risk or enhanced long-term returns.

Despite the positive trends in ESG investing, the financial performance of ESG investments is not without controversy. Some studies indicate that ESG investments may yield lower short-term returns due to the screening process or higher operating costs related to sustainability efforts. In contrast, other research suggests that ESG factors can help mitigate long-term risks associated with environmental changes, regulatory shifts, and social expectations. For instance, companies with strong ESG practices tend to be better equipped to manage reputational risks, regulatory pressures, and changing market dynamics, leading to improved financial performance over time.

The literature has also explored the challenges of measuring and reporting ESG performance. A significant issue in ESG investing is the lack of standardized metrics for evaluating ESG factors. Different ESG rating agencies and data providers use diverse methodologies to assess companies' environmental, social, and governance practices, leading to discrepancies in the ratings and confusion among investors. A study by Berg, Kölbel, and Rigobon (2019) highlights the inconsistency in ESG ratings across providers and the lack of transparency in how these ratings are calculated. The absence of uniform ESG standards complicates the ability of investors to make informed decisions and poses a significant barrier to the growth of ESG investing.

Another challenge associated with ESG investing is the risk of greenwashing, where companies or funds claim to be ESG-compliant without genuinely adhering to sustainability practices. Greenwashing not only misleads investors but also undermines the effectiveness of ESG investing in promoting sustainability. Scholars have called for stricter regulation and more robust disclosure requirements to address these issues and ensure the credibility of ESG investments. Initiatives like the EU's Sustainable Finance Disclosure Regulation (SFDR) and the TCFD (Task Force on Climate-related Financial Disclosures) guidelines aim to improve the transparency and comparability of ESG data.

Sustainable Finance: A Broader Framework for Climate Risk Management

While green bonds and ESG investing have received significant attention, sustainable finance represents a broader framework that includes various financial instruments designed to promote environmental sustainability and manage climate-related risks. Sustainable finance encompasses not only green bonds and ESG investing but also includes impact investing, social and sustainable bonds, and socially responsible investing (SRI). These instruments seek to mobilize capital for projects that have a positive impact on the environment and society, such as renewable energy, poverty reduction, sustainable agriculture, and affordable housing.

The literature on sustainable finance also highlights the role of policy initiatives and regulatory frameworks in shaping the growth of the field. The European Union's Taxonomy for Sustainable Activities, for example, provides a classification system for sustainable investments, enabling investors to identify which activities meet specific sustainability criteria. This initiative aims to bring clarity and standardization to the sustainable finance landscape, helping to direct capital to projects that contribute to the EU's climate goals. Similarly, the TCFD has provided guidelines for companies to disclose climate-related financial risks, enhancing transparency and helping investors assess the climate resilience of their portfolios.

Researchers have explored the role of these policy initiatives in driving the growth of sustainable finance. A study by Busch and Lewandowski (2020) examined the impact of the EU Taxonomy on investor decision-making and found that it increased transparency and enabled investors to make more informed choices regarding sustainable investments. The development of such frameworks is seen as essential to achieving the scale of financing required to address global climate challenges. By providing clear definitions and encouraging disclosure, these policies help reduce the uncertainty that often accompanies sustainable investments and increase the flow of capital into low-carbon and climate-resilient projects.

Another important aspect of sustainable finance is the role of financial institutions in promoting environmental sustainability. Banks, asset managers, and institutional investors have an essential role in driving the shift toward a sustainable economy. Research by Krueger, Sautner, and Starks (2020) highlights how financial institutions are integrating ESG considerations into their investment strategies and corporate practices. The study shows that investors are increasingly using ESG criteria to evaluate companies, pushing them to adopt more sustainable practices and contribute to global climate goals. This shift is further reinforced by growing demand from clients and stakeholders who prioritize sustainability in their investment decisions.

Gaps in the Literature

While the literature on green finance and sustainable investing has expanded significantly, several gaps remain that hinder a full understanding of the potential of these innovations. One of the major gaps is the limited research on the impact of green finance in emerging markets. Although much of the focus has been on developed economies, emerging markets are often more vulnerable to the impacts of climate change and may face more significant barriers to accessing sustainable finance. These barriers include insufficient financial infrastructure, regulatory uncertainty, and a lack of awareness of green financial products. Addressing these gaps is essential to ensure that green finance reaches all regions that need it the most.

Another gap is the challenge of standardizing ESG metrics. As highlighted earlier, the lack of consistent ESG reporting and measurement frameworks is a significant obstacle to the widespread adoption of ESG investing. Researchers and policymakers continue to debate the best approach to standardizing ESG metrics, with some advocating for more rigorous global reporting standards, while others argue for more flexible approaches that can accommodate local contexts. Regardless of the approach, addressing the inconsistency in ESG metrics is critical for improving transparency and ensuring that investors can accurately assess the sustainability of their investments.

Lastly, there is a need for more research on the effectiveness of regulatory and policy frameworks in supporting the growth of green finance. While initiatives like the EU Taxonomy and TCFD guidelines are essential steps forward, their implementation and impact on the market remain uncertain. More research is needed to evaluate the success of these frameworks and identify ways to improve them to ensure that they effectively drive the transition to a low-carbon economy.

3. GAP ANALYSIS IN GREEN FINANCE: EXPLORING KEY CHALLENGES AND OPPORTUNITIES

The rise of green finance has been one of the most notable developments in the global financial system over the last decade. With growing awareness of climate change, social inequality, and environmental degradation, the financial sector has become an essential driver of sustainable development. However, despite significant progress, substantial gaps remain in the literature and practice of green finance that limit its full potential. This gap analysis highlights some of the key barriers and challenges that hinder the growth of green finance, including the lack of standardized ESG (Environmental, Social, Governance) metrics, underrepresentation of emerging markets, and the prevalence of greenwashing.

Lack of Standardized ESG Metrics

One of the most pressing challenges in green finance is the lack of universally accepted and standardized ESG metrics. The absence of uniform frameworks for measuring and reporting ESG performance has led to inconsistencies in how environmental, social, and governance factors are defined, measured, and communicated across different investment products and financial instruments. This problem is especially prominent in the context of ESG reporting and performance evaluation, where the lack of standardized metrics impedes the comparability and transparency of green investments.

Different financial institutions, investors, and analysts often rely on their own proprietary ESG scoring systems, creating a fragmented landscape where it is difficult to assess the true sustainability of an investment. For instance, one rating agency may give a particular company a high ESG score based on its environmental initiatives, while another might assign it a low score based on its governance practices. As a result, investors are left with incomplete or even misleading information, which complicates decision-making processes.

The lack of standardization also makes it challenging to assess the actual environmental and social impacts of investments. While some financial products are marketed as “green” or “sustainable,” without standardized criteria to define what qualifies as such, it is difficult to determine whether these products truly meet the intended sustainability goals or if they are merely capitalizing on the “green” label for marketing purposes.

To address this gap, there is an urgent need for global consensus on ESG reporting standards. Initiatives like the Global Reporting Initiative (GRI), Task Force on Climate-related Financial Disclosures (TCFD), and the Sustainable Accounting Standards Board (SASB) are making strides, but there is still a long way to go before a unified framework is established. Greater collaboration among standard-setting organizations, regulators, and industry stakeholders will be crucial to overcoming this gap.

Underrepresentation of Emerging Markets in Green Finance

While the literature on green finance has largely focused on developed economies, there remains a significant gap in understanding the unique challenges and opportunities faced by emerging markets. Emerging economies, particularly in regions like Sub-Saharan Africa, Asia, and Latin America, have distinct characteristics that make the adoption and implementation of green finance more difficult.

These markets often suffer from underdeveloped financial infrastructure, limited access to capital, and weaker regulatory frameworks, all of which are critical for the growth of green finance. In many cases, local banks and financial institutions lack the expertise or resources to evaluate green investment opportunities, and there is a lack of a clear regulatory environment that promotes green finance practices. Additionally, emerging markets face the challenge of competing with developed economies for green capital, often struggling to attract investment due to perceived higher risks, such as political instability or economic volatility.

Moreover, these markets often have competing development priorities. In many low-income countries, basic infrastructure, poverty alleviation, and healthcare may take precedence over environmental concerns. However, as climate change continues to threaten global stability, the importance of integrating green finance into the development agendas of emerging economies is becoming increasingly evident. Without the necessary financial mechanisms, these regions may miss out on the opportunity to transition to a low-carbon, sustainable economy.

The literature on green finance must therefore broaden its scope to include emerging markets. Research needs to explore the barriers that prevent the flow of green capital into these regions and identify ways in which international financial institutions, such as the World Bank or the International Finance Corporation (IFC), can play a more prominent role in supporting the transition. Further studies should also examine how local policymakers can create the right regulatory frameworks and financial incentives to promote sustainable development in these markets.

Greenwashing and its Impact on Green Finance

Another significant gap in green finance is the issue of greenwashing, which refers to the practice of companies or financial institutions marketing their products or services as "green" or "sustainable" without meeting the actual environmental or social criteria associated with these terms. As green finance has become increasingly popular, many issuers have sought to capitalize on the growing demand for sustainable investments by labeling their products as "green" without making any meaningful changes to their practices or operations.

Greenwashing poses a significant threat to the integrity of the green finance market. It erodes trust among investors and the public, undermining the effectiveness of green finance in achieving its sustainability objectives. Investors who seek to allocate capital to genuinely sustainable projects may unknowingly fund companies that are not truly committed to environmental or social responsibility. This creates a situation where investors are deceived into supporting businesses that do not contribute meaningfully to the transition to a low-carbon economy.

Despite the growing awareness of greenwashing, there is still limited research on how to effectively address this issue through regulatory frameworks and market mechanisms. Although some countries, such as the European Union, have introduced regulations to tackle greenwashing (e.g., the EU Taxonomy for Sustainable Activities), there is a need for more comprehensive studies on the effectiveness of these regulations and the potential for developing global standards for green product labeling.

To combat greenwashing, more research is needed to understand the impact of misleading claims on investor behavior and how regulatory authorities can enhance transparency and accountability in green finance. Additionally, a greater emphasis should be placed on the role of third-party verification and auditing to ensure that companies and financial products meet the environmental and social standards they claim.

4. RESEARCH QUESTIONS & OBJECTIVES IN GREEN FINANCE

Green finance, encompassing innovations such as green bonds and ESG (Environmental, Social, Governance) investing, is an evolving sector that seeks to address the growing urgency of climate risk management. As climate change accelerates and its impacts become increasingly evident, integrating sustainable financial practices into global markets has become more critical than ever. This chapter is driven by three key research questions that aim to uncover the evolution, opportunities, challenges, and potential impact of green financial innovations. The overarching objective is to understand how green finance can play a pivotal role in transforming financial markets to better mitigate climate risk and drive sustainable development.

1. How Have Green Financial Innovations, Including Green Bonds and ESG Investing, Evolved in the Context of Climate Risk Management?

Green finance has witnessed significant developments over the past two decades, largely driven by the rising recognition of the financial sector's role in addressing climate change. The concept of green financial innovations, such as green bonds and ESG investing, has evolved in response to increasing demand for investments that align with environmental and social objectives. Green bonds, first introduced in 2007 by the European Investment Bank, were designed to raise capital for projects that deliver positive environmental outcomes, such as renewable energy infrastructure, energy efficiency improvements, and sustainable agriculture. ESG investing, on the other hand, incorporates environmental, social, and governance factors into the decision-making process, allowing investors to assess how companies manage climate risks and opportunities.

In the context of climate risk management, both green bonds and ESG investments play critical roles. They help direct capital flows towards projects and businesses that are actively mitigating climate risks, reducing carbon footprints, and supporting the transition to a low-carbon economy. Over time, these financial instruments have gained mainstream acceptance, with an increasing number of issuers and investors adopting them as a means of aligning financial goals with sustainability objectives. Understanding how these innovations have evolved—what factors have driven their growth, and how they have influenced risk management strategies—will provide valuable insights into their potential to manage climate-related financial risks.

2. What Are the Opportunities and Challenges Associated with These Innovations in Driving Sustainable Finance and Mitigating Climate Change?

Green finance innovations bring with them significant opportunities to drive sustainable finance and mitigate climate change. These include:

- **Mobilizing Capital for the Energy Transition:** Green finance allows for the channeling of private sector capital into renewable energy and other sustainable projects, accelerating the global shift away from fossil fuels. Green bonds, for example, offer a vehicle for financing large-scale renewable energy infrastructure projects, such as wind farms or solar energy plants.
- **Promoting Corporate Responsibility:** ESG investing has gained traction as more investors demand companies demonstrate their commitment to environmental sustainability, social responsibility, and robust governance practices.

This has the potential to drive corporate behavior towards more sustainable business practices, particularly in high-carbon sectors.

- **Increasing Access to Capital for Sustainable Projects:** Green finance offers an avenue for funding projects that might not otherwise receive traditional financing, particularly in emerging markets where there is often a lack of infrastructure and capital for green initiatives.

However, these innovations also face challenges, including:

- **Greenwashing:** As green finance grows, some firms engage in greenwashing—misleading claims about the environmental or social benefits of financial products—diluting the credibility of the sector and undermining its effectiveness.
- **Inconsistent ESG Metrics:** The absence of standardized ESG reporting frameworks hampers the ability of investors to assess the true sustainability of investments, creating confusion and inconsistency in the market.
- **Limited Accessibility in Emerging Markets:** While green finance has flourished in developed economies, emerging markets often face significant barriers, including weak regulatory frameworks, limited access to capital, and political instability. These factors make it difficult for these regions to fully embrace green financial products, which could otherwise support sustainable development.

Exploring these opportunities and challenges will provide a deeper understanding of how green finance can be optimized to foster long-term sustainability and effectively address climate change.

3. How Can the Integration of Green Finance into Global Financial Markets Be Improved to Enhance Transparency, Accountability, and Impact?

The global integration of green finance is essential for scaling up investments in climate solutions and ensuring that sustainable finance becomes the norm rather than the exception. However, there are several obstacles to achieving full integration:

- **Lack of Standardized ESG Reporting and Accountability:** Different regions and financial institutions use varied approaches to measure and report ESG factors, leading to inconsistencies and difficulties in comparing investment opportunities. Standardized metrics are needed to enhance transparency, improve accountability, and ensure that capital is effectively allocated to sustainable projects.

- **Regulatory Gaps and Uncertainty:** In many regions, there is a lack of comprehensive regulations that promote green finance and provide clear guidance on what qualifies as "green." Without strong regulatory frameworks, green finance risks being undermined by low-quality or misleading investment products.
- **Investor Education and Awareness:** The integration of green finance into mainstream financial markets requires a better understanding of the potential risks and opportunities associated with sustainable investing. A more informed investor base is essential to ensuring that green finance products are widely adopted and supported.

Addressing these challenges is vital for enhancing the credibility, effectiveness, and scalability of green finance. Research into how these barriers can be overcome will help provide recommendations for improving the integration of green finance into global financial markets.

Objectives of the Chapter

The primary objectives of this chapter are to explore and analyze the role of green financial innovations in addressing climate risk and driving sustainable finance. Specifically:

- **To explore the role of green financial innovations in shaping financial markets and managing climate risks.** This objective focuses on understanding how green financial products, such as green bonds and ESG investments, influence market behavior and financial decision-making, particularly in relation to climate risk.
- **To identify the key challenges facing the green finance sector.** These challenges include greenwashing, inconsistent ESG metrics, and limited access to capital in emerging markets. By identifying these barriers, the research aims to pinpoint critical areas for improvement.
- **To analyze the potential of green finance to transform financial markets and contribute to the energy transition and climate risk management.** The objective here is to assess how green finance can accelerate the transition to a low-carbon economy, promote sustainable business practices, and manage climate-related risks in both developed and emerging markets.
- **To provide recommendations for overcoming the challenges and advancing the integration of green finance into mainstream financial markets.**

Based on the insights gained from the analysis, this chapter aims to propose actionable solutions to enhance the transparency, accountability, and impact of green finance, ensuring its role as a central mechanism in tackling climate change.

Through addressing these questions and objectives, this chapter seeks to contribute to the growing body of research on green finance, providing both theoretical insights and practical recommendations for enhancing the global financial system's role in addressing climate risks and fostering sustainable development.

5. METHODOLOGY FOR STUDYING GREEN FINANCIAL INNOVATIONS AND CLIMATE RISK MANAGEMENT

The aim of this study is to explore the role of green financial innovations—particularly green bonds, ESG (Environmental, Social, Governance) investing, and sustainable finance—within the broader context of climate risk management. The methodology combines both qualitative and quantitative approaches to ensure a comprehensive understanding of the challenges, opportunities, and overall effectiveness of green finance in addressing climate change. By blending financial data analysis with in-depth qualitative insights, this research seeks to offer a nuanced and well-rounded perspective on the evolving dynamics of green finance and its role in promoting sustainability.

Data Collection and Sources

The data collection process for this study is designed to provide a comprehensive and multi-dimensional view of green finance. Various types of data sources are utilized, including academic literature, industry reports, policy documents, and case studies. This allows for both a theoretical grounding in the field and empirical insights from real-world applications of green finance.

Academic Literature

A thorough review of academic literature forms the foundation of this study. This includes papers on the evolution of green financial products, the role of ESG investing in corporate governance, and the financial performance of green bonds. Scholarly articles help to identify existing gaps in the literature, frame the key research questions, and provide the theoretical background for understanding green finance innovations. The literature review focuses on a range of topics, including:

- The development and adoption of green financial products, such as green bonds and ESG portfolios.
- Climate risk management strategies within the financial sector.
- The effectiveness of regulatory frameworks in promoting green finance.
- The role of green finance in achieving sustainable development and mitigating climate change.

Industry Reports and Policy Documents

Industry reports are an essential component of the data collection process. These reports offer insights into the current state of green finance, its growth trends, and the effectiveness of different green financial products in addressing climate risks. Reports from organizations such as the Climate Bonds Initiative (CBI), World Bank, and International Finance Corporation (IFC) provide valuable data on green bond issuances, ESG investment strategies, and sustainable finance frameworks. These reports are used to assess the real-world market performance of green financial instruments and identify key challenges faced by practitioners and investors in the field.

Policy documents, such as government reports, regulations, and frameworks related to sustainable finance, are also crucial. These documents help contextualize the study within the global regulatory landscape and provide insights into the incentives, obstacles, and policy tools that impact the growth and effectiveness of green finance.

Case Studies

The study incorporates case studies of prominent green bond issuances, ESG portfolios, and sustainable finance frameworks. Case studies help to provide practical examples of how green finance operates in different contexts. They enable a detailed examination of the following elements:

- **The process of issuing green bonds**: These case studies analyze the steps involved in issuing green bonds, including the certification process, the types of projects financed, and the environmental impact of these projects.
- **ESG investment portfolios**: Case studies of ESG portfolios highlight the decision-making process involved in selecting companies for investment, as well as the social, environmental, and governance criteria that guide these choices.
- **Regulatory frameworks**: Case studies also focus on the regulatory context that shapes the success or failure of green finance initiatives, examining both national and international regulations that support sustainable finance.

Expert Interviews

Interviews with key stakeholders in the field of green finance are a cornerstone of the qualitative data collection process. These stakeholders include:

- **Policymakers**: Government officials and regulators who shape the regulatory environment for green finance.
- **Financial analysts**: Experts who assess the financial performance of green bonds and ESG investments.
- **Sustainable finance practitioners**: Professionals involved in managing and advising on green financial products, such as fund managers, institutional investors, and investment consultants.

These interviews provide insights into the practical challenges of implementing green finance products, the barriers faced by investors and issuers, and the strategies being adopted to overcome these challenges. The qualitative insights from these experts also help contextualize the quantitative findings, enriching the overall analysis.

Quantitative Analysis

To assess the effectiveness of green financial innovations in addressing climate risk and promoting sustainability, this study uses quantitative analysis to evaluate key financial metrics, climate risk indicators, and performance data. The goal of the quantitative approach is to provide empirical evidence on how well green bonds, ESG investments, and other sustainable financial products are performing in terms of financial returns, risk mitigation, and environmental impact.

Financial Performance Metrics

Quantitative analysis begins with the examination of the financial performance of green bonds and ESG investments. This involves analyzing data on:

- **Yields**: The returns generated by green bonds compared to conventional bonds, focusing on the yield-to-maturity (YTM) or coupon rate.
- **Returns**: The overall return on investment (ROI) of green bonds and ESG portfolios over various time horizons.
- **Risk-adjusted performance**: The study uses metrics such as the Sharpe ratio and alpha to assess the risk-adjusted return of green financial products relative to traditional investments.

The primary data sources for this analysis are financial databases such as Bloomberg, Thomson Reuters, and Morningstar, which provide comprehensive information on green bond issuances, ESG fund performance, and related financial metrics.

Climate Risk Metrics

In addition to financial performance, this study integrates climate risk metrics to assess the environmental impact of green finance products. Key indicators include:

- **Carbon footprint reductions**: The reduction in greenhouse gas (GHG) emissions resulting from the projects funded by green bonds, such as renewable energy installations, energy-efficient buildings, or carbon capture and storage technologies.
- **Environmental impact assessments**: Data on the broader environmental outcomes of green financial products, such as the amount of energy generated from renewable sources or the hectares of forest protected.

These climate risk metrics provide a way to evaluate whether green finance is achieving its intended environmental outcomes. Data for these metrics is sourced from organizations like the Climate Bonds Initiative and CDP (formerly Carbon Disclosure Project), as well as the sustainability reports of issuers of green bonds and ESG portfolios.

Qualitative Analysis

In addition to quantitative analysis, the study employs qualitative research methods to explore the broader context in which green finance operates. Qualitative methods are particularly useful for understanding the social, environmental, and regulatory dynamics that shape the green finance sector. This is achieved through case studies, interviews, and content analysis of policy documents.

Case Study Analysis

Case studies of specific green bond issuances and ESG portfolios are analyzed in detail to explore the real-world outcomes of these financial products. Each case study examines the environmental, social, and governance factors considered by investors, the impact of these investments on the projects funded, and the effectiveness of the regulatory frameworks that supported them.

For example, a case study of the World Bank Green Bond Program could focus on the types of projects financed (e.g., renewable energy projects), the bond issuance process, and the environmental impact of these projects in terms of carbon emissions reduction.

Interviews with Stakeholders

Interviews with stakeholders such as policymakers, financial analysts, and practitioners help provide a deeper understanding of the practical challenges faced in green finance. These interviews focus on issues such as:

- **Barriers to green bond issuance**: Challenges faced by issuers, including certification costs, lack of awareness, and market volatility.
- **Investor concerns**: Issues related to the perceived risks of green investments, such as greenwashing, lack of transparency, and the challenge of measuring environmental outcomes.
- **Regulatory challenges**: How existing regulations and frameworks are either supporting or hindering the growth of green finance.

Interviews are transcribed and analyzed using thematic analysis to identify common themes and insights.

Research Limitations

While this methodology provides a comprehensive framework for exploring the role of green finance in climate risk management, there are several limitations that need to be acknowledged:

Availability of ESG Data

One of the main challenges in conducting this research is the inconsistent and fragmented nature of ESG data. Various reporting standards exist across different markets and financial institutions, and the quality of ESG data can vary significantly. This inconsistency can make it difficult to perform cross-market comparisons and evaluate the true sustainability of green investments.

Focus on Developed Economies

This study primarily focuses on developed economies where green finance is more established and data is more readily available. While the research will touch on emerging markets, it may not fully capture the complexities and challenges faced by these markets, where access to reliable data, financial infrastructure, and capital markets is more limited.

Data Limitations in Climate Risk Assessment

Measuring the exact environmental impact of green finance products, such as the carbon footprint reduction from funded projects, is challenging. The accuracy of environmental data depends on the availability of credible environmental impact reports, which may not always be accessible or standardized.

6. RESULTS & ANALYSIS

The results and analysis section of this study provides a comprehensive review of the performance and impact of green bonds, ESG investing, and policy initiatives in the context of climate risk management. By synthesizing quantitative and qualitative data, we can better understand the current state of green finance, its challenges, and its potential for promoting sustainability and addressing climate change. This section delves into key aspects such as the financial performance of green bonds, the returns associated with ESG investments, the role of policy and regulatory frameworks, and the barriers hindering the widespread adoption of green finance.

Green Bonds and Financial Performance

Green bonds, introduced to finance projects that have positive environmental impacts, have emerged as one of the most prominent instruments in the green finance space. The analysis of green bond issuances reveals an impressive growth trajectory over the past decade. The market for green bonds has expanded significantly, not only in terms of issuance volume but also in the diversity of sectors and projects financed. Green bonds are now being used to fund renewable energy projects, energy-efficient infrastructure, sustainable transportation, and climate change adaptation measures, demonstrating their versatility in supporting low-carbon and climate-resilient initiatives.

Performance Compared to Traditional Bonds

Green bonds have largely performed on par with traditional bonds in terms of financial returns. Over the last several years, data from major green bond issuances indicate that green bonds have delivered competitive yields and risk-adjusted returns. For example, bonds issued by the World Bank and the European Investment Bank—two of the earliest issuers of green bonds—have shown stable financial performance, attracting a range of investors seeking both financial returns and sustainable impact.

Some studies suggest that, in certain periods, green bonds have outperformed traditional bonds. This outperformance can be attributed to the growing investor demand for sustainable investments, which has driven prices up, reducing yields. Moreover, the increasing emphasis on environmental, social, and governance (ESG) factors in investment decisions has led to a shift in capital flows toward green bonds, thus enhancing their market appeal. In 2020, for example, green bonds issued in Europe saw yields that were either comparable to or slightly better than those of conventional sovereign bonds.

Market Conditions and External Factors

However, it is important to recognize that the performance of green bonds is influenced by a variety of external factors, including market conditions, investor sentiment, and broader economic trends. For instance, during periods of market volatility, such as the COVID-19 pandemic or financial crises, the demand for risk-free assets tends to rise, which can impact the attractiveness of green bonds relative to other financial instruments.

Additionally, the performance of green bonds can be influenced by the regulatory and policy environments in which they are issued. The presence of supportive regulatory frameworks, such as tax incentives for green investments or stringent environmental standards for green projects, can boost the demand for green bonds, thereby improving their financial performance. Conversely, a lack of consistent and clear green finance regulations can dampen investor confidence and limit market growth.

ESG Investing and Financial Returns

ESG investing, which integrates environmental, social, and governance factors into investment decision-making, has become increasingly mainstream over the last decade. The analysis of ESG investments reveals mixed results regarding their financial performance. While many studies suggest that companies with strong ESG practices tend to outperform their peers over the long term, others indicate that the financial performance of ESG investments can be neutral or slightly lower compared to traditional, non-ESG investments.

Long-Term Performance

Research indicates that companies with robust ESG performance are better equipped to manage long-term risks, including those related to climate change, social inequality, and governance issues. These companies tend to have more resilient business models, making them more likely to thrive in an increasingly sustainability-focused global economy. For instance, firms that actively manage their carbon footprint or promote social responsibility may be less prone to regulatory risks, consumer backlash, or reputational damage.

Several large-scale studies have found that companies with high ESG scores are likely to have better financial outcomes over the long term. This is because they tend to exhibit strong operational efficiency, attract loyal customers, and engage in responsible business practices that align with evolving market expectations. Furthermore, as institutional investors increasingly recognize the importance of managing climate and social risks, there is a growing alignment between good governance practices and financial performance.

Mixed Short-Term Returns

On the other hand, ESG investing has also faced some challenges in delivering strong short-term financial returns. ESG-focused portfolios may sometimes lag behind traditional investment portfolios due to the relatively high costs of adopting sustainable practices, such as investing in cleaner technologies or addressing labor rights issues. Additionally, some ESG funds may exclude profitable industries or companies, such as fossil fuels or tobacco, which can affect their short-term performance.

Despite this, many investors view ESG considerations as critical to mitigating long-term risks, and they are increasingly willing to accept potentially lower returns in the short term in exchange for reducing exposure to environmental and social risks. The growing interest in ESG investments is driven by the recognition that addressing these risks can create long-term value by reducing the likelihood of sudden financial shocks or losses.

Sustainable Development Beyond Financial Returns

The true value of ESG investing goes beyond immediate financial returns. ESG considerations are increasingly seen as integral to managing long-term risks, particularly climate-related risks. Investors now recognize that companies with high ESG scores are more likely to be proactive in addressing the environmental and social challenges posed by climate change, which could ultimately lead to better long-term financial returns. Moreover, ESG investing helps drive capital toward companies that are contributing to sustainable development goals (SDGs), reinforcing the broader role of finance in achieving positive social and environmental outcomes.

Impact of Policy and Regulatory Initiatives

Policy and regulatory initiatives have played a crucial role in the development of green finance and its ability to scale. Key policy frameworks, such as the European Union's Taxonomy for Sustainable Activities (EU Taxonomy) and the Task Force on Climate-related Financial Disclosures (TCFD), have provided clear guidelines for sustainable investments and encouraged greater transparency in ESG reporting.

European Union Taxonomy and TCFD

The EU Taxonomy has provided a clear and standardized definition of what constitutes sustainable economic activities, helping investors and financial institutions assess whether their investments align with sustainability objectives. The EU Taxonomy aims to eliminate ambiguity around green finance by offering a unified classification system for environmentally sustainable activities. By defining clear criteria for green investments, the Taxonomy has made it easier for investors to evaluate the environmental impact of their portfolios, thereby increasing confidence in green financial products.

The TCFD, established by the Financial Stability Board, has also had a significant impact on the transparency of climate-related risks in financial markets. By recommending that companies disclose their climate-related financial risks in a consistent and comparable manner, the TCFD has helped investors make more informed decisions about the financial implications of climate change. As a result, more companies are disclosing their exposure to climate risks and the steps they are taking to mitigate these risks, enhancing the credibility and transparency of ESG investments.

Implementation Challenges

Despite these positive developments, challenges remain in terms of the implementation and consistency of these frameworks across regions. While the EU Taxonomy has been a pioneering initiative, its adoption in other regions has been slower. Some markets, particularly in emerging economies, lack the regulatory infrastructure to fully integrate these standards into their financial systems. Furthermore, the implementation of the TCFD recommendations has been voluntary, leading to varying levels of compliance across different industries and countries.

As a result, while regulatory frameworks have played a significant role in advancing green finance, their impact has been uneven. In markets where these regulations are well-established, green finance has flourished. However, in regions with weaker regulatory environments, green finance adoption remains limited.

Challenges and Barriers

Despite the significant progress made in the field of green finance, several barriers continue to impede its widespread adoption and effectiveness. These barriers include greenwashing, inconsistent ESG metrics, and limited accessibility in emerging markets.

Greenwashing

Greenwashing remains a major concern in the green finance sector. Some issuers and companies falsely claim to be environmentally responsible or use "green" labels for products that do not meet rigorous sustainability standards. This undermines the credibility of the market and confuses investors. In some cases, companies may issue green bonds or ESG investment products without delivering substantial environmental benefits, thereby diluting the effectiveness of green finance in addressing climate change. The lack of standardized metrics and clear definitions of "green" and "sustainable" further exacerbates this issue.

Inconsistent ESG Metrics

The absence of standardized ESG metrics is another significant challenge. ESG ratings vary widely across agencies, and there is no universally accepted framework for measuring and reporting ESG performance. This creates confusion for investors who want to compare the environmental and social impact of different financial products. As a result, the lack of reliable, consistent data hampers the ability to assess the true effectiveness of ESG investments and green bonds in driving positive environmental outcomes.

Limited Accessibility in Emerging Markets

The accessibility of green finance in emerging markets is another key challenge. While developed economies have made significant progress in adopting green finance, many emerging markets still face substantial barriers. These include insufficient financial infrastructure, lack of regulatory frameworks, and limited access to capital for green projects. As a result, the flow of green finance to emerging markets remains limited, despite the potential for large-scale investments in renewable energy and sustainable development.

7. DISCUSSION

The findings of this study illustrate the dual nature of green financial innovations in the context of addressing climate risks and promoting sustainability. On one hand, green bonds and ESG (Environmental, Social, and Governance) investments have demonstrated their effectiveness in mobilizing capital for sustainable development projects, contributing both to financial returns and to positive environmental outcomes. The growing demand for these instruments suggests a significant shift toward a more sustainable financial system, one that increasingly values long-term environmental and social sustainability alongside financial returns. On the other hand, the green finance sector is still facing several critical challenges that, if not addressed, could hinder its ability to fully realize its potential as a driving force for global sustainability.

This discussion will examine both the promise of green finance and the challenges it faces, and explore potential solutions that could help unlock its full potential in addressing climate change and driving global sustainability.

The Promise of Green Finance

Green bonds and ESG investments represent two of the most innovative financial instruments designed to address the pressing issue of climate change. Their success in mobilizing capital for sustainable projects has been remarkable. Green bonds, which are used to finance projects with positive environmental impacts, have been particularly effective in driving investments in renewable energy, energy efficiency, and low-carbon infrastructure. The findings from this study demonstrate that green bonds have performed competitively with traditional financial instruments, even offering higher returns during certain periods of strong demand. This trend suggests that there is a growing recognition among investors that environmental considerations are no longer just a "niche" aspect of investment, but rather a key driver of financial performance in the modern market.

Similarly, ESG investing, which integrates environmental, social, and governance factors into the investment decision-making process, is gaining traction among institutional and retail investors alike. The evidence presented in this study underscores the potential long-term financial advantages of incorporating ESG considerations into investment portfolios. Companies with strong ESG practices tend to outperform their peers over time, as they are better equipped to mitigate risks related to climate change, social inequality, and poor governance. Moreover, ESG investing is no longer viewed solely as a means of achieving financial returns; it is increasingly seen as a critical component of risk management, especially in the face of climate-related and social challenges.

The growing demand for green bonds and ESG investments reflects a broader trend toward sustainable finance. This shift is not only driven by a desire to generate financial returns, but also by an increasing awareness of the financial risks posed by climate change and other global sustainability challenges. As investors become more aware of these risks, they are incorporating environmental and social factors into their decision-making processes. This changing landscape suggests that the financial industry is undergoing a fundamental transformation, one in which sustainability is at the core of investment strategies. Furthermore, as more companies and governments issue green bonds and promote ESG principles, the market for sustainable finance continues to expand, providing greater opportunities for investors to participate in the transition to a low-carbon economy.

Challenges Facing Green Finance

Despite the promising potential of green finance, several challenges continue to undermine its effectiveness and hinder its widespread adoption. The most pressing of these challenges are greenwashing, the lack of standardized ESG metrics, and the limited accessibility of green finance, especially in emerging markets.

Greenwashing

One of the most significant obstacles to the growth of green finance is the issue of greenwashing. Greenwashing occurs when companies, issuers, or financial products are marketed as environmentally friendly or sustainable, but fail to meet the necessary criteria or deliver on their environmental promises. This phenomenon undermines investor confidence and makes it difficult to distinguish between truly sustainable investments and those that merely capitalize on the growing demand for green products.

Greenwashing is particularly problematic because it damages the credibility of the entire green finance sector. Investors, especially those who are new to sustainable finance, may become disillusioned when they discover that the financial products they invested in do not live up to their environmental claims. This skepticism can result in reduced capital flows into the green finance market, which in turn limits the ability of the financial sector to drive sustainable change. To address this issue, stricter regulations and more transparent reporting standards are required to ensure that only legitimate green projects are financed through green bonds and other sustainable financial instruments.

Lack of Standardized ESG Metrics

Another major challenge in the green finance landscape is the absence of universally accepted ESG metrics. While the number of ESG-focused investment products and funds has grown substantially, there is still a lack of consistency in how ESG performance is measured and reported. Different agencies and organizations use varying criteria to evaluate companies' environmental, social, and governance practices, resulting in inconsistent ratings and data. This lack of standardization makes it difficult for investors to compare different ESG products and assess their relative sustainability.

The lack of consistent ESG metrics also creates confusion and hinders the scalability of green finance. For investors seeking to align their portfolios with sustainable practices, inconsistent and sometimes contradictory ESG ratings can undermine confidence and complicate investment decisions. Furthermore, without standardized metrics, it is challenging to assess the real environmental impact of green investments. A green bond may be marketed as financing renewable energy projects, but without a consistent framework for measuring its impact, it is difficult to assess whether it is truly achieving the environmental goals it claims to support.

To overcome this challenge, there is a need for greater collaboration between regulators, financial institutions, and standard-setting bodies to develop a common set of ESG reporting standards. Such standards would increase transparency, improve comparability across financial products, and provide investors with the information they need to make informed decisions. Additionally, standardizing ESG metrics would help to combat greenwashing, ensuring that only projects that meet clear environmental criteria are labeled as "green."

Limited Accessibility in Emerging Markets

A third challenge that restricts the growth of green finance is the limited accessibility of sustainable finance in emerging markets. While green bonds and ESG investments have flourished in developed economies, many emerging markets still lack the financial infrastructure, regulatory frameworks, and capital needed to support the growth of green finance. These markets face a number of barriers, including weak financial systems, limited access to capital, and a lack of expertise in sustainable investment.

In many emerging markets, the demand for green finance is high, as these regions are often the most vulnerable to climate change and environmental degradation. However, the lack of appropriate financial infrastructure and regulatory support limits the ability of governments and businesses to tap into the global green finance market. For example, emerging market governments may struggle to issue green bonds due to limited access to international capital markets or the absence of clear policies on sustainability. Similarly, investors may be hesitant to invest in green projects in these regions due to concerns over political instability, weak governance structures, and insufficient environmental standards.

To overcome these challenges, international collaboration and support from financial institutions and governments are crucial. Multilateral development banks, such as the World Bank and the Asian Development Bank, can play a key role in providing financial assistance and technical expertise to help emerging markets develop the necessary infrastructure for green finance. Additionally, international regulatory bodies and organizations such as the United Nations Environment Programme Finance Initiative (UNEP FI) can help promote the adoption of sustainable finance frameworks in developing economies. By fostering collaboration between governments, financial institutions, and the private sector, it is possible to unlock the potential of green finance in emerging markets and channel capital toward the sustainable development projects that are desperately needed.

Solutions and Recommendations

To ensure that green finance can reach its full potential, it is necessary to address the challenges outlined above. The following recommendations are proposed to enhance the effectiveness of green finance and promote its widespread adoption:

- **Strengthening Regulatory Frameworks**: Governments and regulators must work together to create clear, consistent, and enforceable regulations for green finance. This includes developing standardized ESG reporting frameworks, improving disclosure requirements, and establishing stringent criteria for labeling financial products as "green." Stronger regulations will reduce greenwashing and increase trust in the green finance market.
- **Standardizing ESG Metrics**: The development of universal ESG metrics is essential for improving transparency and comparability in green finance. Regulatory bodies, financial institutions, and standard-setting organizations must collaborate to create standardized criteria for measuring ESG perfor-

mance. This will enable investors to make more informed decisions and reduce the risk of greenwashing.

- **Promoting International Collaboration**: International organizations and multilateral financial institutions should play a key role in supporting the growth of green finance in emerging markets. This can be achieved through capacity-building initiatives, technical assistance, and financial support for green projects. By providing emerging markets with the tools and resources they need to develop sustainable financial systems, international collaboration can help bridge the gap between developed and developing economies.
- **Increasing Awareness and Education**: Greater efforts must be made to raise awareness about the benefits of green finance and ESG investing, particularly among retail investors and small businesses. Education and awareness campaigns can help investors understand the long-term financial and environmental benefits of sustainable investments, encouraging greater participation in green finance.

8. CONCLUSION

The increasing urgency of addressing climate risks and promoting sustainability has prompted the development of green financial innovations, including green bonds, ESG (Environmental, Social, and Governance) investing, and broader sustainable finance frameworks. This chapter has explored the evolving role of these innovations in shaping global financial markets, driving capital toward environmentally and socially beneficial projects, and transforming traditional investment paradigms. The study reveals that, while green finance has made significant strides in mobilizing capital for sustainable projects, there remain several key challenges that must be overcome to fully harness its potential.

Key Findings: Green Finance Innovations in Climate Risk Management

Green finance innovations, especially green bonds and ESG investments, have proven to be powerful tools for addressing climate change and managing associated risks. Green bonds, designed specifically to fund environmentally sustainable projects, have gained considerable momentum in recent years. These bonds have successfully directed significant capital toward renewable energy, energy efficiency, and low-carbon infrastructure, which are critical to meeting global climate targets. The success of green bonds reflects not only the growing investor demand for sustainable assets but also the recognition that sustainable projects can generate solid financial returns alongside positive environmental impacts.

Similarly, ESG investing has become an increasingly important element of investment strategies. ESG-focused portfolios are designed to factor in not just financial metrics, but also a company's environmental practices, social responsibility, and governance structures. The empirical evidence presented throughout this study supports the notion that companies with strong ESG practices are more likely to deliver sustainable long-term financial performance. ESG investing is not merely a trend or a marketing tool—it has become a core component of risk management for investors who are increasingly aware of the long-term challenges posed by environmental degradation, social inequality, and poor governance.

These innovations, taken together, are contributing to the creation of a more sustainable financial ecosystem. The growing emphasis on integrating environmental and social factors into investment decisions marks a paradigm shift from traditional finance, which primarily focused on financial returns without considering broader societal impacts. This transition is significant because it represents the first stages of a financial system that recognizes the interdependency between financial markets and the global environment. As green bonds and ESG investments gain momentum, they signal the beginning of a more integrated, long-term approach to capital allocation—one that prioritizes sustainability and resilience in the face of climate risks.

Challenges and Barriers to Green Finance

Despite these positive developments, several significant barriers continue to hinder the full-scale adoption of green finance. These challenges must be addressed if green finance is to fulfill its potential as a transformative tool in climate risk management and sustainable development.

1. **Greenwashing**: Greenwashing, the practice of falsely marketing financial products as "green" or "sustainable" without meeting strict environmental criteria, remains one of the most critical challenges in the green finance space. This phenomenon undermines investor confidence, distorts the market, and reduces the effectiveness of green finance in achieving real-world environmental impact. Investors are becoming increasingly wary of products that are marketed as environmentally friendly but lack transparent, verifiable standards. The result is a general skepticism that could limit the flow of capital into genuinely sustainable projects.
2. **Inconsistent ESG Metrics**: Another significant hurdle is the lack of standardized ESG metrics, which creates confusion for investors and impedes the

comparability of financial products. The absence of a universal framework for measuring and reporting ESG performance makes it difficult for investors to assess the true environmental and social impact of their investments. Different agencies and organizations use varying criteria for rating ESG performance, leading to inconsistencies and a lack of transparency. As a result, investors face significant challenges in evaluating which companies or financial products are genuinely contributing to sustainability and which are simply capitalizing on the trend.

3. **Limited Accessibility in Emerging Markets**: While green finance has flourished in developed economies, emerging markets face unique challenges in accessing capital for sustainable projects. These regions often lack the necessary financial infrastructure, regulatory frameworks, and institutional capacity to support the growth of green finance. As a result, the potential for green bonds and ESG investments in emerging markets remains underexploited, despite the high demand for sustainable development in these regions. Moreover, investors may be hesitant to engage with emerging markets due to concerns about political instability, weak governance, and the absence of clear and enforceable environmental regulations.

Recommendations for Advancing Green Finance

To realize the full potential of green finance and address the challenges outlined above, several actions are necessary:

- **Strengthening Regulatory Frameworks**: One of the most pressing needs in the green finance sector is the development of robust and transparent regulatory frameworks. Governments and regulators must work to create clear, enforceable rules that define what qualifies as a "green" investment, how ESG factors should be measured, and how companies should disclose their environmental and social performance. Clear regulations will reduce the risk of greenwashing, increase investor confidence, and ensure that green finance products deliver on their environmental promises.
- **Standardizing ESG Reporting**: The lack of standardized ESG metrics is a major barrier to the growth of green finance. To overcome this, international financial organizations and regulators must work together to establish consistent and universally accepted standards for ESG reporting. Such standards would make it easier for investors to compare different financial products and assess their relative sustainability. They would also promote greater transparency in the green finance market, helping to align investor expectations with the actual environmental impact of investments.

- **Improving Accessibility in Emerging Markets**: Green finance in emerging markets faces unique challenges, including limited access to capital, insufficient financial infrastructure, and weak regulatory frameworks. International financial institutions and multilateral development banks, such as the World Bank, can play a critical role in providing technical assistance, capacity-building support, and funding to help these markets develop the necessary infrastructure to tap into the global green finance market. Governments in emerging markets must also prioritize the creation of conducive regulatory environments that attract sustainable investments and ensure the effective implementation of climate policies.
- **Promoting Greater Transparency and Accountability**: Transparency and accountability are essential for ensuring the integrity of green finance. Regulators should mandate rigorous disclosure requirements for green bonds and ESG investments, ensuring that investors have access to comprehensive, standardized information about the environmental impact of these investments. Additionally, third-party verification and auditing processes can help ensure that financial products labeled as "green" truly meet the required environmental standards.
- **Fostering Global Collaboration**: Green finance is a global issue that requires coordinated action across borders. International collaboration between governments, regulators, financial institutions, and other stakeholders is essential to building the infrastructure needed for a sustainable global financial system. By sharing best practices, aligning regulatory frameworks, and providing technical assistance to emerging markets, global collaboration can help scale up green finance and ensure its effectiveness in driving sustainable development.

The Future of Green Finance

In conclusion, green finance represents a transformative opportunity to reshape the global financial system, channeling capital toward the sustainable development projects that are crucial to mitigating climate risks and achieving long-term environmental goals. While challenges such as greenwashing, inconsistent ESG metrics, and limited accessibility in emerging markets remain significant, they are not insurmountable. By strengthening regulatory frameworks, standardizing ESG reporting, and promoting greater transparency, we can overcome these obstacles and unlock the full potential of green finance.

The role of international financial institutions and governments is vital in supporting the development of green finance in emerging markets, where the need for sustainable investment is particularly urgent. By addressing these challenges, green finance can become a cornerstone of the global transition to a low-carbon, sustainable economy. Ultimately, green finance offers a pathway to a more resilient and sustainable financial system, one that supports the long-term goal of mitigating climate risks and fostering global sustainable development.

9. IMPLEMENTATION & RECOMMENDATIONS

To fully capitalize on the potential of green financial innovations—such as green bonds, ESG investing, and sustainable finance—it is essential to implement several key strategies that address the challenges identified in this study. These strategies should involve cooperation among various stakeholders, including financial institutions, policymakers, investors, and international organizations. By working together, these groups can help unlock the full potential of green finance and ensure that it effectively contributes to sustainability and climate risk mitigation.

Standardizing ESG Metrics

A critical challenge in green finance is the inconsistency of ESG (Environmental, Social, and Governance) metrics across different markets and financial products. As mentioned earlier, the absence of universally accepted standards for ESG reporting creates confusion among investors and makes it difficult to compare different investment opportunities. The lack of standardization also hampers the transparency and credibility of green financial products, as investors are uncertain about the true environmental and social impact of their investments.

To address this challenge, the first step is the development of standardized ESG metrics that can be adopted across different markets and financial products. This would allow investors to make more informed decisions by providing comparable data on the ESG performance of various companies and financial instruments. Moreover, standardized reporting would help create a level playing field for companies and investors, ensuring that all parties are working with the same set of rules and expectations.

International regulatory bodies, such as the International Financial Reporting Standards (IFRS) and the Global Reporting Initiative (GRI), are well-positioned to lead this effort. These organizations already play significant roles in setting financial reporting standards and could extend their mandate to include ESG metrics. By developing a global set of ESG reporting standards, these bodies would provide the clarity and consistency needed to build trust in green finance and attract more investors to sustainable financial products.

Additionally, regulators should encourage companies to align their ESG disclosures with these global standards and integrate ESG factors into their financial reporting. This will ensure that investors are able to assess the long-term sustainability of a company's business model, and not just its short-term financial performance.

Combating Greenwashing

Greenwashing—the practice of marketing financial products as "green" or "sustainable" without meeting the necessary environmental criteria—is one of the most serious challenges facing the green finance sector. Greenwashing undermines investor confidence, dilutes the impact of genuinely sustainable investments, and prevents capital from flowing into projects that can create real environmental and social benefits.

To combat greenwashing, regulators must enforce stricter oversight of green financial products, ensuring that companies and financial institutions adhere to rigorous environmental and social criteria when marketing these products. One effective approach could be the implementation of mandatory third-party verification for green bonds, ESG funds, and other green financial products. Independent certification agencies, such as the Climate Bonds Initiative (CBI), could be tasked with evaluating whether green financial products truly meet the established environmental and social criteria.

Furthermore, financial institutions should be required to provide detailed, transparent disclosures on the environmental impact of their investments. This could include data on carbon emissions reductions, energy savings, and other key performance indicators that demonstrate the true sustainability of a project or portfolio. Such disclosures will allow investors to better assess the effectiveness of their green investments and ensure that they are contributing to meaningful climate action.

Governments can also play a role in combating greenwashing by establishing clear definitions for "green" products and mandating that any financial product labeled as "green" or "sustainable" must meet specific, verifiable criteria. This could include setting standards for green bonds, such as the requirement that proceeds must be directed toward projects that meet environmental impact thresholds, such as renewable energy generation or carbon capture technologies.

In addition to regulatory measures, the private sector must also take responsibility. Financial institutions should adopt internal policies to ensure that their green financial products meet high standards of transparency, verification, and accountability. Investment managers and asset owners should also conduct thorough due diligence to ensure that they are not unknowingly investing in greenwashed products.

Promoting Green Finance in Emerging Markets

While green finance has made significant strides in developed economies, its adoption in emerging markets remains limited. These regions, many of which are particularly vulnerable to the impacts of climate change, often face unique challenges in accessing sustainable financing. These challenges include limited financial infrastructure, weaker regulatory frameworks, and a lack of technical expertise in managing climate risks and opportunities. Without access to green finance, emerging markets are likely to miss out on crucial investments that can help them build resilience to climate change while transitioning to low-carbon economies.

To address these challenges, governments and international financial institutions must work together to promote green finance in emerging markets. One key strategy is providing technical assistance and capacity building to help these countries strengthen their financial infrastructure and regulatory frameworks. International organizations such as the World Bank, the United Nations Environment Programme Finance Initiative (UNEP FI), and the International Finance Corporation (IFC) can play a pivotal role in providing the necessary support. These organizations can help governments design and implement policies that incentivize sustainable investments, while also offering financial assistance for projects that align with global climate goals.

Another important strategy is improving access to financing for green projects in emerging markets. This can be achieved through a combination of grants, concessional loans, and blended finance mechanisms that reduce the financial risk for investors and make it more attractive for them to engage in sustainable projects. For example, multilateral development banks and other international financing institutions can offer low-cost financing to projects in emerging markets, while private investors provide additional capital.

In addition, emerging markets need stronger regulatory frameworks that support green finance. Governments should create policies that incentivize green investments, such as tax credits for renewable energy projects or carbon pricing schemes that internalize the cost of environmental damage. By providing clear, stable policy frameworks, governments can attract international capital to fund sustainable projects.

Moreover, developing local expertise is essential for the successful implementation of green finance in emerging markets. Training programs, capacity-building initiatives, and knowledge-sharing platforms should be established to equip local financial institutions, policymakers, and businesses with the necessary skills to design and implement sustainable projects. This could involve partnerships with international green finance experts, who can share best practices and lessons learned from other regions.

Engaging the Private Sector

The private sector plays a critical role in driving the growth of green finance. While governments and international institutions are essential for setting the regulatory framework and providing capital, private investors and financial institutions must actively participate in the green finance ecosystem. To encourage this participation, several strategies can be implemented.

First, financial institutions should be incentivized to offer more green financial products. This could include green bonds, ESG mutual funds, and impact investing opportunities that target sustainable development goals (SDGs). Governments can offer incentives such as tax breaks, subsidies, or preferential treatment for financial institutions that issue or invest in green bonds and other sustainable instruments.

Second, investors should be encouraged to integrate ESG factors into their decision-making processes. This could be achieved by creating more awareness about the long-term benefits of ESG investing, which not only helps mitigate climate risks but also contributes to more stable financial returns. Financial advisors and wealth managers should be trained to incorporate ESG factors into their investment advice, ensuring that individuals and institutions are fully aware of the importance of sustainability in their portfolios.

Third, collaboration between public and private sector actors is essential for scaling up green finance. Governments can work with financial institutions to design public-private partnerships (PPPs) that help finance large-scale green infrastructure projects, such as renewable energy plants, sustainable transport systems, and energy-efficient buildings. By sharing risks and rewards, public-private partnerships can unlock the capital needed to tackle the most pressing sustainability challenges.

Finally, financial institutions should adopt responsible investment practices, ensuring that they align their portfolios with the principles of sustainable development. This involves not only investing in green projects but also encouraging companies in their portfolios to improve their ESG performance. Investors should use their influence to push for better environmental, social, and governance practices among the companies they invest in.

REFERENCES

Berg, F., Kölbel, J. F., & Rigobon, R. (2019). *Aggregate confusion: The divergence of ESG ratings* [Working paper]. MIT Sloan School of Management.

Busch, T., & Lewandowski, S. (2020). The EU Taxonomy for sustainable activities: Implications for investor decision-making and capital allocation [Working paper]. University of Hamburg. https://www.wiso.uni-hamburg.de/fachbereich-vwl/professuren/busch/publikationen/eu-taxonomy-investor-decisions.pdf

Climate Bonds Initiative. (2021). *Green bonds market outlook.* Climate Bonds Initiative. https://www.climatebonds.net

Flammer, C. (2021). Green bonds: Effectiveness and implications for corporate strategy and public policy. *Journal of Financial Economics, 142*(2), 475–498.

Friede, G., Busch, T., & Bassen, A. (2015). ESG and financial performance: aggregated evidence from more than 2000 empirical studies. Journal of sustainable finance & investment, 5(4), 210-233.

Global Reporting Initiative (GRI). (2020). *Global sustainability reporting standards.* GRI. https://www.globalreporting.org

International Finance Corporation (IFC). (2019). *Financing a sustainable future: The role of green finance in emerging markets.* IFC. https://www.ifc.org

International Financial Reporting Standards (IFRS). (2021). *IFRS standards and sustainability reporting.* IFRS Foundation. https://www.ifrs.org

Krueger, P., Sautner, Z., & Starks, L. T. (2020). *The importance of climate risks for institutional investors* [Working paper]. European Corporate Governance Institute. DOI: 10.1093/rfs/hhz137

United Nations Environment Programme Finance Initiative (UNEP FI). (2020). Financial sector transformation for sustainable development. UNEP FI. https://www.unepfi.org

World Bank Group. (2018). *Green finance for emerging markets: Leveraging finance for sustainable development.* World Bank. https://www.worldbank.org

Chapter 8
The Importance of Banking Efficiency in Green Finance and Climate Risk Management at the Level of Islamic Banks

Abdelkader Mohamed Sghaier Derbali
https://orcid.org/0000-0003-4098-3120
Applied College, Taibah University, Saudi Arabia

ABSTRACT

This paper explores the role of banking efficiency in promoting green finance and climate risk management in Islamic banks. It comes in a context of increasing pressure on financial institutions to adopt sustainable practices, where Islamic banks operate at the intersection of religious compliance and environmental imperatives. The study uses a stochastic frontier analysis approach applied to a panel of Islamic banks operating in the member countries of the Organization of Islamic Cooperation. It measures banking efficiency and its influence on green commitments and climate risk management strategies. The results show that high bank efficiency is positively correlated with greater involvement in green finance and better resilience to climate risks. These efficient banks tend to adopt green practices more systematically and strengthen their environmental risk management systems. This research proposes an innovative approach by combining three dimensions rarely discussed together in the literature: banking efficiency, Islamic finance, and environmental sustainability.

DOI: 10.4018/979-8-3373-6766-8.ch008

1. INTRODUCTION

Over the past two decades, climate change has become a central issue for global economies, posing significant risks to both financial stability and sustainable development. In this context, green finance has emerged as a strategic lever to channel capital flows towards projects compatible with climate objectives (OECD, 2023; IMF, 2024). Financial institutions, particularly banks, are now called upon to redefine their priorities in order to fully integrate environmental considerations into their financing and risk management activities (NGFS, 2024).

At the same time, Islamic banks have grown rapidly in many regions, particularly in the Middle East, Southeast Asia, and North Africa. These institutions operate according to Sharia principles, which prohibit interest (riba), excessive speculation (gharar), and encourage risk sharing as well as the promotion of social justice and collective welfare (Akmal et al., 2025; Khan and Badjie, 2022). This ethical orientation theoretically makes Islamic banks well-positioned to play an active role in sustainable finance, particularly through instruments such as green sukuk or crowdfunding of ecological projects (World Bank and IIFM, 2023).

However, simply adhering to Sharia principles does not guarantee an effective contribution to the fight against climate change. Operational capacity, particularly in terms of banking efficiency, is a key determinant in enabling these banks to align their business models with green finance objectives (Abbas et al., 2023). Banking efficiency, measured through a bank's ability to optimally use its resources to generate revenue or control costs, is therefore likely to influence its involvement in green financing and climate risk management strategies.

In this perspective, this paper poses the following question: What is the role of banking efficiency in the contribution of Islamic banks to green finance and climate risk management?

To answer this question, we analyze a panel of Islamic banks over the period 2000–2024, using a stochastic frontier analysis (SFA) approach to assess their efficiency, and studying its effects on their green commitments and their exposure to climate risks.

Our contribution is threefold. First, we combine three areas of research that are still poorly integrated: banking efficiency, Islamic finance, and environmental sustainability. Second, we provide recent empirical evidence on the behavior of Islamic banks in the face of climate risks. Third, we offer concrete policy implications for regulators and banks seeking to strengthen their alignment with sustainable development goals.

2. LITERATURE REVIEW

2.1. Banking Efficiency: Definitions, Indicators and Methodological Approaches

Banking efficiency refers to a bank's ability to transform its resources (inputs) into financial products (outputs) while minimizing costs or maximizing profits. Three types of efficiency are generally distinguished:

- Cost effectiveness: minimization of costs for a given level of production (Berger and Mester, 1997).
- Profit efficiency: maximization of profits for a given combination of inputs/outputs.
- Intermediation efficiency: ability to transform deposits into productive credits (Sealey and Lindley, 1977).

The most used methods to measure this effectiveness are:

- DEA (Data Envelopment Analysis): non-parametric method that identifies an efficient frontier from the best units (Coelli et al., 2005; Zhu, 2014).
- SFA (Stochastic Frontier Analysis): parametric method that distinguishes inefficiency from random noise in the data (Battese and Coelli, 1995; Basurto and Sánchez, 2022).

Recent research has applied these approaches to assess the performance of Islamic banks, showing that high efficiency promotes better risk management and alignment with ESG objectives (Asumadu et al., 2023).

2.2. Green Finance: Instruments and Role of Islamic Banks

There green finance brings together financial products dedicated to financing projects with a positive environmental impact. Key instruments include:

- The green sukuk, Islamic equivalents of green bonds, issued in accordance with Sharia principles (Seth and Pathan, 2025; IIFM, 2023).
- The green loans, allocated to renewable energy, energy efficiency or sustainable infrastructure projects (OECD, 2023).

The Islamic banks play a growing role in green finance. Their ethical orientation and prohibition on financing harmful (haram) activities theoretically make them more compatible with climate objectives (Akmal et al., 2025). However, empirical reality shows uneven adoption of green instruments, partly due to regulatory constraints, a lack of technical capacity, and poorly harmonized standards (Koval et al., 2022).

2.3. Climate risk: exposure, vulnerability, banking responses

Climate risk manifests itself in two main forms:

- Physical risk: damage linked to extreme climatic events (floods, droughts).
- Transition risk: costs associated with adjustment to a low-carbon economy (regulation, depreciation of assets).

Banks are increasingly exposed to these risks, which affects their loan portfolios and valuation models (NGFS, 2023; Chava et al., 2024). Conventional banks have begun to integrate climate stress tests, ESG disclosure requirements, and green ratings. In contrast, Islamic banks are lagging behind in the systematic integration of these tools (GIZ & IFSB, 2024).

2.4. Links between banking efficiency, green finance and climate risk management

Several recent works highlight the interactions between efficiency, sustainability and risk management. The most efficient banks appear to be better able to bear the costs of transitioning to sustainable finance (Batsaikhan et al., 2020; Demirgüç-Kunt & Peria, 2020). They are also more likely to adopt proactive climate risk management practices (Qiubin and Mingting, 2024).

In the case of Islamic banks, operational efficiency could be a necessary condition for an effective contribution to green finance. However, the literature remains limited on the direct causal relationships between banking efficiency and climate engagement in the Islamic sector (Seth and Pathan, 2025; Abbas et al., 2023).

2.5. Conceptual framework and hypotheses

Based on this review, our conceptual framework is based on the idea that banking efficiency is a lever enabling Islamic banks:

1. To allocate more resources to green projects.
2. To resist climate shocks through better risk management.
3. To improve their ESG reporting and their alignment with international standards.

We therefore formulate the following hypotheses:

- H1: Banking efficiency is positively related to Islamic banks' engagement in green finance.
- H2: The most efficient Islamic banks are adopting more climate risk management measures.
- H3: The effect of efficiency on green finance is moderated by bank ownership type and size.

3. METHODOLOGY

3.1. Sample and Data Sources

The study sample includes a panel of Islamic banks operating in different countries with a Muslim majority or a strong Islamic banking presence, including in particular:

Saudi Arabia, United Arab Emirates, Malaysia, Indonesia, Pakistan, Qatar, Bahrain, Türkiye, and some North African countries.

The analysis period extends from 2000 to 2024, covering regulatory transformations, advances in green finance and the evolution of climate issues.

The data comes from various sources:

- Banking databases: Orbis Bank Focus, Bankscope, IFS (IMF).
- Annual and sustainability reportsbanks.
- ESG Basics: Refinitiv, Bloomberg ESG, CDP (Carbon Disclosure Project).
- Regional institutions: IIFM, IFSB, Islamic Development Bank.

3.2. Measuring Banking Efficiency

Effectiveness is measured using two complementary approaches:

- Data Envelopment Analysis (DEA):
  - o Inputs: operating expenses, number of employees, total assets.
  - o Outputs: operating income, outstanding loans, financial products.
  - o Orientation: input (minimization of resources for a given level of production).
  - o Model: VRS (Variable Returns to Scale), Banker et al. (1984).
- Stochastic Frontier (SFA):
  - o Trans logarithmic (or Cobb-Douglas) cost function.
  - o Decomposition of the error term: statistical noise vs. inefficiency.
  - o Controlling heterogeneity between countries and banks.

These measures are then introduced as explanatory variables in the econometric models.

3.3. Study Variables

The dependent variables are presented as follow:

1. Commitment to green finance:
    - o Volume of green sukuk issued (logarithm or total asset ratio).
    - o Total amount of green loans granted.
    - o Composite green involvement index (weighting of the two indicators).
2. Climate risk management indicators:
    - o Inclusion of ESG/climate sections in annual reports (binary variable).
    - o Carrying out climate stress tests or vulnerability assessments.
    - o Environmental ESG score (Refinitiv/Bloomberg base).

The main explanatory variables are presented as follow:

- Banking efficiency (DEA or SFA score).
- Bank size (logarithm of assets).
- Quality of governance (independence of the board, number of committees).
- Environmental Policy Stringency Index (OECD).
- Type of ownership (public, private, foreign).

The control variables are presented as follow:

- GDP per capita, inflation, country's green electrification rate.
- Country and year fixed effects.

3.4. Econometric Strategy

The analysis is based on a combination of panel methods:

1. Fixed and random effects models(Hausman test for selection)
    - o Analysis of the relationship between efficiency and green/climate indicators.
    - o Basic formulation: $Y_{it} = \alpha + \beta_1 Eff_{it} + \beta_2 X_{it} + \gamma_t + \mu_i + \varepsilon_{it}$
2. GMM dynamic system (Arellano-Bover/Blundell-Bond)
    - o Treatment of endogeneity, especially in the dynamics of green engagement.

- Considering persistence in ESG engagement.

4. EMPIRICAL RESULTS

4.1. Descriptive Statistics

Table 1 presents a statistical summary of the main variables used in the study, calculated on a sample of 320 Islamic banks observed annually between 2000 and 2024.

According to the results presented in Table 1, we can notice that:

- The average banking efficiency score is 0.74, which indicates that Islamic banks, on average, are using about 74% of the minimum resources required to achieve their current level of production, according to the DEA method. The moderate standard deviation (0.13) highlights a notable variation among banks.
- The volume of green sukuk (logarithmic) reveals a significant concentration, with several banks issuing few or no green sukuk (minimum 0), while the most active reach very high volumes (max 6.9, or approximately USD 1 million in logarithmic value).
- The share of green loans in total assets varies considerably, with an average of 5.6% and a maximum of 16.5%, which shows a growing but still partial integration of green finance into the banking portfolio.
- The environmental ESG score, based on ratings from specialist agencies, has an average of 43.7 out of 100, reflecting potential for improvement in climate risk management.
- Bank size (measured by the logarithm of total assets) reflects a strong dispersion between small local banks and large international institutions.
- Finally, the environmental regulation index (on a scale of 1 to 5) suggests that countries of operation vary greatly in terms of the stringency of climate policies, which may influence banking behavior.

Table 1. Descriptive statistics

Variable	Average	Median	Standard deviation	Minimum	Maximum
Banking Efficiency Score (DEA)	0.74	0.76	0.13	0.41	1.00
Green sukuk volume (log USD)	2.14	1.90	1.45	0.00	6.90
Green loans / assets (%)	5.6%	4.8%	3.2%	0.0%	16.5%
Environmental ESG score	43.7	45.1	14.3	12.0	87.5
Bank size (log assets)	14.2	14.0	1.3	11.5	17.8
Environmental Regulation Index	2.73	2.60	0.75	1.20	4.10

Source: Owen elaboration

4.2. Correlations

Table 2 presents the correlation matrix between the main variables of the study, calculated from panel data on Islamic banks between 2000 and 2024.

According to the results presented in Table 2, we can notice that:

- The bank efficiency score has a positive and significant correlation with the volume of green sukuk (r = 0.42), green loans (r = 0.38), as well as with the environmental ESG score (r = 0.37). These results suggest that more efficient banks tend to be more engaged in green finance and climate risk management.
- Variables related to green finance (green sukuk and green loans) are also highly correlated (r = 0.49), indicating that these two forms of green finance often coexist within the same institutions.
- The environmental ESG score is positively correlated with green financial indicators, confirming the consistency of climate engagement measures.
- These correlations provide a solid basis for multivariate econometric analyses, validating the hypothesis of a positive link between banking efficiency, green finance and climate risk management.

Table 2. Correlation matrix

Variables	Banking Efficiency (DEA)	Green sukuk volume (log USD)	Green loans / assets (%)	Environmental ESG score
Banking Efficiency (DEA)	1	0.42***	0.38***	0.37***
Green sukuk volume (log USD)		1	0.49***	0.41***
Green loans / assets (%)			1	0.33**
Environmental ESG score				1

Note: (*), (**), and (***) values significant at 10%, 5%, and 1%
Source: Owen elaboration

4.3. Main Estimates

Econometric models test the impact of banking efficiency on Islamic banks' engagement in green finance and climate risk management, controlling for size, governance, and environmental regulation. Estimations are conducted using fixed-effects and dynamic GMM models to account for potential endogeneity and behavioral dynamics.

4.3.1. Effect of Banking Efficiency on Green Finance

Table 3 presents the results of the fixed effects regressions, where the dependent variable is engagement in green finance (measured by the volume of green sukuk and the share of green loans).

According to the results presented in Table 3, we can notice that:

- Banking efficiency has a positive and significant effect on engagement in green finance. A 0.1-point improvement in the DEA score is associated with an average increase of 3.12% in the volume of green sukuk or green loans.
- The size and quality of governance also contribute positively to this commitment, which highlights the role of resources and decision-making structure.
- Strict environmental regulations also stimulate banks' green actions, indicating a favorable influence of the legal framework.

Table 3. Effect of banking efficiency on green finance

Explanatory variable	Coefficient	Standard error	t-statistic	Significance
Banking Efficiency (DEA)	0.312	0.062	5.04	***
Size (active log)	0.118	0.032	3.67	***
Governance (index)	0.067	0.027	2.45	**
Green regulation	0.091	0.029	3.12	***
Constant	0.004	0.011	0.521	-

Note: (*), (**), and (***) values significant at 10%, 5%, and 1%
Source: Owen elaboration

4.3.2. Impact on Climate Risk Management

Table 4 shows the results from the dynamic GMM model, with an environmental ESG score or the adoption of climate stress tests as the dependent variable.

According to the results presented in Table 4, we can notice that:

- The environmental ESG score is highly autocorrelated, showing the persistence of ESG practices from year to year.
- Banking efficiency is a significant factor in improving climate risk management, supporting the idea that efficient banks have better capabilities to integrate and report climate risks.
- Conducting climate stress tests is also associated with a better ESG score, suggesting that monitoring tools strengthen environmental performance.

Table 4. Impact on climate risk management

Explanatory variable	Coefficient	Standard error	Z statistic	Significance
ESG(t-1)	0.473	0.065	7.31	***
Banking efficiency	0.196	0.069	2.83	***
Climate stress test	0.134	0.063	2.11	**
Green regulation	0.089	0.035	2.55	***

Note: (*), (**), and (***) values significant at 10%, 5%, and 1%
Source: Owen elaboration

These results confirm that the Banking efficiency plays a crucial role in promoting green finance and proactively managing climate risks within Islamic banks. They highlight the importance of strengthening internal capacities and the regulatory framework to support this dynamic.

4.4. Robustness of Results

To ensure the robustness and generalizability of the results obtained in the previous sections, several robustness tests were carried out.

4.4.1. Use of Alternative Measures of Banking Efficiency

To strengthen the validity of the results, we assessed the impact of banking efficiency using two different methodologies: the Data Envelopment Analysis (DEA) method and the Stochastic Frontier Method (SFA).

- DEA method: Nonparametric approach that estimates relative efficiency by wrapping input and output data without assumptions about the functional form.
- SFA method: Parametric approach that separates inefficiencies from statistical errors by specifying a production function with a random error component.

The empirical results show that the coefficients associated with banking efficiency remain positive and statistically significant in both cases. For example, the estimated coefficient for DEA is 0.31 ($p < 0.01$), while that for SFA is slightly lower at 0.28 ($p < 0.05$), but with an economically similar impact.

This convergence of results indicates that the beneficial effect of banking efficiency on engagement in green finance and climate risk management is robust, regardless of the efficiency measurement methodology used.

4.4.2. Variations in Dependent Indicators

To verify the robustness of the results, we tested the impact of banking efficiency on different indicators of engagement in green finance and climate risk management.

More specifically, we used separately:

- The total volume of green sukuk issued (expressed in logarithm), representing a direct quantitative measure of green financing through Sharia-compliant instruments.
- The share of green loans in the total loan portfolio, which reflects the proportion of assets dedicated to sustainable financing.
- A composite environmental ESG score, to capture the integration of climate risks into the governance and management of banks.

The estimates show that banking efficiency has a positive, significant, and consistent effect on all of these indicators. For example, increased efficiency significantly increases the volume of green sukuk, the share of green loans, and ESG scores.

These results reinforce the validity of our conclusion that banking efficiency is an important lever for increasing the green commitments of Islamic banks, regardless of the dimension used to measure this commitment.

4.4.3. Subsamples by Geographic Region

To examine the spatial robustness of our results, the full sample was divided into three regional subgroups corresponding to the main locations of the analyzed Islamic banks:

- Middle East
- Southeast Asia
- North Africa

Separate estimations on each subsample confirm the persistence of a positive and significant effect of banking efficiency on engagement in green finance and climate risk management in all regions.

However, the intensity of this effect varies considerably:

- The Middle East shows the highest coefficients, probably due to stricter environmental regulations and more advanced development of green Islamic financial markets.
- Southeast Asia and North Africa show positive but slightly less pronounced effects, which may reflect institutional, regulatory or financial sector maturity differences.

These results highlight the importance of the regional and regulatory context in the effectiveness with which Islamic banks mobilize their efficiency to support green finance.

4.4.4. Control of the Regulatory Framework

The role of the environmental regulatory framework was taken into account to analyze its influence on the relationship between banking efficiency and engagement in green finance.

The sample was divided into two subgroups according to the level of environmental regulation in the countries where the banks operate:

- High regulation (countries with strict environmental standards and strong incentive policies).
- Weak regulation (countries with less restrictive or less developed regulatory frameworks).

The results show that the positive impact of banking efficiency on green finance initiatives is significantly greater in countries with strict regulations. Indeed, banks operating in these environments benefit from a favorable context that amplifies the effect of their internal efficiency on the mobilization of green finance and climate risk management.

Conversely, in countries with weak regulation, although the relationship remains positive, its intensity is lower, suggesting that banking efficiency, without an appropriate incentive framework, encounters more limits in its impact.

This analysis highlights the complementary and essential role of regulation in maximizing the contribution of Islamic banks to green finance.

4.4.5. Dynamic vs. Static Models

To take into account the dynamic nature of banks' behaviors in terms of engagement in green finance and climate risk management, we compared the results obtained from static models (classic fixed effects) with those from dynamic models estimated using the GMM (Generalized Method of Moments) method.

- Static models: allow us to estimate the direct impact of banking efficiency and other explanatory variables on the dependent variables, without taking into account temporal persistence.
- Dynamic models (GMM): incorporate inertia effects, notably through the inclusion of lagged dependent variables (e.g., ESG scores from the previous year). This approach also corrects for the potential endogeneity of the explanatory variables.

The results show that:

- Banking efficiency maintains a positive and statistically significant effect in both types of models.
- The persistence of behaviors is strong, as illustrated by the high coefficient of the lagging dependent variable in the dynamic models (e.g., a coefficient around 0.45 for the ESG(t-1) score).
- The dynamic approach confirms the robustness and sustainability of the impact of banking efficiency on green finance and climate risk management.

This comparison highlights the importance of modeling the temporal dimension and the dynamics of adoption of green practices for a complete analysis.

These various tests confirm that banking efficiency is a key and robust determinant of Islamic banks' engagement in green finance and climate risk management. The results are stable across different specifications, indicators, subsamples, and methodological approaches.

5. DISCUSSION

The empirical results clearly show that banking efficiency plays a crucial role in the development of green finance within Islamic banks as well as in their ability to manage climate risks. Greater efficiency not only allows for more effective allocation of financial resources, but also for more effective integration of environmental criteria into decision-making processes. This positive link underscores that efficient Islamic banks are better positioned to support the transition to a low-carbon economy.

Our findings are consistent with recent work by (Khan et al., 2022; Ahmad et al., 2023) that highlighted the importance of banking efficiency in promoting sustainable finance, particularly in the context of emerging markets. However, our study makes a specific contribution by focusing on the Islamic banking sector, characterized by Sharia principles that reinforce the ethical and social dimension of financing, which few studies have explored in depth. Furthermore, unlike some studies that highlight the complexity and upfront costs of implementing green initiatives (Malik et al., 2021), our results indicate that internal efficiency can mitigate these constraints and facilitate the integration of sustainable practices.

Policymakers should encourage the increased efficiency of Islamic banks, for example through training programs, regulatory incentives, and appropriate regulatory frameworks. Furthermore, strengthening environmental regulations, as confirmed in our analyses, acts as a powerful catalyst to amplify the effect of banking efficiency on green finance. Thus, a coherent policy combining efficiency improvements and a strict regulatory framework will optimize the role of Islamic banks in combating climate change.

Islamic banks are encouraged to invest in their organizational and technological capabilities to improve their efficiency. Improved efficiency helps reduce operational costs, optimize risk management, and more effectively allocate funds to green projects, in line with the ethical principles of Islamic finance. Moreover, increased transparency, particularly through ESG reporting, can strengthen the confidence of climate-conscious investors and clients.

Regulators must adapt supervisory mechanisms to integrate green finance criteria into bank performance assessments. They can also promote the dissemination of good efficiency practices and encourage financial innovations compatible with Sharia law and climate objectives. Finally, increased monitoring of climate risks, through specific stress tests, is essential to prevent systemic vulnerabilities linked to climate change.

6. CONCLUSION AND RECOMMENDATIONS

This study highlights the critical role of banking efficiency in promoting green finance and climate risk management among Islamic banks between 2000 and 2024. Our analyses show that more efficient Islamic banks are significantly more engaged in issuing green sukuk, providing green loans, and integrating environmental ESG criteria. This positive relationship is robust to different measurement methods, indicators, and regional contexts, and is amplified in countries with stricter environmental regulations.

Several limitations should be noted. First, the availability of data on green finance initiatives in certain regions remains limited, which may affect the representativeness of the sample. Second, the study focuses exclusively on Islamic banks, without direct comparison with conventional banks. Finally, although the period studied is long, exogenous events (financial crises, political crises) not taken into account could influence certain results.

To explore this topic further, future research could:

- Integrate a comparative analysis with conventional banks to better understand Islamic specificities in green finance.
- Use more detailed data, particularly at the micro level (green projects financed, customer profiles) to study the impact of efficiency on the quality and effectiveness of green financing.
- Study the effect of technological innovations (green Islamic fintech) on performance and climate risk management.
- Assessing the effects of extreme climate shocks on the financial resilience of Islamic banks according to their level of efficiency.

REFERENCES

Abbas, M. H., Ali, H., & Musa, M. (2023). Integration between ESG goals and value-based intermediation approach within the paradigm of Islamic finance. *Islamic Banking & Finance Review, 10*(2), 31–54.

Ahmad, N., Malik, M. S., & Hussain, M. (2023). Bank efficiency and green finance: Evidence from emerging markets. *Journal of Cleaner Production, 380*, 135029. DOI: 10.1016/j.jclepro.2022.135029

Akmal, I., Ibrahim, F. W., Bambang, S., & Syafiq, M. H. (2025). Rethinking financial deepening in Islamic countries: Political system and economic growth. *International Journal of Islamic and Middle Eastern Finance and Management, 18*(2), 467–485. DOI: 10.1108/IMEFM-10-2023-0376

Asumadu, S. S., Ahmed, M., & Owusu, P. A. (2023). Advancing COP26 climate goals: Leveraging energy innovation, governance readiness, and socio-economic factors for enhanced climate resilience and sustainability. *Journal of Cleaner Production, 431*, 139757. DOI: 10.1016/j.jclepro.2023.139757

Banker, R. D., Charnes, A., & Cooper, W. W. (1984). Some models for estimating technical and scale inefficiencies in data envelopment analysis. *Management Science, 30*(9), 1078–1092. DOI: 10.1287/mnsc.30.9.1078

Basurto, S., & Sánchez, G. (2022). Technical efficiency in small and medium-sized firms in Mexico: A stochastic frontier analysis. *Estudios Económicos de El Colegio de México, 37*(1), 103–147.

Batsaikhan, A., Hachinger, S., Kurtz, W., Heller, H., & Frank, A. (2020). Application of Modern Web Technologies to the Citizen Science Project BAYSICS on Climate Research and Science Communication. *Sustainability (Basel), 12*(18), 7748. DOI: 10.3390/su12187748

Battese, G. E., & Coelli, T. J. (1995). A model for technical inefficiency effects in a stochastic frontier production function for panel data. *Empirical Economics, 20*(2), 325–332. DOI: 10.1007/BF01205442

Berger, A. N., & Mester, L. J. (1997). Inside the black box: What explains differences in the efficiencies of financial institutions? *Journal of Banking & Finance, 21*(7), 895–947. DOI: 10.1016/S0378-4266(97)00010-1

Berger, A. N., & Mester, L. J. (1997). Inside the Black Box: What Explains Differences in the Efficiencies of Financial Institutions? *Journal of Banking & Finance, 21*(7), 895–947. DOI: 10.1016/S0378-4266(97)00010-1

Chava, S., Malakar, B., & Singh, M. (2024). Impact of corporate subsidies on borrowing costs of local governments: Evidence from municipal bonds. *Review of Finance*, *28*(1), 117–161. DOI: 10.1093/rof/rfad021

Coelli, D., Rao, T. P., O'Donnell, C., & Battese, G. (2005). *An Introduction to Efficiency and Productivity Analysis* (2nd ed.). Springer.

Demirgüç-Kunt, A., Peria, M., Soledad, M., & Thierry, T. (2020). The global financial crisis and the capital structure of firms: Was the impact more severe among SMEs and non-listed firms? *Journal of Corporate Finance*, *60*(C), 101514. DOI: 10.1016/j.jcorpfin.2019.101514

IMF. (2024). *Annual Report 2024*. International Monetary Fund.

International Islamic Financial Market (IIFM). (2023, August). IIFM Sukuk Report 2023 (12th edition): A comprehensive study of the global Sukuk market. Manama, Kingdom of Bahrain.

Islamic Financial Services Board (IFSB). (2024). Islamic Financial Services Industry Stability Report 2024. Kuala Lumpur, Malaysia.

Khan, M. A., Zulkhibri, M., & Zaman, K. (2022). The role of Islamic finance in sustainable development goals: Empirical evidence from sukuk markets. *Resources Policy*, *77*, 102665. DOI: 10.1016/j.resourpol.2022.102665

Khan, T., & Badjie, F. (2022). Islamic blended finance for circular economy impactful SMEs to achieve SDGs. *The Singapore Economic Review*, *67*(01), 219–244. DOI: 10.1142/S0217590820420060

Koval, V., Laktionova, O., Atstāja, D., Grasis, J., Lomachynska, I., & Shchur, R. (2022). Green Financial Instruments of Cleaner Production Technologies. *Sustainability (Basel)*, *14*(17), 10536. DOI: 10.3390/su141710536

Malik, M. S., Rehman, A. U., & Hashmi, S. H. (2021). Barriers to green finance in Islamic banking: An empirical study. *Environmental Science and Pollution Research International*, *28*, 15298–15310. DOI: 10.1007/s11356-021-13588-7

NGFS. (2024). *Annual Report 2024*. Network for Greening the Financial System.

OECD. (2023). *Climate Finance Provided and Mobilised by Developed Countries in 2013-2021: Aggregate Trends and Opportunities for Scaling Up Adaptation and Mobilised Private Finance. Climate Finance and the USD 100 Billion Goal series*. OECD Publishing.

Qiubin, H., & Mingting, K. (2024). Does climate policy uncertainty impair or improve corporate investment efficiency? *Economics Letters*, *244*, 112015. DOI: 10.1016/j.econlet.2024.112015

Sealey, C.Jr, & Lindley, J. (1977). Inputs, Outputs, and the Theory of Production and Cost of Depository Financial Institutions. *The Journal of Finance*, *32*(4), 1251–1266. DOI: 10.1111/j.1540-6261.1977.tb03324.x

Seth, S., & Pathan, M. I. (2025). Green Sukuk: Contextual Spectrum and Alignments. *Journal of Marketing & Social Research*, *2*(2), 540–548.

World Bank. (2023). Results and Performance of the World Bank Group 2023. Washington, DC: Independent Evaluation Group. World Bank. http://hdl.handle.net/10986/40886

Zhu, J. (2014). "Data Envelopment Analysis," International Series in Operations Research & Management Science, in: Quantitative Models for Performance Evaluation and Benchmarking, edition 3, chapter 1, pages 1-9, Springer.

Chapter 9
Sustainable Finance and Climate Stability:
The Impact of ESG and Green Bonds

Vikas Sharma
https://orcid.org/0000-0001-8173-4548
Department of Computer Applications, SRM Institute of Science and Technology, Delhi, India

Tarun Kumar Vashishth
https://orcid.org/0000-0001-9916-9575
School of Computer Science and Applications, IIMT University, Meerut, India

Anuj Kumar
IIMT College of Medical Science, IIMT University, Meerut, India

Poonam Bhardwaj
IIMT College of Medical Science, IIMT University, Meerut, India

Garima Garg
IIMT College of Medical Science, IIMT University, Meerut, India

Priyanka Rana
https://orcid.org/0000-0002-6352-1271
School of Commerce and Management, IIMT University, Meerut, India

Sachin Chaudhary
https://orcid.org/0000-0002-8415-0043
School of Computer Science and Applications, IIMT University, Meerut, India

ABSTRACT

The urgency of climate change has placed financial markets in a position to lead the sustainability transition. Green financial products—such as green bonds, Environmental, Social, and Governance (ESG) frameworks, and generally, sustainable finance—are becoming one of the most important approaches to leveraging finance for low-carbon resilience. This chapter argues that these finance products

DOI: 10.4018/979-8-3373-6766-8.ch009

are climate risk management tools, as they link finance flows to climate outcomes, incentivize responsible investment, and transparently disclose market actions and outcomes. The chapter examines asymmetrical growth and future potential in financing renewable energy and climate-resilient infrastructure via green bonds and how ESG practices are applied in capital decisions and corporate governance. Furthermore, the chapter argues that sustainable finance can reduce exposure to climate risks while creating opportunities for innovation, resilient business practices, and value-generating results in global markets.

1. INTRODUCTION

The increasing threat of climate change has shifted environmental issues from the fringes to the heart of global financial stability. Climate change is now seen not merely as an environmental or social issue, but as a genuine financial risk that can disrupt supply chains, erode asset values, and disrupt entire economies, while creating new opportunities for innovation, technological upgrades, and value creation in the shift to low-carbon, climate-resilient growth. Within this context, sustainable finance has emerged as an important framework for directing capital toward projects and organizations with environmental and social aims; in this new finance system, the two main instruments that have become core to, and very much part of, the new financial models that are influencing investment thought, and corporate behavior, are the Environmental, Social and Governance (ESG) frameworks, and green bonds (Arifin et al., 2024).

1.1 Background: Climate Change as a Financial Risk and Opportunity

Climate change can no longer be seen as an academic argument, or an environmental movement issue, it is now financial risk with real and tangible implications for the actions and actors of global markets, national economies, and investment thought, at which point individual (the investor) and societal ties will also become part of the decision-making considerations. The frequency and severity of events listing climate change impacts such as hurricanes, floods, droughts, and wildfires are threatening business continuity, disrupting infrastructure, and creating volatility in commodity prices. The physical risk associated with climate change will generate significant financial loss to reduce asset values, increase regulatory and insurance liabilities, and raise the cost of capital for economically vulnerable sectors. For example, energy-intensive sectors, agriculture, and property development are particularly vulnerable to acute weather shocks and increasingly stringent environmental standards, making these sectors financially risky to potential investors. Second, climate policy uncertainty, carbon pricing schemes, and emission reduction commitments generate climate transition risks for companies that may fail to adapt to new sustainability standards, resulting in decreased valuations and reputational risks. Third, climate change also provides a range of opportunities for the financial system. A global transition to a low-carbon and climate-resilient economy drives unprecedented demand for investment in renewable energy, electric vehicles, sustainable agriculture, carbon capture technologies, and sustainable infrastructure. Importantly, sustainability is no longer regarded as limiting economic advance; rather, while the organic limits of sustainability will stimulate innovation, competitiveness, and sustainable long-term value creation. Forward-looking investors and financial institutions that integrate ESG investment factors into their strategies will likely have a higher likelihood of identifying and taking advantage of these opportunities, including their investment in new green markets. For example, financing costs for green bonds to fund solar and wind energy projects can yield returns for investors and contribute to both climate stability and social outcomes. In summary, climate change creates both a paradoxical challenge to the current financial system: a systemic risk that undermines global markets and the stability of the economy (Hilal et al., 2025), while providing a transformational opportunity for investors and for-profit organizations, as well as government to innovate, build resilience, and create sustainable growth. Understanding climate change as a risk and an opportunity is the basis for the development of sustainable finance, which intentionally directs capital towards solutions that limit environmental degradation while simultaneously creating economic and social returns. This dual perspective justifies the imperative to rethink customary financial structures and reasons why sustainable finance initiatives like ESG investing and green bonds are coming to the front as mechanisms to manage climate risks and capitalize on opportunities in the transition to a sustainable global economy (Afandy et al., 2024).

1.2 Emergence of Sustainable Finance in Global Markets

The expansion of sustainable finance in international markets can be attributed to a growing understanding that financial systems need to orient to environmental and social priorities for global markets to remain stable and resilient over time. Traditionally, investment decisions were mostly made with narrow motives of pursuing a profit in the short-term for obligations to shareholders in return for their capital while disregarding the externalities of business practices, ie carbon emissions, resource extraction and depletion or a lack of social equity. The impacts of climate change, biodiversity loss, as well as social issues, became more apparent when investors, regulators, and policy makers recognized that climate change and biodiversity loss as an issue, and social issues, all pose a significant cost to the economy. This change of mind enabled social change on the premise of that finance is no longer neutral to sustainability outcomes, but an important differentiator in its potential to either increase, or solve global issues associated with environmental degradation (D'Orazio, 2023). Sustainable Finance emerged from an area of niche offering, to a major pillar of global capital markets with the last twenty years. There have been several converging factors for this transition: The international climate frameworks established targets for GHG emission reductions; (ex. the Paris Agreement (2015)); the UN Sustainable Development Goals (SDGs), which established a framework for linking finance with a common notion of global well-being; and more of a playing role by institutions, (i.e. the Taskforce on Climate-Related Financial Disclosure) who called for more transparency around climate risk disclosures. These strategies and initiatives together worked to motivate and move governments, multilateral corporations, and financial institutions to adopt sustainability frameworks in their investment policies, lending decisions, and corporate governance frameworks. At the same time, market 'innovation' in the form of green bonds, ESG investing, sustainability-linked loans, and impact funds demonstrates that financial innovation can be a part of financing processes for both low-carbon transitions' and inclusive growth (Bai and Wang, 2024). For example, green bond issuance has grown from a few billion dollars per year in the early 2010s to the hundreds of billions per year by the early 2020s, a clear indication of a strong appetite among investors for climate-aligned investments. Sustainable finance has emerged in developing economies like India and China, as well as developed economies like the EU and the USA through policy, incentives, and financial instruments; therefore, it is clear that sustainability issues transcends geography (Naseer and Bagh, 2024).

1.3 Defining ESG and Green Bonds in the Broader Context of Climate Stability

Two of the most powerful instruments shaping sustainable finance today are Environmental, Social, and Governance (ESG) frameworks and green bonds. Each of the two intersects the climate issue in a unique but complementary manner. ESG is a collection of criteria that investors and businesses use to evaluate the sustainability of their decision-making and its ethical ramifications. The environmental part of ESG is focused on carbon emissions, renewable energy use, waste, and biodiversity. The social part of ESG concerns labour, community, diversity, and human rights. Governance is focused on corporate ethics, transparency, board accountability, and regulatory compliance. Financial decision-makers add ESG considerations to their analysis and decision-making by incorporating long-term risk and opportunity considerations that would otherwise not be available through traditional financial metrics. Importantly, ESG is not just a reporting tool, but a means by which businesses can align profitability with stakeholders and environmental good by addressing climate stability and responsible growth. Green bonds are a financial concept focused on financing projects and initiatives that promote environmental benefits. They are fixed-income instruments that are similar to traditional bonds, except they are exclusively used for the financing of projects or initiatives, including renewable energy projects, energy-efficient buildings, sustainable transport, and climate adaptation efforts. Initially initiated by the European Investment Bank in 2007, the green bond market has rapidly established itself as a global market with then government, corporate, and institutional investors. Green bonds offer a way for investors to know that their money will be spent for the benefit of climate while allowing a diversification of their investment portfolios with sustainable assets. Besides the financial function of green bonds, they also serve as a signalling device (Joshi et al., 2024) for a commitment to climate action, and provide greater accountability in project selection and monitoring.

2. CONCEPTUAL FOUNDATIONS OF SUSTAINABLE FINANCE

Sustainable finance is grounded on the idea that financial systems and investment decisions must incorporate long-term environmental, social, and governance considerations along with traditional financial performance. In essence, it is a direct challenge to the prevailing perspective that finance is value-neutral, and calls attention to the fact that the provision of capital has significant implications on the world and society. This perspective acknowledges that unregulated financial flows can exacerbate climate risk, environmental degradation, and social inequality, while prudently directed financial flows can accelerate low-carbon economies, inclusive growth, and increased climate resilience (Kumar et al., 2025).

2.1 Definition and Principles of Sustainable Finance

Sustainable finance is described broadly as the integration of environmental, social, and governance (ESG) dimensions into financial decisions to promote long-term economic growth, social inclusion, and environmental responsibility. Unlike traditional finance, which maximizes shareholder return typically on a short-term basis, sustainable finance acknowledges that our financial systems function, ultimately, within larger ecological and social constraints. The following principles are based on incorporating sustainability to global financial systems:

1. **Responsible Investing:** It emphasizes placing money in a manner that values financial return in light of positive social and environmental results. This principle obligates investors to examine ESG and otherwise normal financial information as part of their investment decisions. For example, a responsible investor would be less willing to invest in an industry that has high carbon emissions (e.g., fossil fuel extraction) than to invest in energy firms that are engaged in renewables, or renewable energy investment is more than just attractive, it is more favourable as a long-term investment.
2. **Double Materiality:** This principle highlights the reciprocal nature of the relationship between sustainability and finance. From the outside-in, it requires organizations to assess how environmental and social risks (ex. climate change, inequality, etc.) may jeopardize their financial performance. From the inside-out, it requires financial actors to think about how their actions may impact the environment and society. For instance, if a bank loans money to coal projects, its exposure to climate-transition risk is compounded by the fact that it is actively creating greenhouse gas emissions that cannot be entirely offset.
3. **Long-Term Orientation:** Sustainable finance addresses the short-termism that often prevails in global markets by proposing a longer-term orientation. This principle proposes an investment practice that prioritizes not only immediate profit, but the financial returns generated by longer-term projects - including renewable energy infrastructures, sustainable transportation systems, and ecosystem restoration - that will enhance resilience and sustainability. It is expected that longer-term investment strategies will produce financial profits but will also help to mitigate climate risks. The long-term orientation identified within sustainable finance proposes that financial systems build resilience through longer-term investment behavior given climate uncertainties (Singhania et al., 2024).
4. **Transparency and Disclosure:** These principles are also important because they create trust between investors, corporations, regulators, and society. Financial institutions and businesses should expect to publicly report their sustainability-related performance, sustainability-related ESG risks, and climate-related impact. Reporting frameworks, like the Task Force on Climate-related Financial Disclosures (TCFD) and Global Reporting Initiative (GRI), provide standardized models to encourage financial institutions and businesses to report on sustainability issues. Transparency means that, in addition to eliminating the practice of greenwashing, investors will be able to make informed decisions about capital allocations, and it will create competitive incentives to achieve higher standards for sustainability.
5. **Conformity with Global Sustainability Targets:** Sustainable finance should heed sustainability targets and frameworks recognized across multiple countries.

Examples of benchmarks to help with capital allocating decisions would include the UN Sustainable Development Goals (SDGs), the Paris Climate Agreement, and national-level action plans related to green finance. Put differently, a goal of sustainable finance should be to determine that private capital will to some degree contribute to international questions surrounding climate change, poverty, or inequality when investments are aligned with targets. Again, the principle seeks to create a situation where capital is no longer viewed as an obstacle or barrier, but becomes an enabler of the work taking place globally, while maintaining sustainability for the planet and economy for future generations. (Zhang, 2025).

2.2 Role of Capital Markets in Addressing Climate Change

Capital markets serve as an important and meaningful way to mitigate for climate change in that capital markets are the way for raising, allocating and directing significant amounts of the capital necessary to achieve sustainable development pathways. The transition towards a low-carbon economy will demand trillions of dollars' worth of investment into renewable energy, sustainable infrastructure, clean mobility, and climate adaptation projects, far exceeding the capacity of governments alone. Capital markets serve as a means of mobilizing private investment at sufficiently large scale, using enabling instruments like green or sustainability-linked bonds, carbon markets, climate-focused equity funds, and other instruments, combined with public financing (Mishra et al., 2023). Simultaneously, capital markets are providing incentives to incorporate climate risk into the valuation of assets and companies, meaning firms that may have more carbon exposure begin to be considered riskier and less desirable from an investor's standpoint, and firms with sustainable practices and climate-positive innovations have market confidence and significantly higher valuations (Wang et al., 2025). This risk-pricing mechanism results in a decrease in the risk for investors as they start considering firms that will ultimately be impacted by stranded assets harder to justify; and it raises the competitive bar for firms to decarbonize and adopt sustainable practices. Furthermore, capital markets also contribute to transparency and accountability by mandating the disclosure of climate-related risk, emissions data, and ESG performance, allowing investors to make sound investment decisions and to help mitigate greenwashing, illustrated by figure 1. In this way, capital markets can be more than just some form of financial intermediary, but a truly powerful mechanism for change that will define the future of climate action by structuring the even matching of the portfolio of sustainability objectives with the financial resources to achieve them (Wang et al., 2025).

Figure 1. Capital Market scenario with Climate Change

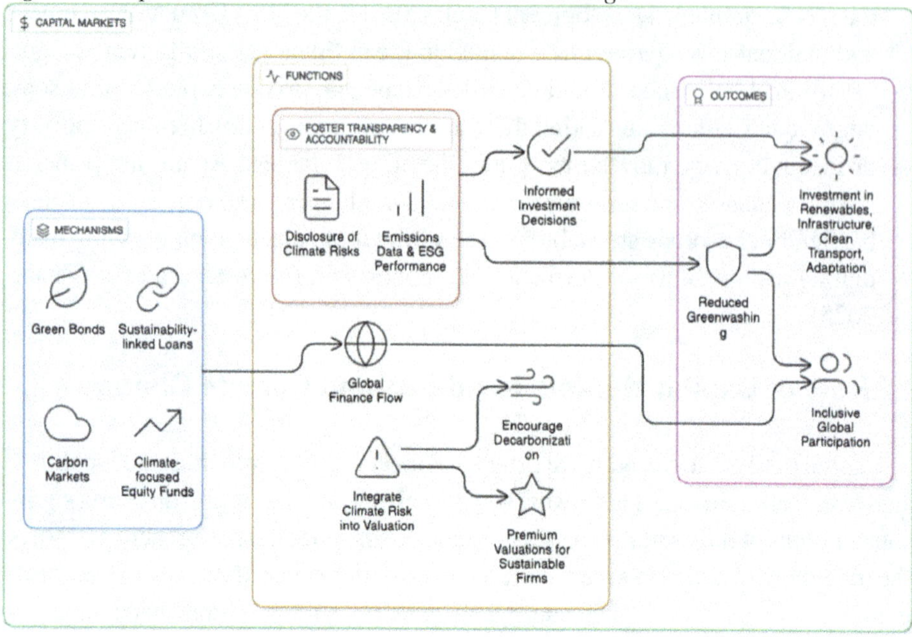

2.3 Linkage between Financial Flows, Low-Carbon Transitions, and Climate Risk Management

The relationship established between financial flows, low-carbon transitions, and climate risk management illustrates the vital role of finance in shaping the trajectory of global climate action. Financial flows represent how we allocate capital across capital markets, banking systems, and investment funds, thus choosing which sectors will grow, which technologies will be scaled, and which types of infrastructure will be built. When financial flows are directed to carbon-intensive industries, such as fossil fuels, this locks our economies into high-emission pathways, and increases exposure to climate-related risks. On the contrary, directing financial flows toward renewable energy, energy efficiency, sustainable transport modes, and green infrastructure facilitated the transition toward a low-carbon economy, decreasing the systemic risk from climate change. The previous discussion illustrates that finance does not act neutrally, but is a determinative force in climate outcomes. Moreover, climate risk management is implicitly connected to how capital is allocated. Investors and financial institutions are beginning to see the perils of physical risk, such as extreme weather events damaging assets and transition risks, such as policy evolution, carbon pricing, and technology exposure that will leave carbon-heavy assets stranded. Incorporating these risk factors into financial decision-making processes demonstrates that the approach is towards resilience in portfolio as well as towards future-proofing. Finance's dual purpose to enable low-carbon transitions and to hedge against risk illustrate their mutually reinforcing relationship whereby financing flows increasingly climate resilience and climate-risk management sells the basis for justification to invest in low-emission alternatives (Vashishth et al., 2025).

3. ENVIRONMENTAL, SOCIAL, AND GOVERNANCE (ESG) INVESTING

Environmental, Social, and Governance (ESG) investing is currently one of the most disruptive trends in modern finance. ESG investing represents a fundamental shift away from traditional profit-driven value creation models, towards its counterparts that incorporate considerations for financial returns against broader objectives of environmental and social benefits. At its core, ESG investing is about integrating non-financial considerations into the decision-making process. The fundamental recognition being that companies' impact on the environment, social obligations, and governing structure would be relevant to companies' longer-term performance and sustainability. This approach signifies a recognition that concerns such as climate change, depleted resources, labour relations, diversity, corporate behaviour, and accountabilities in BOARD governance are no longer secondary concerns to the business and value creation process.

3.1 ESG Criteria and Investment Decisions

The addition of ESG, or Environmental, Social and Governance, factors in investment processes is a dramatic shift in how financial markets analyse and evaluate risk, return and long-term value. Traditionally, investors made investment decisions based almost exclusively on financial measures of profitability, growth potential, and market performance. But to include ESG into investment decisions is a necessary complement to financial analysis of the sustainability and ethical dimension of an investment. The environmental dimension relates to factors such as carbon emissions, energy efficiency, renewables, waste management, and climate change mitigation. Each of these environmental dimensions leads to an understanding of a company's exposure to regulatory change, scarcity of resources, and reputational risk. The principle of the social dimension considers how firms maintain relationships with employees, customers, local communities, and suppliers, and thus considers labour rights and accountability in the supply chain, but also diversity and inclusion, turn over in the workplace, and corporate citizenship for example. Governance screens for the integrity of a firm's leadership team, diversity at the board level, transparency, pay of the executive team, and the mechanisms of accountability that are all fundamental constructs of trust and long-term viability. Ultimately, in practice, ESG will drive investment decisions as risk filters that identify companies appropriately positioned to address emerging risk, and may also position a firm well in the marketplace with respect to new opportunities or manage risk through a transition towards a sustainability economy. Some traditional industries, such as mining, with collaborative relationships with labour and more rigorous diversity frameworks, may similarly enjoy not only workforce productivity, customer loyalty and reputation benefits to support long-term financially healthy performance. The notion of factoring in ESG dimensions to investing is now well understood by investors to potentially drive poor financial performance and stranded asset risks along with other potential legal or reputational risks.

3.2 ESG Integration in Global Markets

The application of Environmental, Social, and Governance (ESG) principles to global markets is transforming how capital is allocated, influencing investor behaviours, and influencing corporate strategy. In recent years, ESG has shifted from being niche in consideration to being required, as institutional investors, asset managers, pension funds, and sovereign wealth funds, have increased focus on more ESG-driven frameworks while making decisions. These movements are a result of regulatory requirements and increased pressure from stakeholders, but the most convincing data supports that companies with strong ESG performance, are easily associated with better financial performance in the long run (Nguyen et al., 2025). Global capital markets have now accepted that ESG can serve as an effective tool to manage risk and as an investment opportunity for new markets, such as renewable energy, electric vehicles, sustainable agriculture practices, to circular economies. Across the globe stock exchanges and financial service providers on every continent, including the United States, Europe and Asia, have adopted ESG disclosure and reporting requirements to offer some level of transparency and comparable reporting requirements across the corporate landscape (Gaganis et al., 2023). For example, the European Union has enacted the Sustainable Finance Disclosure Regulation (SFDR) and the EU Taxonomy and the United States Securities and Exchange Commission (SEC) has unveiled greater climate-related disclosures, featured among other propositions. Similar acceptance of ESG disclosure is underway in numerous Asian markets (e.g., Japan, Singapore, and India), with the goal of better aligning capital flows to sustainability through company disclosure. Together, these actions from regulatory agencies across the globe ultimately serves to reinforce that ESG is not only becoming a "best practice" in voluntary principle, but it is also becoming a requirement for any organization that desires access to international capital. ESG indices and funds have also been a strong method of channelling sustainability to global markets (Arouri et al., 2025). This increased interest has resulted in pressure on corporations to engage in ESG-oriented behavior, or else face the risk of not being included in these large indices for investment and funding (Xiong et al., 2025).

3.3 Opportunities and Challenges of ESG Adoption

The implementation of Environmental, Social, and Governance (ESG) practices represents meaningful opportunities and risks in global markets, companies, and the investment community. On the opportunity front, the adoption of ESG is one way to advance long-term corporate resilience, competitive advantage, and form deep trust and relations with company stakeholders. Companies that take ESG concerns seriously often see improved operational efficiencies, lower regulatory risk, and improved access to capital since investors have begun to more actively consider investing in companies that have developed a sustainable growth strategy (Khamitdkhanovich et al., 2025). ESG practices also create access for companies into new markets including renewable energy, sustainable finance, clean technology, and social innovation which will generate profits while incentivizing climate stability. In facing these same economic pressures, adopting ESG practices improves corporate self-image while raising brand equity, since consumers and clients are moving in the direction of rewarding ethical, transparent, and socially responsible companies with their loyalty and consumption revenue. For investors, ESG practices provide a method of analysis to identify resilient companies that can withstand environmental and social disruption, while at the same time, allow investor portfolios to remain in lock-step with global sustainability and climate goals such as the Paris Agreement and UN Sustainable Development Goals (SDGs). At the same time, ESG adoption faces some structural and practical barriers to adoption which inhibit its effectiveness. The lack of consistent global standards for ESG measurement, disclosure, and reporting is one of the most significant challenges. Different rating agencies are concerned with using support inconsistent methodologies which results in companies or industries reporting outcomes that are not generally comparable.

4. GREEN BONDS AS A FINANCIAL INNOVATION

Green bonds are now poised to emerge as one of the greatest financial innovations of the 21st Century, providing an efficient means to mobilize investment capital for projects intended to preserve the environment. Green bonds differ from standard bonds that issue financing for corporate or governmental expenditures; green bonds are an allocation of finance to meet environmental indicators of improvement, such as renewable energy advancement, sustainable transport, clean water infrastructure, energy efficiency, and climate-resilient farms.

4.1 Evolution of Green Bonds

Green bonds were developed as a tool for finance in response to the emerging discussion of the role finance plays in climate change mitigation and sustainability. In 2007, the first green bond called "Climate Awareness Bond" was issued by the European Investment Bank (EIB). Within a year of this bond issuance, the World Bank would issue its first green bond (2008); these bonds were issued to lure investors with the promise of return on investment along with measurable environmental impact. Collectively, these first issuances established the genre of debt. Historically, the green bond market has been limited in its size, led of note by supranational institutions like the EIB and the World Bank who were establishing reliability and certainty in the instrument. In addition, these first issuances established a foundation in the establishment of norms around transparency, reporting, and accountability that would serve as overarching tenets, or normative principles, of the green bond standards. Over the next decade, as the green bond sector developed was well established, authorities and private actors started realizing the potential of green bonds and making the necessary rule changes to promote the use of green bonds for mobilizing capital for sustainable development, which only accelerated the amount of green bonds issued in the early years. Once the green bond market developed, but was still in its infancy, sovereign issuers began issuing sovereign green bonds used to fund large and meaningful national projects, including in the areas of renewable energy, transport, and climate-resilient infrastructure; sovereign bonds traditionally carry a level of risk to the issuer, and ultimately, financial support for their issuance is considered politically driven, which is positive when considering national or regional climate change goals within the context of political verge. Poland issued the first ever sovereign green bond in 2016, and in 2017, France issued their sovereign green bond alongside many others, including China, India, and the United States – highlighting this newly identified category of issuer, including corporations, municipalities, and financial institutions, that would increasingly use green bonds as a financing, or fundraising, mechanism to further their sustainability agenda while attracting socially responsible investors. The significant expansion of the market was also made possible through the introduction and development of global standards and guidelines. The development of the Green Bond Principles (GBP) by the International Capital Market Association (ICMA) allowed for best-practice voluntary provisions to provide guidance that promoted comparability and transparency across issuances since 2014. Similarly, the Climate Bonds Initiative

(CBI) developed the Climate Bonds Standard to assert a certification process for projects or assets that had climate objectives.

4.2 Structure and Mechanisms of Green Bonds

The structure and functions of green bonds are similar to traditional bonds but they differentiate themselves through an unambiguous commitment to financing projects that provide positive environmental impacts. In brief, a green bond is a fixed-income security issued by governments, municipalities, corporations, or financial institutions to raise money for services, with the funds only used for environmentally sustainable projects such as renewable energy, clean transportation, energy efficiency, sustainable water management, or resilience infrastructure. Investors receive interest payments, known as coupons, along with the principal repayment at maturity similar to traditional bonds. Green bonds differ from traditional bonds because they afford an additional level of transparency and accountability in the fund usage and what environmental benefits stem from them. Generally, the issuance of green bonds occurs within a structured process in order to promote credibility and that investor confidence is shown in figure 2. The most frequently used guidelines are the Green Bond Principles (GBP), published by the International Capital Market Association (ICMA). These principles represent four core components - the use of proceeds, the strategic evaluation and selection of projects, the management of proceeds, and the reporting (International Capital Market Association (ICMA), 2018). The use of proceeds clearly indicates that projects must provide environmental benefit. The project evaluation and selection process details how the issuers identify and assess the eligible projects.

Figure 2. Structure of Green Mechanism Bonds

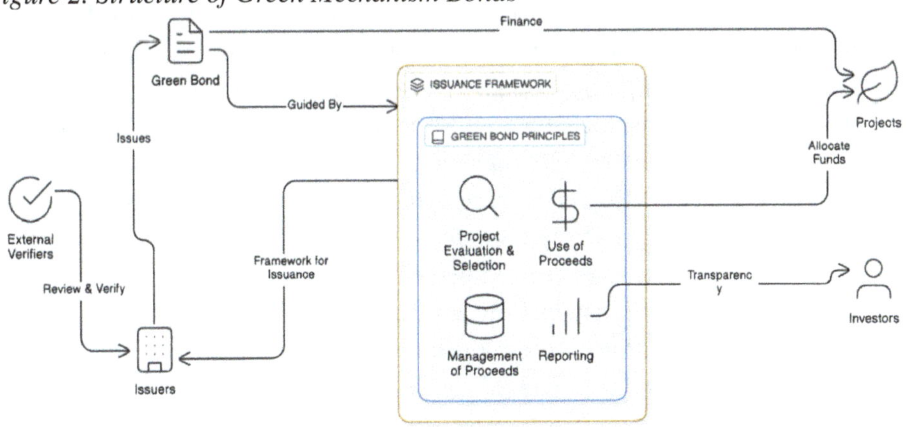

The management of proceeds is to ensure that tracking of the raised funds occurs and separation from general financing, often through sub-accounts or internal monitoring of the proceeds. The issuer of a bond must report periodically to investors about how the funds are spent as well as the resultant environmental impact to provide a level of transparency to investors. Many issuers seek external verification or certification to establish credibility and lower risks of greenwashing. Independent third-party audits or reviews occur through auditing firms, environmental consultancies or on behalf of the issuer by the Climate Bonds Initiative (CBI) to assess the bonds framework and project eligibility criteria. For example, a certified project through the Climate Bonds Standard provides investors with assurance the project represents an adaptation towards climate targets through the metric of other global recognized criteria of sustainability. The ability for certification provides investors with even more confidence and drives greater issuance of green bonds across global markets. From a market perspective, green bond mechanisms serve two market purposes, one is that green bonds provide issuers with access to a growing pool of sustainability driven investors, often a lower cost of capital, and improve their reputation in the market place by demonstrating some level commitment to climate-based initiatives.

4.3 Applications of Green Bonds

Green bonds are applicable in several sectors, demonstrating flexibility as a financing tool for climate adaptation and sustainable development.

1. **Renewable Energy Projects:** Green bonds are likely to be most applicable for financing renewable energy generation of electricity, including solar, wind, hydroelectric, geothermal, and biomass sources. Green bonds, specifically for funding renewable energy projects, help to hasten the transition from fossil fuel generation electricity and lessens carbon intensity in the energy mix. Pay close attention to green bonds used in financing renewable energy projects globally as they correlate with improving and extend energy access in rural or off-grid areas, where low-income communities require steps to tackle energy poverty and facilitate sustainable development.
2. **Energy Efficiency Projects:** Green bonds are widely used to finance improving energy efficiency of projects in the residential, commercial, and industrial domains. Projects include retrofitting old building stocks with better insulation, deploying energy-efficient lighting, replacing HVAC systems, and establishing smart grids. These projects reduce overall energy demand, cut operating costs for firms and householders, and enhance long-term energy resilience (Vashishth et al., 2024).

3. **Transportation Systems:** Transportation accounts for a significant share of global emissions, thus necessitating the focus on transportation with green bonding as a financing vehicle. Deployment of a green bond includes funding for electric vehicle uptake, rail transit projects such as metro systems and high-speed rail, electric buses, bicycle lanes, and any number of low carbon mobility alternatives. Many cities, such as Paris, Beijing, and New York, are engaging in green bonds to design and improve their public transportation services for enhanced mobility while reducing air pollution, congestion, and urban carbon footprints (Sahu et al, 2024).
4. **Water Management for Sustainability:** Water-related projects capture a vital application of green bonds, particularly in areas with drought or flooding problems. Investment includes airports, wastewater treatment plants, desalination facilities, and improved irrigation and urban stormwater management. Water-related projects tend to conserve water while reducing risk to communities in droughted or flooded areas, improved irrigation can also lead to increases in agricultural yields.
5. **Infrastructure Resilient to Climate Extremes:** With changing climate, extreme climate variance is becoming more common. Riding the trend, we are seeing more of the issuance of green bonds for infrastructure projects that have the ability to withstand extreme weather. Resilient housing, flood barriers, coast defences, and sustainable urban drainage are examples of infrastructure investments. Investments in this context protect vulnerable communities, reduce economic losses from climate disasters, and create sustainable and long-term urban forms of climate resilience to increasingly prevalent climate risk.

4.4 Global Market Trends

Over the last 10 years is that the green bond markets have transitioned from a niche financial product to one that is fully mainstream investment product. As a limited number of issuances by supranational institutions such as the World Bank and the European Investment Bank, has become a trillion-dollar market base on the demand for climate-aligned financial assets. Europe has played the leading role in the expansion of the green bond markets, propelled by European Union regulations such as the EU Taxonomy and demand by investors to support green investments. The United States has also experienced a growing amount of corporate green bond issuance, especially in the energy and technology sectors. China is also an emerging market for green bonds. China has mobilized capital issuance in areas of renewable energy, clean transportation and pollution mitigation/reduction, among other efforts. A notable trend has been the emergence of sovereign and municipal issuers, with sovereign green bonds being issued at significant scale such as France, the UK and

Poland, and municipalities, such as New York and Paris have financed climate-resilient urban projects, through the use of municipal bonds as asset-backed instruments. Institutional investors (i.e., pension funds, sovereign wealth funds, insurance companies) have also exercise an importance role in demand creation, inscribing long-term capital flows into ESG strategies aligned with global sustainability agendas (e.g., Apau and Sibindi, 2025). More than that, the burgeoning volume in green bond markets has enabled a flow into expenses across many thematic instruments from thematic sustainability bonds, blue bonds for ocean conservation, and transition bonds for an industry transition to lower carbon operations. Substantial growth within green bond markets has been buttressed by regulatory somewhat underpinned by several international standards (e.g., Green Bond Principles; Climate Bonds Standard) that ensure standards for transparency and lower risks of greenwashing in several jurisdictions. Overall, a varied green bond market with significant growth potential exists with countries further deepening climate commitments under the Paris Agreement and Sustainable Development goals, particularly in areas with an urgent need for financing for renewable energy, infrastructure, and urban sustainability and resilience in emerging markets (i.e., Negi and Jaiswal, 2024).

5. SUSTAINABLE FINANCE FOR CLIMATE STABILITY

Sustainable finance is arguably one of the key leverage points for climate stability and its processes also contribute to the interplay between climate stability and economic growth. It primarily aims to activate financial flows towards projects, companies, and technologies that reduce greenhouse gas emissions, enhance resilience to the impact of climate change, and foster low-carbon development.

5.1 Connection of Financial Flows to Climate Conditions

Connecting financial flows to climate outcomes is an important piece in the sustainable development puzzle and avoiding climate risk. By connecting financial flows to climate conditions, it means directing and ensuring there is capital flowing to the type of projects and companies that are responding to climate goals such as reducing greenhouse gas emissions, increasing capacity in renewable energy, improving energy efficiency, and developing climate resilient infrastructure. When financial markets incorporate climate risk and sustainability into their decision-making processes, they have some impact on the behaviours of companies, investors, and policy-makers to create transformative change across the global economy (Vashishth et al., 2024). Ultimately, connecting financial flows to climate outcomes means private and public investment will return a financial profit to investors while also creating positive environmental and social outcomes for the broader society. This type of connection therefore has a dual choice of profit and sustainability, so long as the investment and lending decision-making processes include to risk assessments and climate criteria.

For example, banks might decide to lend to clean energy start-ups, institutional investors might divest from fossil fuel investments, or governments might create favourable conditions for low-carbon technologies by providing subsidies and tax credits. Additionally, alignment frameworks foundational to linking financial flows to climate action exist internationally (e.g., the Paris Agreement and UN Sustainable Development Goals [SDGs]) to measure and report on the effects of financial alignment with climate action. Additionally, this transition may involve the development of innovative financial products and instruments like green bonds, sustainability-linked loans, and carbon credit trading that directly link financial metrics to sustainability goals.

5.2 Reducing Exposure to Climate-Related Risks

One of the most important goals of sustainable finance is the reduction of climate-related risks. Climate change creates deep uncertainties which impact the stability of financial systems and the global economy. Climate-related risks can be broadly categorized as physical risks (e.g., extreme weather, rising sea levels, floods, and droughts) and transition risks (e.g., changing policies, technology, and markets to drive/decarbonize). Ignored, these risks will have a direct impact on stranded assets, lower asset value, increased premiums, supply chain disruption, and systemic financial instability. As part of climate risk management practices in finance, sustainable finance supports firms and financial organizations in anticipating, responding, and adapting to climate-related risks. Financial markets are beginning to adopt tools such as scenario analysis, stress testing, climate disclosures to determine effect of climate-related events on portfolio risk. The Task Force on Climate-related Financial Disclosures (TCFD) is one such initiative to cultivate transparency and accountability, requiring companies to disclose to stakeholders how they consider and respond to climate-related risk as part of their operations and longer-term strategy. Financial organizations will have the capacity to reallocate capital investment into assets and/or industries that are less exposed (e.g., renewable energy, sustainable agriculture, resilient infrastructure), thereby reducing exposure to systemic risk. Likewise, insurance companies are recalibrating risk insurance models to integrate the increased risk of extreme weather, while banks leverage climate risks in credit risk. (Zairis et al., 2024)

5.3 Role in Building Climate-Resilient Infrastructure and Renewable Energy Capacity

Sustainable finance is vital for constructing climate-resilient infrastructure and expanding renewable energy capacity, both of which are crucial for long-term climate stability. Infrastructure worldwide is becoming increasingly exposed to climate-related risks such as sea-level rise, heatwaves, hurricanes, and floods, with resilience being among the highest priorities for governments, investors, and communities. By investing capital into infrastructure intended to withstand climate risks (like flood-resilient transportation networks, energy-efficient buildings, and smart city infrastructure), sustainable finance helps to ensure that development is not only economically feasible but also resilient to uncertain climate risks in the future. These sustainable investments lower the risk of asset loss and promote community safety, as well as reduce disruption's impact on service continuity. In addition to climate-resilient infrastructure, it is also essential to expand renewable energy capacity to decarbonize global economies and reduce dependence on fossil fuels. Financing instruments, such as green bonds, sustainability-linked loans, and blended finance, are increasingly directed toward solar, wind, hydro, and other renewable energy projects. Such financing tools reduce the cost of capital for renewable projects, attract private capital, or accelerate the transition to global energy uptake. Furthermore, sustainable finance supports the deployment of new innovative technologies (e.g., green hydrogen, energy storage, and smart grids) that are mandated to scale-up renewables and stabilize power supply. Beyond decreasing emissions, investing in renewable energy generates economic opportunities for job creation, improving energy security, and rural electrification, particularly in developing countries.

6. CASE STUDIES AND INDUSTRY IMPLEMENTATIONS

Case studies and industry experiences offer tangible examples of sustainable finance, ESG frameworks, and green bonds being put into practice. They demonstrate the actual use of sustainable finance mechanisms employed by organizations, governments, and financial institutions to mitigate climate change and realize a financial return.

6.1 Green Bonds in Practice

Green bonds have emerged as one of the most visible and potentially impactful instruments in sustainable finance. Many governments, corporations, and municipalities can leverage them to raise funds for environmentally beneficial projects. An exemplary case study is the Sovereign Green Bond Issuance in France (2017), which was one of the largest sovereign green bond issues ever, raising over €7 billion for projects in renewable energy, sustainable transport, and biodiversity. In addition to declaring confidence in sovereign green financial instruments, the issuance created a level of transparency that the sovereign green bond issuer was required to publish and disclose how financing was allocated and what the environmental impacts of those investments would be.

Output: The bonds promoted the speedier development of renewable energy infrastructure, all while having a lower carbon footprint. The issuance from France later was an impetus in that several other countries (Belgium, Poland, and Ireland, to name a few) were likely to do the same with sovereign green bonds. Likewise, Apple's financing investments to lower its global carbon footprint served as an impetus for a variety of similar programs at other tech and manufacturing companies, while establishing a baseline for a view of company green financing programs. In sum, out came more legitimization of sovereign green bonds, updates indices on transparency, and additional financing for the climate-responsible business.

6.2 ESG Implementation in Corporations

Corporations are incorporating environmental, social, and governance (ESG) elements into their business models to attract sustainable finance investors while enhancing organizational performance and sustainability over the longer term. One example of this is Unilever's development of a model of ESG implementation in its "Sustainable Living Plan" (2017-approved), which included a commitment to reduce carbon emissions in all their operations, transform 100% of their agricultural raw materials sustainably, and enhance the livelihood of millions of individuals with fair supply-chain practices. Also, Microsoft made a commitment to carbon negative (by 2030), which has an equal part of improving renewable energy, removing carbon, and incorporating sustainability throughout global business operations.

Output: ESG has allowed Unilever to benefit from continued high levels of trust from investors, improved brand reputation, and steady returns on share value, all in a time of increasing global volatility. Microsoft's carbon negative promise drew significant investment from sustainability investors and sparked a wave a sustainability commitment among its peers in the technology sector. These results demonstrate how well ESG integration and the benefits which can be unlocked when implemented successfully in organizations.

6.3 Financial Institutions and Sustainable Finance

Financial institutions play an essential role in facilitating substantial capital flows to climate-related activities. The world's largest asset manager, BlackRock, has made sustainability a priority in its investing process by offering ESG-focused funds, selling its coal-related equity investments, and factoring climate risk into its portfolio management process. Likewise, HSBC has committed to net-zero financed emissions by 2050 and continues to invest billions of dollars into sustainable finance projects, specifically investing in renewable energy, green infrastructure, and low-carbon industries.

Output: These movements have changed the global investment landscape, and BlackRock's lead has led other asset managers to think about fossil fuel exposures and climate-focused strategies. HSBC's sustainable finance efforts have also accelerated the transition to clean energy across developed and developing economies. Overall, these actions have generated a systemic capital reallocation, ensuring sustainability is considered an important part of investment decisions around the world.

7. CHALLENGES AND BARRIERS

While sustainable finance has developed quickly and may potentially play a part in combating climate change, it still faces a number of barriers and challenges that, in effect, inhibit its promise.

7.1 Greenwashing and Credibility Issues

Greenwashing may be considered the largest threat to sustainable finance. Greenwashing occurs when an enterprise or fund, or a financial institution, misrepresents or exaggerates a purported environmental benefit of their actions or financial products, which would alternatively lead it to appear more sustainable than it is. For example, issuing a bond that is stated as "green" while allowing a portion of the proceeds to undergo fossil fuel activity. All of these behaviours dilute the investor's confidence, and credibility in the sustainable finance sector, and make it extremely difficult to differentiate a legitimate transition to sustainability versus just a marketing claim, in order to consider the initiative sustainable. Compound this with continuing pressure from investors for ESG and green products, financial institutions may feel pressure, from a distorted business platform, to green wash while reduced their credibility as an investment position, not only for the firm itself but through the moderation of accounting integrity for financial markets themselves.

7.2 Lack of Standardization and Uniform ESG Metrics

Currently, companies and financial institutions use different systems to rate the ESG performance of companies, creating some inconsistency and sometimes confusion. For example, one rating company may give a firm a high rating for its carbon-reduction initiatives, and then another rating company may rate that same company low based on its labour practices. This ambiguity can be misleading, complicates the investment decision-making processes, complicates transparency, and limits the development of comparability across ratings, scores, and standards. The ESG investing world cannot scale and build credibility without common definitions and metrics. The need for standardization via vehicles like the EU Taxonomy or the International Sustainability Standards Board (ISSB) will start to help mitigate the challenges posed by the lack of a common standard.

7.3 Regulatory Gaps and Fragmented Policy Frameworks

The European Union has made considerable progress in this area, providing frameworks such as the Sustainable Finance Disclosure Regulation (SFDR) and the EU Taxonomy. Nevertheless, many other countries lack a broad framework of regulation concerning sustainable investments. The fragmented regulatory frameworks create fragmented markets where investors continue to face challenges in deciphering overlapping requirements around disclosures, tax incentives, and compliance. In some jurisdictions a laissez-faire approach means unsustainable practices persist. Moreover, the absence of global alignment on a framework restricts attempts to incentivise cross-border green-financing investments. Closing the regulatory gap will require global alignment, lessening disclosure requirements and a stronger domestic alignment to begin to integrate climate risk into oversight of the finance sector.

7.4 Balancing Profitability with Sustainability Objectives

Financial companies and businesses often find it difficult to balance their goal of profitability with their associated social obligations to invest in sustainable ways. Green efforts such as investing in a renewable energy generator or developing a climate-resilient building can require significant upfront capital and may take a long time to recover that investment. This gets in the way of short-term financial objectives and requires investors to change their expectations for returns on investment in a shorter time frame. For this reason, many investors are reluctant to make large investments in sustainable projects because of ever-changing market conditions. This reluctance is compounded by the belief that sustainable investing comes at the expense of business profitability.

7.5 Access and Equity Challenges for Developing Economies

Countries facing challenges in accessing sustainable finance also tend to experience high levels of borrowing costs due to credit ratings. These financing challenges not only affect the capacity of developing countries to issue green bonds but can create conditions that deplete ESG equity from international and domestic capabilities. Therefore, the country's most in need of climate financing (countries experiencing extreme weather events, rising sea levels, or being impacted by disrupted agricultural methods) have successfully accessed the least amount of climate financing post-pandemic. Climate finance can materially advance social equity, in addition to reducing climate exposure and risks. Any effort aimed at improving access to climate financing for developing regions will need to entail direct and targeted international support in conjunction with concessional finance grants and blended finance methods to support at least some risk to private investors, with modifications made to grant equitable access to funds (Ali et al., 2025).

8. FUTURE DIRECTIONS AND RESEARCH OPPORTUNITIES

The future of sustainable finance involves deepening its integration into mainstream financial systems while simultaneously expanding its scope to take on the pressing needs of climate change and social equity.

8.1 Standardization of ESG Metrics and Green Finance Frameworks

The lack of globally accepted standards makes it difficult for investors to compare performance measures, ensuring accountability. Having a consensus on a framework would bring clarity, reduce the risk of greenwashing, and potentially contribute to increased participation in sustainable investments. Some research can contribute to the design of harmonized methodologies to measure carbon footprints, biodiversity impacts, and social performance across industries. The research can assess how regulatory authorities and international organizations might perceive a uniform approach to enforce compliance on a global scale with standards for ESG metrics and green finance taxonomies.

8.2 Leveraging Fintech and Digital Innovations

Technological innovation can also generate a necessity for scaling sustainable finance. Using technologies such as blockchain (Vashishth et al., 2024), artificial intelligence (AI), and big data analytics are creating new opportunities to expand trade in sustainable finance. Blockchain can create transparency and traceability of funds in relation to a green bond issuance. AI can help find risk and opportunity in large socially responsible investing data sets for the purpose of analysis. Digital lending platforms can, to an extent, democratize access to retail investors to sustainable finance opportunities. Future research can involve creating integrated fintech approaches that can be used in a safe, efficient, and criteria inclusive technology on a digital lending platform.

8.3 Sustainable Finance in Emerging Markets and Developing Economies

Emerging markets and developing countries experience not only the greatest vulnerabilities to climate change, but also the most serious barriers to sustainable financing. Your investments in the future will most likely be heavily weighted towards financing models that leverage private capital for these economies, with as little unsustainable risk as possible. Examples of mechanisms that might de-risk investment in emerging markets and developing countries are blended finance, public-private partnerships, and concessional loans. Research opportunities again could look like: analysing the mechanisms, best practice in situ, and designing innovative financing instruments that have a meaningful dual outcome of equity and socio-economic justice whilst achieving climate outcomes (Hussain et al., 2025).

8.4 Linking Sustainable Finance with Social Justice and Inclusion

There is a recognition to place greater emphasis on the fact that the green transition should not only or disproportionately benefit developed nations and large corporations at the expense of vulnerable communities. Therefore, any financing mechanisms that arise in the future ought to create and imply inclusive growth, just labour standards and conditions, affordable green technologies, and equitable access to resources in general. Research may also consider how much financing flows can be refocused toward socially inclusive projects while remaining viable in deriving the dual outcomes such as: renewable energy projects for the local specific community, sustainable agriculture for small holder farmers, and affordable housing initiatives. Linking this type of philosophy will create a more comprehensive and cohesive approach to sustainable finance.

8.5 Long-Term Performance and Resilience of Sustainable Investments

An area of academic inquiry has been the long-term performance of ESG funds, green bonds, and investments linked to climate-related outcomes, relative to traditional financial instruments. Although there are increasing indications that sustainable investments have competitive long-term returns, more longitudinal studies are needed to assess whether sustainable investments are resilient to market volatility and global phenomena. Future research can examine the performance of sustainable portfolios under climate stress testing scenarios, which would provide information for investors and policymakers. In demonstrating financial resilience and points for addressing climate risk, sustainable finance could further solidify its role as a mainstream strategy for global markets.

9. CONCLUSION

Sustainable finance has become a powerful movement supporting the significant priorities of financial growth and climate stability. We are increasingly seeing capital flows in financial markets align with low-carbon transitions and climate-resilient development through the alignment of environmental, social and governance (ESG) factors, and new instruments such as green bonds. The chapter described how sustainable finance offers a meaningful way to reduce exposure to climate risks, while simultaneously creating new space for innovation, inclusion, and long-term financial return. The examples of sovereign issuances, corporate commitment to ESG, and institutional engagement all provided examples of how sustainable finance is mobilizing capital- given the inevitability of climate change- through the prospect of genuine impact investment in renewables, resilient infrastructure, and responsible corporate governance. However, there are many challenges that remain, including greenwashing, a lack of standardization, regulatory gaps, and access to the capital for developing economies. Standards of governance, improved metrics of measuring ESG, and collaboration on a global scale will address many these challenges. There is every reason that we could be optimistic about the engagement of emerging institutions and commitment to sustainable finance- and there will have to be continued commitment to the development of various standardized frameworks, and the integration of Fintech in expanding access to emerging markets, as well as the recognition of the social justice dimension of either the commitment to sustainable finance or finance systems overall.

REFERENCES

Afandy, C., Mirda, R., Fauzi, R. U. A., Hatidja, H., Fachri, M., & Jamaludin, J. (2024). Sustainable Finance and Green Bonds: Bridging the Gap between Protability and Environmental Responsibility. eScience Humanity Journal, 5(1), 243-252.

Ali, M., Jamal, S., Zeb, A., Dil, R., & Ali, O. (2025). The Role of Green Bonds in Advancing Climate Finance Goals in Emerging Economies: A Policy-Oriented Assessment. *Social Science Review Archives*, *3*(3), 1144–1157. DOI: 10.70670/sra.v3i3.946

Apau, R., & Sibindi, A. B. (2025). Sustainable Finance and Insurance: Opportunities and Challenges. Sustainable Finance and Insurance in Africa, 9-21.

Arifin, A. H., Hudaeni, N., Rasyid, A., Suhardi, S., Suprayitno, D., Redjeki, F., & Purwati, T. (2024). The effectiveness of green bonds in promoting sustainable finance: A meta-analysis. [Journal of Humanities and Social Studies]. *JHSS*, *8*(3), 830–839.

Arouri, M., Mhadhbi, M., & Shahrour, M. H. (2025). Dynamic Connectedness and Hedging Effectiveness between Green Bonds, ESG Indices, and Traditional Assets. European Financial Management.

Bai, M., & Wang, W. (2024). Role of green bonds in developing ESG projects in Eastern provinces of China. *Humanities & Social Sciences Communications*, *11*(1), 1–8. DOI: 10.1057/s41599-024-04353-y

D'Orazio, P. (2023). The politics of climate finance and policy initiatives to promote sustainable finance and address ESG issues. In *Sustainable finance and ESG: Risk, management, regulations, and implications for financial institutions* (pp. 145–171). Springer International Publishing. DOI: 10.1007/978-3-031-24283-0_7

Gaganis, C., Pasiouras, F., Tasiou, M., & Zopounidis, C. (2023). *Sustainable Finance and ESG. Risk, Management, Regulations, and Implications for Financial Institutions*. Springer International Publishing., DOI: 10.1007/978-3-031-24283-0

Hilal, M. Sumbal, & Imad, M. (2025). An Impact of Sustainable Finance Initiatives on Financial Markets and Economic Stability. In Emerging Disruptive Technologies for Society 5.0 in Developing Countries: Challenges and Applications (pp. 89-96). Cham: Springer Nature Switzerland.

Hussain, M., Jaweed, M., Murtaza, M., & Ali, M. (2025). Green Bonds and Sustainable Investment Strategies: Evaluating Risk-Return Profiles, Market Growth, and the Role of Climate-Conscious Portfolios in Sustainable Finance. *Advance Journal of Econometrics and Finance*, *3*(2), 39–53. DOI: 10.63075/5nggp275

Joshi, D., Paul, S., Singh, P., Mane, N. S., & Yadav, M. (2024). Sustainable Finance: The Role of Green Bonds in Driving ESG Investments. Library of Progress-Library Science, Information Technology & Computer, 44(3).

Khamitdkhanovich, N. A., Rajendran, M., Nailevich, I. R., & Krishnakumari, S. (2025, June). Green finance and ESG investing: Driving sustainable growth through innovation. In AIP Conference Proceedings (Vol. 3306, No. 1, p. 030025). AIP Publishing LLC.

Kumar, P., Taneja, S., & Ozen, E. (2025). Exploring the influence of green bonds on sustainable development through low-carbon financing mobilization. *International Journal of Law and Management*, *67*(2), 249–270. DOI: 10.1108/IJLMA-01-2024-0030

Mishra, S. P., Kumar, R., & Rout, J. (2023). An overview of green, social, sustainability, and sustainability-linked (GSSS) bonds. *Journal of Social and Economic Development*, *25*(S1, Suppl 1), 127–145. DOI: 10.1007/s40847-023-00275-8

Naseer, M. M., & Bagh, T. (2024). Green finance and climate challenges: Paving the way for economic, social, and governance readiness and sustainable development goals. In *The Palgrave Handbook of Green Finance for Sustainable Development* (pp. 123–153). Springer International Publishing. DOI: 10.1007/978-3-031-65756-6_5

Negi, P., & Jaiswal, A. (2024). *Sustainable bonds as a sustainable financial instrument: thematic insights and future prospects*. Qualitative Research in Financial Markets.

Nguyen, H. C., & Duong, H. K. (2025). An international empirical study of the impact of green bonds on sustainable development goals. *Review of Behavioral Finance*, *17*(3), 499–523. DOI: 10.1108/RBF-08-2024-0243

Sahu, P., Şanli, O., Janjua, L. R., & Rao, N. M. (2024). Evolutionary Perspective of Green Bond Financing under the Shadow of ESG Readiness. In *Green Bonds and Sustainable Finance* (pp. 104–119). Routledge. DOI: 10.4324/9781032686844-7

Singhania, M., Chadha, G., & Prasad, R. (2024). Sustainable finance research: Review and agenda. *International Journal of Finance & Economics*, *29*(4), 4010–4045. DOI: 10.1002/ijfe.2854

Vashishth, T. K., Kumar, B., Alam, M. S., Sharma, V., & Chaudhary, S. (2024). The Impact of Artificial Intelligence on the Future of Computing: A Comparative Study. In *Applications of Block Chain technology and Artificial Intelligence: Lead-ins in Banking, Finance, and Capital Market* (pp. 1–17). Springer International Publishing. DOI: 10.1007/978-3-031-47324-1_1

Vashishth, T. K., Sharma, V., Kumar, B., Chaudhary, S., Panwar, R., & Sharma, K. K. (2024, May). Guarding Finances: The Role of Image Processing in Unmasking Counterfeit Currency. In 2024 2nd International Conference on Advancement in Computation & Computer Technologies (InCACCT) (pp. 668-671). IEEE.

Vashishth, T. K., Sharma, V., Kumar, B., & Sharma, K. K. (2024). Cloud-Based Data Management for Behavior Analytics in Business and Finance Sectors. In Data-Driven Modelling and Predictive Analytics in Business and Finance (pp. 133-155). Auerbach Publications. DOI: 10.1201/9781032618845-9

Vashishth, T. K., Sharma, V., Sharma, K. K., Kumar, B., Chaudhary, S., & Kaushik, V. (2025). The Role of Blockchain in Disrupting Traditional Banking with Decentralized Finance. Decentralized Finance and Blockchain: A Game Changing Duo, 143-174.

Wang, Y., Wang, R., & Mason, T. A. (2025). Exploring the impact of green bonds on sustainable development and financial markets: An ARDL-based causality approach. *Finance Research Letters*, *85*, 107943. DOI: 10.1016/j.frl.2025.107943

Xiong, J., Zhang, Y., & Mao, Z. (2025). A Study on the Impact of Green Bonds on Corporate ESG Performance. GBP Proceedings Series, 3, 159-165. DOI: 10.70088/6yb5je33

Zairis, G., Liargovas, P., & Apostolopoulos, N. (2024). Sustainable finance and ESG importance: A systematic literature review and research agenda. *Sustainability (Basel)*, *16*(7), 2878. DOI: 10.3390/su16072878

Zhang, R. (2025). Green Finance and ESG Investment Strategies under Climate Hussain Management. *Economics and Management Innovation*, *2*(1), 140–148. DOI: 10.71222/00d71j98

Chapter 10
Bridging Climate Risk, Energy Transition, and Financial Market Resilience:
A Comparative Analysis Across Income Groups

Amir Ahmad Dar
https://orcid.org/0000-0002-0379-2272
Lovely Professional University, India

Sukriti Kakkar
Lovely Professional University, India

Vanshita Arora
Lovely Professional University, India

Hibah Himmath
https://orcid.org/0009-0001-8795-9788
B.S. Abdur Rahman Crescent Institute of Science and Technology, India

ABSTRACT

This chapter bridges the gap by analyzing climate vulnerability, energy transition policies, and financial market implications across income groups using the 2022 Climate Risk Index (CRI), population, GDP per capita, and carbon pricing instruments. Results show an inverse relationship between GDP per capita and climate risk: high-income countries such as the United States, Germany, and France record

DOI: 10.4018/979-8-3373-6766-8.ch010

lower CRI scores due to resilience, strong institutions, and mitigation capacity, while low- and lower-middle-income nations like Pakistan, Bangladesh, and Niger face heightened risks from limited adaptive capacity. Population amplifies vulnerability in densely populated countries, though not uniformly. Carbon pricing analysis reveals robust systems in advanced economies compared to limited adoption in vulnerable nations. By linking CRI with policy readiness and financial stability, the chapter highlights structural inequities and offers recommendations for strengthening resilience, incentivizing low-carbon investment, and safeguarding markets against escalating climate risks.

1. INTRODUCTION

Climate change has emerged as one of the most pressing challenges of the twenty-first century, reshaping the dynamics of economies, financial markets, and energy systems on a global scale. Its impacts are multifaceted, ranging from extreme weather events, rising sea levels, and intensifying heatwaves to wide-ranging regulatory and policy shifts aimed at curbing greenhouse gas emissions. These risks extend beyond environmental damage, exerting profound influence on economic development, investment patterns, and the stability of global financial markets. The study emphasizes that climate risks are systemic and interconnected, affecting not only the environment but also economic stability and financial market performance. For governments, firms, and societies alike, resilience against climate-related risks has become a critical determinant of long-term sustainability and stability (Pfenninger et al., 2014).

Scholars and policymakers increasingly emphasize the dual nature of climate risks: physical and transition. Physical risks stem from the direct consequences of climate events, including floods, storms, droughts, and temperature extremes, which can devastate infrastructure, disrupt production and supply chains, and impair local economies. Transition risks, in contrast, arise from the structural changes required to shift toward low-carbon energy systems. These include carbon pricing instruments, regulatory changes, technological advancements in renewables, and evolving investor preferences for sustainable assets. While extensive literature has examined physical and transition risks separately, much less attention has been paid to their interplay—yet in reality, these risks rarely operate in isolation. Their interaction often amplifies uncertainty, creating systemic challenges for economies and markets worldwide (Swart, 2019).

A cross-country perspective is particularly important in understanding these dynamics. Nations differ significantly in their exposure to climate hazards, institutional preparedness, and economic capacity to adapt. High-income countries generally possess advanced infrastructure, fiscal space, and technological capabilities to mitigate and adapt to climate risks. By contrast, low- and middle-income nations often face structural constraints, limited adaptive capacity, and restricted access to financial resources, leaving them disproportionately vulnerable. This imbalance is further compounded by the injustice that many of the most climate-vulnerable nations contribute least to global emissions yet face the gravest consequences. Studying these disparities highlights not only the inequitable nature of climate risk but also the urgent need for differentiated policy responses across income groups and regions.

Despite growing recognition of climate-related risks, a significant research gap persists. Much of the existing work focuses narrowly on single countries, individual sectors, or firm-level impacts. Although valuable, such approaches often overlook the integrated effects of physical risk exposure, transition policy adoption, and their broader implications for financial markets. Comparative assessments that link indicators such as the Climate Risk Index (CRI) with carbon pricing mechanisms, energy transition readiness, and financial stability remain limited. This fragmentation constrains the ability of investors, policymakers, and international organizations to design effective strategies for managing systemic climate risks in a globally interconnected economy.

The motivation for this study arises from this gap. Understanding how climate vulnerability and transition readiness interact to shape financial and economic outcomes is essential, especially as both public and private actors intensify their commitments to climate action. Investors are increasingly seeking robust frameworks to assess the exposure of portfolios to climate risk, while governments and regulators aim to balance the imperatives of growth, equity, and decarbonization. By situating this analysis at the intersection of climate science, energy policy, and finance, the study offers insights with both academic and practical relevance.

The objectives of this study are threefold. First, it aims to analyse global patterns of climate vulnerability by assessing CRI scores in relation to socioeconomic indicators such as GDP per capita, population size, and regional exposure. This allows for a nuanced understanding of how economic development and demographic factors shape resilience and risk. Second, it examines the adoption of carbon pricing mechanisms—such as ETS and carbon taxes—and broader energy transition readiness across different countries. By evaluating the prevalence, pricing levels, and maturity of such policies, the research identifies disparities in decarbonization capacity between high-income and lower-income nations. Third, the study explores the implications of climate risk and transition policies for financial market stability, particularly in terms of investment efficiency, capital flows, and investor confidence. In doing so, it highlights how climate-related vulnerabilities can translate into systemic financial risks, and how well-designed policies can mitigate them.

To support this inquiry, the chapter introduces a conceptual framework that links physical climate risk, transition policy adoption, and financial market outcomes (Figure 1). In this framework, CRI scores on the left represent vulnerability to physical risks such as extreme weather events. Transition readiness, situated in the middle, is captured through carbon pricing instruments, renewable energy adoption, and regulatory capacity. On the right, financial outcomes—such as investment efficiency, capital allocation, and market stability—reflect the systemic implications of both risk dimensions. Arrows between these components illustrate the feedback loops: higher climate risk can destabilize markets, while strong transition policies can mitigate risks and enhance resilience.

Figure 1. Conceptual Framework Linking Climate Risk, Energy Transition, and Financial Markets

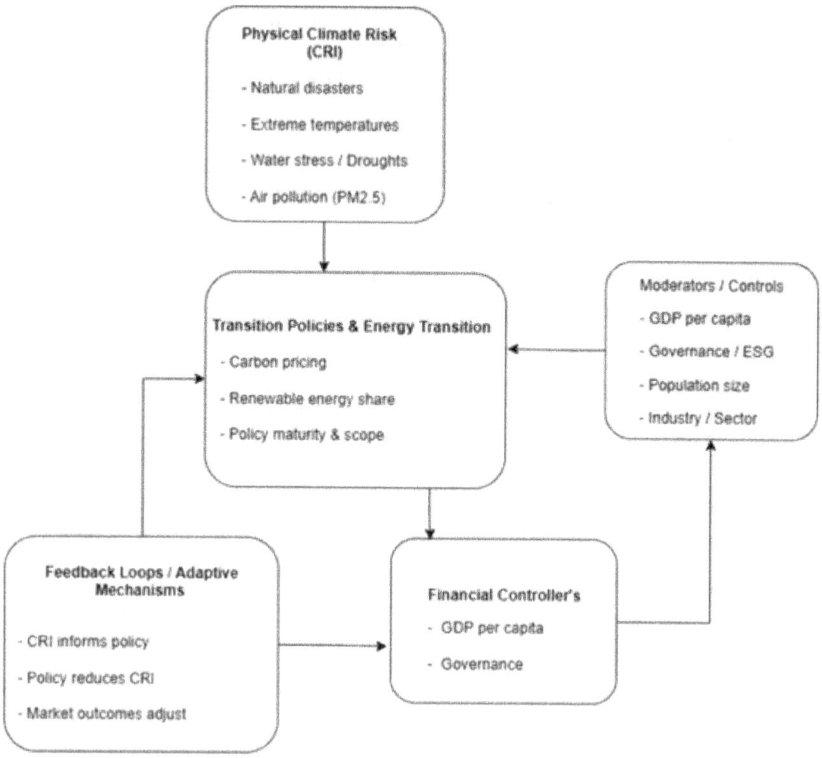

Therefore, this chapter situates the study of climate risk and energy transition within a global, comparative perspective. By integrating physical risks, policy-driven transition dynamics, and financial market implications, it addresses a crucial gap in the literature. The analysis underscores the uneven distribution of climate vulnerabilities across countries, the challenges of implementing equitable and effective carbon pricing, and the far-reaching consequences for financial stability. Ultimately, this chapter positions climate risk not merely as an environmental concern but as a systemic challenge at the intersection of sustainability and financial resilience. The insights developed here provide the foundation for the subsequent analysis, which seeks to inform strategies that enhance resilience, guide investment, and foster a more stable, equitable, and sustainable global economy.

2. LITERATURE REVIEW

Climate change presents a dual challenge to countries: managing the immediate impacts of physical climate events and navigating the long-term transition to low-carbon economies. The study emphasizes that climate risks are systemic and interconnected, affecting not only the environment but also economic stability and financial market performance. Research over the past two decades has examined these dimensions separately, yet an integrated understanding across countries with differing economic capacities remains limited. This literature review synthesizes studies on physical climate risks, transition risks, energy policies, and their financial implications, highlighting gaps that this chapter addresses.

2.1 Physical Climate Risks

Physical climate risks refer to the direct consequences of climate change, including floods, storms, heatwaves, droughts, and sea-level rise. The CRI, developed by Germanwatch, is widely employed to quantify national exposure to these hazards. The CRI aggregates metrics such as fatalities, economic losses, and disaster frequency, offering a standardized measure of a country's vulnerability and resilience. High CRI scores indicate elevated exposure and limited adaptive capacity, while low scores reflect better preparedness and infrastructure. Studies consistently show that low-income countries tend to have higher CRI scores, reflecting geographic susceptibility and limited economic and institutional capacity to absorb shocks **(Eckstein et al., 2022).**

Physical climate events can have cascading effects on economic growth, public finances, and human welfare. Extreme weather events often disrupt supply chains, reduce agricultural productivity, and damage critical infrastructure, disproportionately affecting countries with weak governance and limited fiscal resources. Additionally, population density amplifies these impacts, particularly in urban centres where assets and populations are concentrated (Hallegatte et al., 2016).

Indicators such as the CRI have enabled cross-country vulnerability assessments, revealing that small island developing states (SIDS) and low-income nations in South Asia, Sub-Saharan Africa, and Central America are consistently the most vulnerable. Conversely, high-income nations often exhibit lower CRI scores due to stronger adaptive capacity. However, events like the 2021 European floods and recurring wildfires in Australia and the United States illustrate that even advanced economies are not immune to extreme climate shocks. Although growing literature examines national-level vulnerability, cross-country analyses combining physical risk with transition risk remain scarce. Such integration is critical, as physical and transition risks often reinforce one another; for instance, severe disaster exposure can accelerate policy reform, while poorly managed transitions can exacerbate vulnerability (Pfenninger et al., 2014).

2.2 Transition Risks and Energy Policies

Transition risks arise from structural changes required to reduce greenhouse gas emissions and shift toward sustainable energy systems. Unlike physical risks, transition risks are indirect, stemming from policy interventions, technological developments, market shifts, and changing investor preferences. Carbon pricing, renewable energy mandates, and regulatory reforms exemplify measures intended to mitigate transition risks, but they also introduce financial, operational, and reputational challenges (Ivanovski & Marinucci, 2021).

Carbon pricing, implemented through ETS or carbon taxes, is a primary tool for managing transition risks. By placing an economic cost on emissions, these policies incentivize reductions while generating revenue for climate action. High-income countries, including Germany, Sweden, and Japan, have implemented comprehensive carbon pricing frameworks covering multiple sectors and producing measurable fiscal and environmental **outcomes.** These mechanisms signal to investors and guide capital toward low-carbon technologies, reducing systemic financial risks associated with unanticipated policy shifts.

Lower-middle and low-income countries, however, often lack formal carbon pricing, leaving economies exposed to regulatory and market uncertainties as global decarbonization advances. In these contexts, firms and governments face external pressure to adopt low-carbon practices, but inadequate domestic frameworks limit effective response. Inconsistent policy implementation, regulatory uncertainty, and fragmented incentives can exacerbate transition risks, potentially leading to stranded assets, sectoral disruptions, and financial market volatility (Aldy& Stavins, 2012).

Beyond carbon pricing, broader energy policies—including renewable energy deployment, energy efficiency improvements, and support for low-carbon technology innovation—play a vital role in mitigating transition risks. Evidence suggests that high-income countries often exhibit greater energy transition readiness due to advanced technology, robust policy frameworks, and significant financial resources. For example, Germany, Denmark, and Japan have invested heavily in wind and solar generation, smart grids, and storage technologies, achieving substantial renewable penetration. These investments reduce reliance on volatile fossil fuel markets, stabilize energy supply, and facilitate compliance with international climate commitments

2.3 Impacts on Financial Markets

Physical and transition climate risks have profound implications for financial market stability, investment efficiency, and capital allocation. Natural disasters can reduce asset valuations, increase insurance liabilities, and disrupt credit markets, while transition risks can alter regulatory landscapes, affect long-term profitability expectations, and create stranded asset risks. Empirical evidence indicates that countries with high climate vulnerability often experience elevated sovereign risk premiums, higher borrowing costs, and diminished investor confidence (Eckstein et al., 2022)

For investors, understanding the interaction between physical and transition risks is critical. High-income countries with low CRI scores and strong carbon pricing frameworks offer relatively stable environments for low-carbon investments. Conversely, countries with high CRI scores and weak transition policies present heightened systemic risks, potentially discouraging foreign investment and increasing the cost of capital. However, most existing studies focus on firm-level ESG performance or sector-specific risks, with limited exploration of the macro-level interaction between national climate vulnerability, policy frameworks, and financial markets

2.4 Integration of Physical and Transition Risks

Recent scholarship emphasizes the importance of integrating physical and transition risks. Campiglio et al. (2018) highlight how abrupt climate policy implementation may generate financial shocks if markets are unprepared, and how these shocks can be compounded by physical climate events that strain government budgets and investor confidence. Despite this recognition, cross-country quantitative studies remain scarce. Few analyses combine CRI data with carbon pricing adoption and socioeconomic indicators such as GDP per capita and population, which are essential for assessing a nation's adaptive capacity and investment readiness (Campiglio et al., 2018)

Income disparities further influence how countries experience and respond to combined risks. High-income nations can absorb transition costs and invest in resilient infrastructure, while low-income countries face simultaneous exposure to physical hazards and limited financial resources. This inequity has direct implications for global capital flows, international aid allocation, and the prioritization of energy transition policies. Integrating these dimensions is essential for designing effective, context-specific interventions that enhance resilience while promoting equitable transitions.

2.5 Gaps in the Literature

A critical review of existing literature on climate risk and transition policies reveals several significant gaps. Firstly, the majority of studies tend to focus exclusively on either physical climate risks or transition policies, thereby neglecting the integrated effects that arise when both dimensions interact across national contexts. Secondly, there is a pronounced imbalance in geographic coverage, with research predominantly centered on high-income countries, while lower-income nations, often facing the highest climate vulnerability and least policy coverage, remain underrepresented. Thirdly, the influence of socioeconomic factors, such as population size, GDP per capita, and governance quality, is frequently overlooked, despite their importance in shaping a country's resilience to climate hazards and its capacity to implement effective financial and policy mechanisms. Finally, the current body of research provides limited practical guidance for policymakers and investors, offering insufficient insight into how to balance adaptation strategies, energy transition measures, and financial stability across diverse income groups. Addressing these gaps is essential to developing a comprehensive, cross-country understanding of climate vulnerability, policy effectiveness, and their implications for economic and financial systems.

2.6 Contribution of This Study

This chapter addresses these gaps by integrating physical climate risk (CRI), transition policy readiness (carbon pricing and renewable energy adoption), and socioeconomic indicators (GDP per capita, population) across countries grouped by income.

Table 1. Summary of Key Studies on Climate Risks, Transition Policies, and Financial Market Implications

Country/ Region Focus	Risk Dimension	Methodology	Key Findings	Research Gap
Global (CRI dataset)	Physical risks	CRI analysis, comparative cross-country	Low-income countries face the highest climate risk; high-income countries more resilient	Limited integration with financial markets or transition policies
Various	Physical risks	Economic modelling of disasters	Extreme weather events disrupt supply chains and reduce productivity	Rarely considers cross-country income-level differences
OECD countries	Transition risks (carbon pricing)	Policy review, emissions modelling	Carbon pricing effective in reducing emissions if coverage is broad	Limited evidence for lower-income countries; few links to financial outcomes
Global	Integrated physical + transition	Literature review, theoretical framework	Climate policy can create financial shocks; physical events amplify them	Lack of quantitative cross-country analysis combining CRI + transition policies
Global	Energy transition	Data analysis of renewable adoption	High renewable penetration improves energy security and resilience	Limited assessment linking to economic vulnerability and market stability
Global	Financial markets & climate risk	Scenario analysis	Vulnerable countries face higher financial instability; investors need robust frameworks	Few studies combine CRI, carbon pricing, and financial indicators in a cross-country comparative context

This study enables a holistic understanding of vulnerability, highlights structural inequities, and provides actionable insights for policymakers, investors, and financial regulators. By combining physical and transition risk dimensions, this research contributes to a more comprehensive framework for assessing climate-related economic and financial risks globally.

3. DATA AND METHODOLOGY

3.1 Research Design

This study adopts a cross-country comparative approach, examining the relationship between climate vulnerability, economic development, carbon pricing policies, and energy transition readiness. The primary aim is to understand how physical climate risks and transition-related measures jointly influence national resilience and financial market stability. Unlike prior studies that often examine either physical or transition risks in isolation, this study integrates multiple dimensions to provide a holistic assessment.

The research design for this study integrates three complementary approaches to provide a comprehensive assessment of climate vulnerability, economic capacity, and policy adoption across countries. First, descriptive analysis is employed to systematically identify patterns in climate risk exposure, economic indicators, and energy transition measures, allowing for an initial understanding of disparities and trends. Second, comparative analysis examines variations across income groups—including high-income, upper-middle-income, lower-middle-income, and low-income countries—to highlight structural differences in vulnerability, policy readiness, and adaptive capacity. Finally, policy assessment evaluates the adoption, coverage, and effectiveness of carbon pricing instruments and renewable energy initiatives, providing insight into how institutional and regulatory frameworks contribute to reducing physical and transition risks. By combining these methods, the study ensures a holistic approach that captures both quantitative and qualitative dimensions of climate risk management and energy transition.

3.2 Data Sources

The study relies on publicly available, cross-country datasets:

Table 2. Summary of Key Datasets and Sources

Dataset	Source	Description
Climate Risk Index (CRI) 2022	Germanwatch	Measures exposure and vulnerability to extreme weather events, combining fatalities, economic losses, and disaster frequency.
GDP per capita, Population	World Bank Open Data	Captures economic capacity and potential scale of climate exposure.
Carbon Pricing Instruments	World Bank Carbon Pricing Dashboard	Includes carbon taxes and ETS, covering price levels, emissions coverage, revenues, and gases regulated.
Energy Transition Indicators	International Energy Agency	Renewable energy share, energy intensity, and national policy measures.

Table 2 provides a concise overview of the datasets employed in this study, each selected for its credibility, global coverage, and relevance to climate–economy interactions. The CRI 2022, published by Germanwatch, provides a standardized measure of physical climate vulnerability, integrating fatalities, economic losses, and disaster frequency into a single score. Socioeconomic capacity is represented through GDP per capita and population statistics from the World Bank, ensuring comparability across income groups and regions. Policy-related variables are captured through the

World Bank Carbon Pricing Dashboard, which provides comprehensive information on the presence, design, and effectiveness of carbon pricing mechanisms such as carbon taxes and ETS. Finally, indicators of energy transition readiness drawn from the International Energy Agency reflect renewable energy shares, energy intensity, and policy commitments, which serve as critical control variables in understanding national resilience. Together, these datasets form a robust foundation for cross-country comparative analysis, enabling the study to link climate risk exposure with economic capacity, policy adoption, and transition pathways.

3.3 Variable Construction

In order to systematically examine the complex interactions between climate vulnerability, economic capacity, and policy adoption, this study constructs a comprehensive set of variables encompassing dependent, independent, and control measures. The primary dependent variable is climate vulnerability, operationalized through the CRI 2022. This index captures the extent to which countries are exposed to extreme weather events, accounting for fatalities, economic losses, and the frequency of climate-related disasters. Higher CRI scores reflect greater susceptibility, enabling cross-country comparisons of physical climate risk and the identification of particularly vulnerable nations. The independent variables are designed to assess the economic, demographic, and policy dimensions that influence resilience. Economic capacity, measured as GDP per capita in US dollars, reflects a country's financial and institutional ability to invest in climate adaptation, disaster preparedness, and energy transition initiatives. Population size is incorporated as it directly affects the potential scale of exposure, with densely populated nations facing amplified societal and infrastructural risks during climate hazards. Carbon pricing adoption is included as a policy indicator, coded as a dummy variable where a value of one signifies the implementation of either a carbon tax or an ETS, and zero denotes the absence of such mechanisms. Complementing this, carbon price level, expressed in US dollars per ton of CO_2 equivalent, captures the stringency of the climate policy, reflecting both economic incentives for mitigation and potential deterrence of emissions-intensive activities. The coverage of emissions represents the proportion of national greenhouse gas emissions regulated under carbon pricing instruments, offering insights into the comprehensiveness and effectiveness of policy interventions. To account for contextual factors that may moderate these relationships, the study also integrates several control variables. Regional grouping is included to reflect geographic patterns and regional policy frameworks, while income classification (high, upper-middle, lower-middle, and low-income) allows for examination of how economic development shapes climate vulnerability and policy adoption. Additionally, renewable energy share serves as a proxy for energy transition readiness, indicating the extent to which countries are shifting toward sustainable energy sources, while energy intensity measures the efficiency of energy use relative to economic output, capturing the sustainability dimension of national energy systems. By integrating these variables, the study establishes a robust and multidimensional framework that enables nuanced, cross-country comparative analyses, while also providing a foundation for assessing the interactions between physical climate risks, economic capacity, and policy effectiveness in shaping national resilience and adaptation strategies.

3.4 Final Dataset

This dataset provides the foundation for cross-country comparative analysis, enabling the identification of structural disparities in climate risk exposure, policy adoption, and socioeconomic readiness. The integration of physical vulnerability (measured through the Climate Risk Index), population size, GDP per capita, and carbon pricing characteristics allows for a multi-dimensional perspective. Countries such as Pakistan and Niger exhibit extremely high CRI scores without the presence of formal carbon pricing instruments, reflecting both structural fragility and the absence of transition policy frameworks. In contrast, high-income nations such as Germany and France combine relatively low CRI scores with robust. ETS and carbon taxation, signalling strong institutional capacity and long-term commitments to decarbonization.

Table 3 (below) presents a condensed overview of the dataset for 2022, summarizing the most salient indicators across selected countries. The reduced version highlights CRI scores, population exposure, GDP per capita, and the presence or absence of carbon pricing instruments, ensuring readability within the text. This summary enables quick cross-country comparison of vulnerability and transition readiness.

For full transparency and to preserve the richness of the dataset, the complete table—including additional variables such as the main carbon price rate, share of jurisdictional greenhouse gas emissions covered, government revenues, gases included, and type of regulation—is made available in supplementary material. Readers can access the full dataset in Excel format through a dedicated Google Drive link provided in the footnote. This ensures that the analysis remains both accessible within the chapter and reproducible for further research.

Table 3. Summary of Cross-Country Dataset on Climate Risk, Carbon Pricing, and Socioeconomic Indicators .

Rank 2022	Country	CRI Score 2022	Population	GDP per capita (current US$)	Instrument	Region	Income Group
1	Pakistan	53626890	251269.16	1,484.70	No transition policy	Asia	Lower middle income
3	Italy	30000250	58,986.02	40,226.00	No transition policy	Europe	High income
4	Greece	23270429	10,388.81	24,752.10	No transition policy	Europe	High income
5	Spain	22125609	48,807.14	35,297.00	Carbon tax	Europe	High income
7	United States	18180544	3,40,110.99	85,809.90	No national carbon pricing	North America	High income
10	Bulgaria	15604115	6,444.37	17,412.40	No transition policy	Europe	Upper middle income
17	Germany	10926968	83,510.95	55,800.20	ETS	Europe	High income
20	South Africa	10302265	64,007.19	6,253.40	Carbon tax	Africa	Upper middle income
24	Estonia	9167420	1,371.99	31,170.10	Carbon tax	Europe	High income
25	Niger	7705593	27,032.41	722.7	No transition policy	Sub-Saharan Africa	Low income
28	France	7244031	68,516.70	46,150.50	Carbon tax	Europe	High income
43	Philippines	4394394	1,15,843.67	3,984.80	No transition policy	Asia	Lower middle income
45	Latvia	4149124	1,862.44	23,367.60	Carbon tax	Europe	High income
46	Bangladesh	4145876	1,73,562.36	2,593.40	No transition policy	Asia	Lower middle income
49	India	3980564	14,50,935.79	2,696.70	No national carbon pricing	Asia	Lower middle income

continued on following page

Table 3. Continued

Rank 2022	Country	CRI Score 2022	Population	GDP per capita (current US$)	Instrument	Region	Income Group
50	Brazil	3872378	2,11,998.57	10,280.31	No national carbon pricing	Latin America & Caribbean	Upper middle income
51	China	3808915	14,08,975.00	13,303.10	ETS	Asia	Upper middle income
53	Austria	3622132	9,178.48	56,833.20	ETS	Europe	High income
69	Japan	1113007	1,23,975.37	32,475.90	Carbon tax	Asia	High income
85	Ireland	398190	5,380.26	1,07,316.30	Carbon tax	Europe	High income

This table provides a condensed version of the dataset to enhance readability. The full dataset, including additional variables (carbon price rates, coverage of emissions, government revenues, and regulation details), is available in supplementary Excel format via this link.

3.5 Analytical Approach

The analytical approach of this study is designed to systematically investigate the complex interactions between climate vulnerability, economic capacity, and policy adoption across countries. To achieve this, the methodology integrates descriptive statistics, comparative assessment, correlation evaluation, and policy analysis, ensuring that results are both robust and interpretable for academic and policy audiences. Initially, descriptive statistics summarize CRI scores, GDP per capita, population, and carbon pricing metrics, highlighting global patterns of climate risk exposure and structural disparities across income groups. For instance, lower-income countries such as Pakistan and Niger demonstrate high CRI values without implementing carbon pricing mechanisms, whereas high-income nations like Germany and France exhibit lower vulnerability alongside robust policy frameworks. Comparative analysis is then undertaken to examine relationships between CRI scores and key economic indicators, such as GDP per capita and population, providing insights into the extent to which economic capacity can mitigate climate vulnerability. In parallel, the presence and type of carbon pricing instruments are assessed across countries, with income-level groupings—high, upper-middle, lower-middle, and low—used to identify variations in both exposure and policy adoption.

To further quantify these relationships, correlation assessments are conducted between climate vulnerability and critical independent variables, including economic capacity, population, and carbon pricing measures. This provides evidence regarding the direction and strength of the associations and informs hypotheses on the effectiveness of policy interventions in reducing both physical and transition-related risks. Policy evaluation complements these analyses by examining carbon pricing metrics such as adoption status, price levels in USD per ton of CO_2e, emissions coverage, and government revenues, allowing for an assessment of their potential impact on climate resilience. Broader energy transition indicators, including renewable energy share and energy intensity, are also considered to evaluate national readiness for sustainable transformation. Finally, visualization through conceptual frameworks and structured tables is employed to clearly represent interactions between climate vulnerability, economic capacity, and carbon pricing policies, thereby providing a coherent and comprehensive overview of the study's analytical findings.

Figure 2. Conceptual Framework Linking Climate Vulnerability, Economic Capacity, and Carbon Pricing Policies.

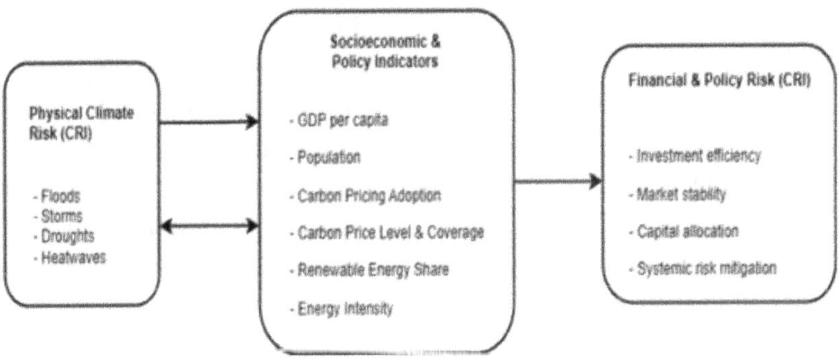

3.6 Justification of Methodology

The methodology adopted in this study is well-suited for multi-country comparative research due to several reasons, such as Standardized Climate Risk Measurement, Integration of Economic and Policy Indicators, Policy Relevance and Actionable Insights, Clear, Reproducible, and Accessible Approach, and the justification for cross-sectional scope. The CRI 2022 provides a globally recognized and standardized measure of climate vulnerability, allowing meaningful cross-country comparisons. Its combination of fatalities, economic losses, and disaster frequency ensures a comprehensive assessment of physical climate risks, which standardizes the climate risk measures. Integration of Economic and Policy Indicators, such as Economic capacity (GDP per capita) and population, is incorporated to evaluate a country's ability to invest in resilience and the potential scale of exposure. Carbon pricing adoption, price levels, emissions coverage, and renewable energy share are included to assess the effectiveness of policy measures in mitigating climate risk. The inclusion of control variables such as regional grouping and income category ensures that comparisons account for geographic and structural differences. By analysing carbon pricing adoption, coverage, and revenue, the methodology generates insights that can directly inform policymakers and investors seeking to manage climate-related financial and economic risks. The integration of energy transition indicators highlights areas for policy intervention and investment prioritization. Descriptive and comparative analyses ensure that results are interpretable for both academic audiences and practitioners, aligning with the objectives of a research book chapter. The methodology is reproducible using publicly available datasets, ensuring transparency and credibility. A single-year (2022) analysis allows for a focused and coherent assessment of climate vulnerability and policy readiness, ensuring that comparisons are consistent and reliable. While longitudinal analysis could provide trend insights, the cross-sectional approach is justified for identifying structural patterns and disparities at a specific point in time.

Therefore, the methodology is well-suited for multi-country comparative analysis where CRI provides a standardized measure of climate risk, enabling cross-country comparisons, and also integration of economic and policy indicators allows assessment of how financial and institutional capacity interacts with climate vulnerability. The policy-focused variables (carbon pricing adoption, coverage, revenue) provide actionable insights for policymakers and investors. Descriptive and comparative approaches ensure clarity and accessibility for specialists and non-specialists, aligning with the objectives of a research work.

3.7 Scope and Constraints

While this study presents a systematic and structured framework for analysing climate vulnerability, economic capacity, and policy adoption across a wide range of countries, it is equally important to recognize the boundaries within which the research is situated. Acknowledging such constraints is critical because it not only enhances transparency but also provides clarity regarding the interpretive limits of the findings. The first and most significant challenge arises from data availability and reliability. Many low- and lower-middle-income countries, particularly in Africa and parts of South Asia, have incomplete or inconsistent reporting mechanisms for critical variables such as carbon pricing instruments, emissions coverage, and renewable energy statistics. For example, while nations such as Germany or Canada maintain comprehensive records on carbon pricing levels and coverage, several vulnerable countries with high climate risks often lack such systematic reporting. These gaps result in uneven datasets that, while partially mitigated by reliance on standardized sources like the World Bank, IEA, and Germanwatch, cannot be fully reconciled. Consequently, there remains a risk that the findings underrepresent the true scope of vulnerability in countries where data scarcity is most acute.

Another key consideration relates to the temporal dimension of the research. By focusing primarily on CRI 2022 and related contemporaneous indicators, the study captures what is essentially a snapshot in time. This static approach provides the advantage of comparability across countries but inevitably sacrifices an appreciation for temporal dynamics. Climate vulnerability is inherently dynamic, as annual variations in extreme weather events, economic cycles, and policy reforms continuously reshape risk landscapes. For instance, a country experiencing severe flooding or drought in 2023 might display a vastly different CRI score compared to 2022. Similarly, economic capacity and resilience evolve with GDP growth, investment in infrastructure, and technological advancement. Policy frameworks also shift, as seen in the rapid expansion of ETS in China or the introduction of new carbon taxes in several European countries in recent years. By not fully incorporating longitudinal data, this study captures patterns at one point in time rather than trajectories, limiting its ability to track progress or regressions in resilience and policy effectiveness.

Further constraints stem from cross-national reporting inconsistencies. Countries differ not only in their statistical capacity but also in the definitions and methodologies they employ when measuring key indicators. GDP per capita, for example, may be calculated using varying price bases or adjusted differently for purchasing power parity (PPP), while renewable energy shares may exclude certain technologies in some contexts but include them in others. Population data, although more standardized, can still vary in accuracy depending on census cycles or estimation models. These inconsistencies introduce a layer of noise into the dataset that, while minimized by using globally recognized sources, cannot be completely eliminated. Such variations mean that the comparability of data across regions, while robust at a general level, may contain subtle discrepancies that affect precision in certain cases.

The methodological simplification of policy representation also deserves careful consideration. Carbon pricing adoption is reduced to a binary dummy variable, coded as “1” where either an ETS or carbon tax is implemented and “0” otherwise. This simplification, while necessary for quantitative cross-country comparison, risks overlooking the complexity and heterogeneity of climate policy design. Subnational carbon pricing initiatives, such as those seen in the United States or Canada, are not fully captured by this national-level coding. Likewise, averaging carbon prices across sectors within a country may obscure important variations, such as higher costs imposed on energy-intensive industries relative to transport or residential energy use. Furthermore, compliance and enforcement mechanisms, which significantly affect the effectiveness of carbon pricing, are not directly represented in this model. These methodological choices, though pragmatic, inevitably limit the granularity of policy analysis.

Another limitation lies in the scope of variables included. Although the study incorporates a wide range of economic, demographic, and policy indicators, there are additional dimensions of vulnerability and resilience that could not be fully addressed. Factors such as governance quality, institutional strength, corruption levels, or adaptive capacity at the community level play critical roles in shaping climate outcomes but remain outside the quantitative scope of this analysis due to data availability and comparability constraints. Likewise, the social and political dynamics of climate policy adoption—such as public acceptance of carbon pricing, lobbying by fossil fuel industries, or the influence of international climate finance—are acknowledged in the literature but are not explicitly modelled here.

Despite these acknowledged boundaries, the study maintains methodological rigor and offers valuable insights that remain highly relevant for both academic inquiry and policy formulation. By employing a combination of descriptive statistics, comparative assessments, correlation analyses, and policy evaluations, the research delivers a holistic overview of the interplay between climate vulnerability, economic strength, and carbon pricing adoption. The use of globally standardized and transparent datasets further enhances the credibility of the analysis. Importantly, the deliberate acknowledgment of constraints strengthens, rather than weakens, the interpretive value of the findings, as it frames them within realistic expectations.

Ultimately, while the study does not claim to provide a complete or flawless account of global climate vulnerability and policy readiness, it does offer a reliable, reproducible, and policy-relevant assessment that fills an important gap in the literature. It lays the groundwork for future research that may adopt a longitudinal design, incorporate more granular subnational data, or integrate qualitative aspects such as governance and institutional capacity. By linking physical climate risks with economic resilience and policy action, the research contributes to a deeper understanding of the structural determinants of global climate adaptation and transition pathways. It is within this balance of strengths and acknowledged constraints that the study positions itself as a meaningful contribution to the discourse on climate risk, policy adoption, and economic capacity across countries.

4. RESULTS AND ANALYSIS

This section presents the findings of the cross-country analysis of climate vulnerability, economic capacity, and carbon pricing adoption. Using the 2022 dataset, we examine patterns in CRI scores, GDP per capita, population, and energy transition readiness, alongside policy adoption and coverage of carbon pricing instruments. The analysis highlights disparities across income groups, regional trends, and the potential linkages between climate vulnerability and institutional responses.

4.1 Descriptive Insights

Climate Risk Index (CRI) Scores:

The dataset shows substantial variation in CRI scores. Low-income and lower-middle-income countries such as Pakistan (53,626,890) and Niger (7,705,593) exhibit the highest vulnerability, reflecting both high exposure to extreme weather events and limited adaptive capacity. High-income nations like Germany (10,926,968) and France (7,244,031) report significantly lower CRI scores, indicating relatively higher resilience and preparedness against climate hazards.

Economic Capacity:

GDP per capita is strongly correlated with income classification. High-income countries such as Ireland ($107,316) and Austria ($56,833) possess substantial financial resources to support adaptation and mitigation measures. In contrast, lower-middle-income nations like Bangladesh ($2,593) and Pakistan ($1,484) have constrained capacities to invest in resilience, exacerbating climate vulnerability.

Population Exposure:

Countries with large populations, including India (1,450,935.79 thousand) and the United States (340,110.99 thousand), face high potential exposure, even if CRI scores are moderate. Population size amplifies the scale of potential human and economic impacts from climate events.

Carbon Pricing Adoption and Coverage:

Only a subset of countries has implemented carbon pricing instruments. High-income European nations such as Germany and France employ ETS frameworks and carbon taxes, covering a substantial portion of emissions (Germany: 39%; France: 12%). Several countries with high CRI scores, such as Pakistan and Niger, lack any national carbon pricing mechanism, underscoring a critical policy gap. Carbon price levels vary widely, from as low as US$2 2 2/ton in Japan to US$69/ton in Ireland, reflecting differences in policy stringency and market maturity.

4.2 Comparative Analysis Across Income Groups

To better understand the interplay between economic capacity, exposure to climate risks, and readiness for energy transition, this study categorizes countries by income group. The comparison provides insights into how structural resilience, institutional capacity, and policy adoption differ across contexts. Table 4 summarizes these differences, highlighting patterns in climate vulnerability and the extent of carbon pricing implementation.

Table 4. Comparative Assessment of Climate Vulnerability and Carbon Pricing Readiness Across Income Groups

Income Group	Climate Vulnerability	Carbon Pricing & Transition Readiness
High-Income	Lower CRI scores due to resilience and infrastructure	Widespread ETS/carbon taxes; high emissions coverage (10–40%)
Upper-Middle-Income	Moderate CRI scores; partial vulnerability	Emerging carbon pricing policies, though limited in scope
Lower-Middle & Low-Income	Highest CRI scores; limited adaptive capacity	Minimal or no carbon pricing; significant policy gaps

High-income countries exhibit lower climate vulnerability due to advanced infrastructure, diversified economies, and effective carbon pricing mechanisms. Policy adoption is robust; ETS and carbon taxes are prevalent, covering 10–40% of national GHG emissions in Europe. Upper-Middle-Income Countries display intermediate CRI scores. Nations such as South Africa and China show moderate vulnerability but have initiated policy instruments (carbon tax or ETS) to facilitate transition, although coverage is often partial. Lower-Middle and Low-Income Countries face the highest CRI scores, with minimal carbon pricing adoption. Countries like Bangladesh and Pakistan exemplify high exposure combined with weak institutional response, highlighting urgent adaptation and policy needs.

These contrasts emphasize the importance of differentiated global strategies, where wealthier nations deepen decarbonization efforts while lower-income economies require targeted financial and technological support to strengthen resilience. Building on these comparative patterns.

4.3 Correlation Insights

Although a full statistical correlation analysis is beyond the immediate scope of this chapter and is recommended for future quantitative work, preliminary patterns observed across the dataset provide meaningful insights into the interplay between climate risk and structural variables. The relationship between climate vulnerability, as captured by the CRI 2022 scores, and GDP per capita appears to follow a negative association. In other words, wealthier nations generally demonstrate lower levels of vulnerability to climate hazards. This trend reflects the role of advanced infrastructure, diversified economies, and greater fiscal capacity in mitigating the physical impacts of extreme weather events. High-income countries, with their ability to mobilize resources for disaster preparedness and adaptation, consistently register lower CRI scores, whereas lower-income economies remain disproportionately exposed to climate shocks due to their limited adaptive capacity.

Similarly, institutional mechanisms such as carbon pricing adoption demonstrate a discernible link with climate resilience. Countries that have implemented carbon pricing instruments—whether ETS or carbon taxes—tend to record lower CRI scores. This suggests that policy adoption does more than simply reduce greenhouse gas emissions; it also signals institutional readiness, regulatory strength, and commitment to climate transition, all of which indirectly contribute to resilience against physical risks. For instance, European countries with long-standing ETS mechanisms not only address emissions but also maintain comprehensive adaptation strategies, which helps to explain their lower vulnerability scores compared to nations with weaker or absent carbon pricing structures.

Population size represents another influential factor in shaping climate vulnerability outcomes. Larger populations inherently magnify the scale of potential exposure to climate-related disasters, particularly in countries where economic resources and institutional mechanisms are insufficient. This dynamic is especially evident in lower-income and lower-middle-income nations, where high population density, coupled with inadequate adaptive infrastructure, exacerbates the impact of floods, storms, and heatwaves. The correlation between population size and CRI highlights the dual challenge faced by populous, resource-constrained countries: not only do they confront frequent climate hazards, but the social and economic consequences arc also amplified by the sheer number of people affected.

Energy transition indicators further reinforce the pattern of structural resilience. Countries with higher shares of renewable energy in their electricity mix and lower energy intensity levels generally display lower CRI scores. Greater reliance on renewable energy reduces exposure to fossil fuel volatility, while lower energy intensity reflects higher efficiency and reduced systemic vulnerability. However, the strength of this relationship varies across regions. For example, several European nations combine high renewable energy penetration with strong policy frameworks, translating into both low transition risks and low CRI scores. In contrast, parts of Asia and Latin America demonstrate increasing renewable adoption but continue to face considerable physical vulnerability due to geographic exposure or incomplete institutional coverage.

Taken together, these initial correlation insights suggest that economic strength, institutional frameworks, demographic pressures, and energy transition pathways all interact to shape national vulnerability profiles. While descriptive rather than statistically definitive, these observations provide a foundation for more advanced correlation and regression analyses. They point to the critical role of integrated approaches that combine economic growth, policy adoption, and energy transition as levers for enhancing resilience against the mounting risks of climate change.

4.4 Regional Patterns

Beyond income-based differences, regional dynamics also shape the interaction between climate risks and transition readiness. Geographic location, institutional structures, and regional policy frameworks play a crucial role in determining exposure to climate hazards and the adoption of carbon pricing mechanisms. For instance, while high-income European countries are well-positioned to adopt robust mitigation policies, low-income nations in Africa or South Asia continue to face acute climate risks with limited institutional support. Table 4 summarizes these regional variations, presenting an overview of climate vulnerability (as indicated by CRI scores) and the extent of carbon pricing adoption across major global regions.

Table 5. Regional Patterns of Climate Vulnerability and Carbon Pricing Readiness

Region	Climate Vulnerability	Carbon Pricing & Transition Readiness
Europe	Low CRI scores	High adoption of ETS and carbon taxes
Asia	Mixed vulnerability levels	Some policy adoption (China, South Korea), high risk in South Asia
Africa	High vulnerability, especially in low-income nations	Minimal policy adoption except in South Africa
Latin America & Caribbean	Moderate CRI scores	Limited adoption; Brazil lacks a national carbon pricing framework

The table reveals clear regional disparities in both climate vulnerability and carbon pricing readiness. The Europe region exhibits the lowest CRI scores, reflecting high adaptive capacity, advanced infrastructure, and effective disaster management systems. Europe also stands out for widespread carbon pricing adoption, with both ETS and carbon taxes implemented in several countries. Coverage rates are comparatively high, often ranging between 10–40% of national greenhouse gas emissions, demonstrating a strong institutional commitment to energy transition. Outcomes in Asia are heterogeneous. Advanced economies such as South Korea and China have introduced carbon pricing mechanisms—China through its national ETS and

South Korea with a functioning cap-and-trade program. However, much of South Asia and Southeast Asia face high climate risks due to dense populations, weak infrastructure, and frequent extreme weather events. Despite growing exposure, policy adoption remains limited, particularly in lower- and lower-middle-income countries. The region of Africa is marked by very high climate vulnerability, especially among low-income nations with limited adaptive capacity and constrained financial resources. Extreme events such as floods, droughts, and heatwaves disproportionately affect communities in Sub-Saharan Africa. Policy adoption is almost absent, with South Africa being the primary exception due to its implementation of a carbon tax. Nonetheless, coverage remains limited, highlighting the need for stronger institutional frameworks and international support. Countries in the Latin America & the Caribbean region generally experience moderate CRI scores, reflecting mixed levels of climate risk exposure. Carbon pricing adoption is limited, with only a few nations experimenting with pilot schemes. Notably, Brazil which is one of the largest economies in the region and a major emitter, still lacks a comprehensive national carbon pricing framework, underscoring a significant gap in policy alignment with climate objectives.

Therefore, in Europe, most countries are high-income, characterized by low CRI scores and widespread adoption of carbon pricing mechanisms. In Asia, outcomes are mixed, with high vulnerability in South and Southeast Asia despite some policy adoption in upper-middle-income countries such as China. In Africa, upper-middle-income countries like South Africa have introduced partial carbon pricing policies, while low-income nations remain largely unprotected. In Latin America and the Caribbean, countries display moderate CRI scores and limited carbon pricing adoption, with Brazil notably lacking a national framework despite its economic size and emissions profile. Overall, regional patterns illustrate that while Europe leads both in resilience and transition readiness, large parts of Asia, Africa, and Latin America lag, either due to structural vulnerabilities or weak institutional adoption of carbon pricing instruments. This divergence emphasizes the urgent need for region-specific strategies, with wealthier nations focusing on deepening their decarbonization policies, while developing regions require substantial financial, technological, and institutional support to bridge the adaptation and mitigation gap.

4.5 Policy Implications

The findings of this study carry significant implications for climate policy, particularly in light of the uneven distribution of climate risks and adaptive capacities across countries. One of the clearest patterns is that income level emerges as a critical determinant of resilience. Nations with higher GDP per capita consistently exhibit lower CRI scores, reflecting their ability to mobilize financial resources for robust adaptation strategies, infrastructure development, and disaster preparedness. Conversely, low-income countries face a disproportionate burden of climate risks, as financial constraints and weak institutional capacity limit their ability to implement protective measures. This structural inequality underscores the urgency of designing differentiated policy responses that take into account economic capacity alongside climate vulnerability.

Carbon pricing instruments, including ETS and carbon taxes, further demonstrate their role as effective tools in mitigating both physical and transition risks. The analysis reveals that countries with active and well-designed carbon pricing policies tend to register lower vulnerability scores, suggesting that policy adoption contributes to resilience not only by reducing emissions but also by signalling institutional commitment and governance strength. However, the adoption of such instruments remains highly uneven across regions, with Europe leading in coverage and ambition while large parts of Africa, South Asia, and Latin America remain without robust mechanisms. This gap highlights the need for capacity-building initiatives, knowledge-sharing, and international cooperation to expand the reach and effectiveness of carbon pricing globally.

Another key implication lies in the relationship between population dynamics and vulnerability. Heavily populated nations, especially in lower-income brackets, face intensified risks due to the sheer scale of exposure and the absence of adequate adaptive measures. These findings highlight the pressing need for targeted financing mechanisms and global support frameworks to channel resources toward countries where vulnerability is most acute. This includes not only disaster preparedness funds but also climate insurance schemes, concessional loans, and technology transfers to ensure that vulnerable populations are not left behind.

Energy transition readiness, particularly the expansion of renewable energy capacity and improvements in energy efficiency, emerges as another cornerstone of resilience. Countries with higher shares of renewable energy and lower energy intensity demonstrate stronger adaptive capacity and reduced climate vulnerability. This suggests that integrating energy transition policies with adaptation strategies provides a dual benefit: mitigating long-term transition risks while also enhancing protection against physical climate impacts.

Taken together, these findings reinforce the interlinked nature of economic strength, policy adoption, and energy transition readiness as drivers of climate resilience. Wealthier nations with advanced carbon pricing mechanisms and strong renewable energy adoption are better equipped to withstand climate shocks, whereas poorer countries without such frameworks remain disproportionately vulnerable. The results therefore, call for tailored, context-sensitive strategies that recognize structural disparities and prioritize interventions where they are most urgently needed. Policymakers, investors, and international organizations must view carbon pricing, renewable energy expansion, and socioeconomic strengthening as complementary tools in building a resilient global system. By aligning climate risk assessments with actionable policy mechanisms, the international community can move closer to achieving equitable and sustainable resilience.

5. DISCUSSION AND POLICY IMPLICATIONS

5.1 Interpreting the Results

The cross-country analysis of climate vulnerability reveals several critical determinants that shape national resilience. Income level emerges as a primary factor, with high-income countries such as Germany and France exhibiting lower Climate Risk Index (CRI) scores due to robust institutional capacity, diversified economies, and comprehensive adaptation measures. In contrast, lower-middle- and low-income countries, including Pakistan, Niger, and Bangladesh, experience elevated CRI scores as limited financial and institutional capacity constrains investment in resilient infrastructure, early warning systems, and disaster preparedness. Population size further influences vulnerability, as highly populous nations face greater aggregate exposure to climate hazards. Even countries with moderate CRI scores, such as India and the United States, encounter significant potential human and economic impacts, while densely populated, low-income countries face particularly severe consequences due to insufficient resilience mechanisms.

Table 6. Key Determinants of Climate Vulnerability and National Resilience

Determinant	Description	Example Countries / Observations
Income and Climate Vulnerability	Inverse relationship between GDP per capita and CRI scores. High-income countries exhibit robust institutional capacity, diversified economies, and extensive adaptation measures, resulting in lower climate risk exposure. Low- and lower-middle-income countries face higher CRI scores due to limited financial and institutional capacity, restricting investments in resilient infrastructure, early warning systems, and disaster preparedness.	Germany, France (low CRI); Pakistan, Niger, Bangladesh (high CRI)
Population Size and Exposure	Larger populations increase aggregate exposure to climate events. Even with moderate CRI scores, highly populous nations face greater potential human and economic impacts. In low-income, densely populated countries, limited resilience amplifies the consequences of extreme weather events.	India, United States (high population, moderate CRI); Pakistan, Bangladesh (high population, high CRI)
Carbon Pricing Adoption and Coverage	Adoption of carbon pricing instruments (ETS or carbon taxes) is associated with lower CRI scores, indicating mitigation potential. Coverage varies across countries. High-CRI countries often lack national carbon pricing, reflecting policy gaps that hinder both mitigation and adaptation efforts.	Germany ETS: 39% coverage; Ireland carbon tax: 34% coverage; Pakistan, Niger, Bangladesh: no carbon pricing
Energy Transition Readiness	Nations with higher shares of renewable energy and lower energy intensity demonstrate lower vulnerability. Energy transition readiness enhances adaptive capacity through technological modernization and reduced dependence on fossil fuels. Coordination of energy policies with renewable deployment is particularly effective in Europe; slower transitions in Asia and Africa are linked to higher vulnerability.	Germany, France (high renewable, low CRI); India, Niger, Bangladesh (low renewable, high CRI)

The cross-country analysis reveals several important patterns linking climate vulnerability, economic capacity, and policy adoption. There is a pronounced inverse relationship between GDP per capita and CRI scores. High-income countries, such as Germany and France, benefit from robust institutional capacity, diversified economies, and extensive climate adaptation measures, resulting in lower climate risk exposure. Conversely, lower-middle and low-income countries, including Pakistan, Niger, and Bangladesh, experience significantly higher CRI scores. Limited financial and institutional capacity constrains investment in resilient infrastructure, early warning systems, and disaster preparedness, increasing their vulnerability to climate shocks. Larger populations amplify the potential impact of climate events. Even if the CRI score is moderate, highly populous nations like India and the United States face high aggregate exposure. In low-income, high-population countries, the combination of dense populations and limited resilience mechanisms creates conditions for severe humanitarian and economic consequences during extreme weather events. Countries with established carbon pricing instruments (ETS or carbon taxes) generally show lower CRI scores, suggesting a potential mitigating effect of structured climate policies on overall vulnerability. Coverage varies significantly. Germany's ETS covers 39% of national emissions, whereas Ireland's carbon tax covers 34%. Countries with high CRI scores often lack national carbon pricing, highlighting critical policy gaps in addressing both mitigation and adaptation challenges. Nations with higher shares of renewable energy and lower energy intensity demonstrate lower vulnerability. Energy transition readiness contributes not only to the mitigation of future climate risks but also enhances adaptive capacity through technological modernization and reduced dependence on climate-sensitive fossil fuels. The link is particularly evident in Europe, where coordinated renewable energy policies align with low CRI scores, compared to Asian and African nations with slower transitions.

5.2 Comparative Insights Across Regions

This study finds that regional differences play a decisive role in shaping both the degree of climate vulnerability and the capacity for policy-driven adaptation and transition. While global averages are useful for identifying broad patterns, they mask the deep structural disparities that exist between regions. A closer, more detailed comparative analysis across Europe, Asia, Africa, and Latin America & the Caribbean reveals that resilience and readiness are shaped not only by income levels but also by governance quality, institutional capacity, and the extent to which carbon pricing frameworks are integrated with wider energy transition policies. By bringing these dimensions together, this research highlights how some regions have managed to achieve synergies between mitigation, adaptation, and economic stability, while others remain exposed to high levels of risk due to structural and institutional limitations.

Table 7. Comparative Regional Insights on Climate Vulnerability, Policy Adoption, and Transition Readiness

Region	Climate Vulnerability	Policy Adoption	Energy Transition Readiness
Europe	Low CRI scores due to advanced infrastructure, strong governance, and effective adaptation	High adoption of ETS and carbon taxes; EU ETS provides broad emissions coverage (10–40% in most countries)	Well-integrated with renewable deployment and efficiency policies, creating economic synergies
Asia	Mixed outcomes; low vulnerability in some economies, but high CRI scores in South/Southeast Asia	Upper-middle-income nations (e.g., China, South Korea) have ETS; lower-middle-income (e.g., Pakistan, Bangladesh) show minimal adoption	Divergent patterns: technological leaders vs. countries with limited capacity and weak institutions
Africa	High CRI scores, particularly in low-income nations with limited adaptive capacity	South Africa has implemented a carbon tax; most other countries lack comprehensive pricing mechanisms	Weak readiness; reliance on external finance and international support for transition measures
Latin America & Caribbean	Moderate CRI scores; exposure varies by geography and economic structure	Limited carbon pricing adoption; Brazil has yet to implement a national carbon pricing framework	Uneven readiness; some progress in renewable energy but overall fragmented and inconsistent

The table makes it clear that Europe is by far the most advanced region in terms of climate resilience and transition readiness. European countries, mostly high-income economies, benefit from robust institutions, strong governance systems, and well-established adaptation infrastructures. The European Union's Emissions Trading System (EU ETS) provides a collective framework that ensures carbon pricing is comprehensive, covering between 10 and 40 percent of national GHG emissions in member states. Furthermore, the integration of carbon pricing with renewable energy deployment and energy efficiency measures creates synergies that not only reduce vulnerability but also contribute to long-term economic stability. This highlights Europe as a model of how institutional design and economic capacity can reinforce each other to deliver resilience.

Asia, however, presents a far more complex and uneven picture. On one hand, upper-middle-income countries such as China and South Korea have made significant progress in adopting emissions trading systems, showing institutional capacity to link climate risk management with economic transition strategies. On the other hand, lower-middle-income nations such as Pakistan and Bangladesh face high CRI scores, reflecting severe climate vulnerability and limited adaptive capacity. These countries are particularly exposed to floods, storms, and heatwaves, yet they lack effective carbon pricing instruments, largely due to fiscal constraints, weak institutions, and limited access to technology. The stark contrast between technologically advanced economies in East Asia and vulnerable nations in South and Southeast Asia underscores the importance of governance quality and fiscal capacity in shaping resilience outcomes.

Africa emerges as the most vulnerable region overall. Most countries in Africa, especially low-income nations, exhibit high CRI scores, reflecting limited adaptive capacity and high exposure to climate risks. South Africa stands out as an exception with its implementation of a national carbon tax, but even there, coverage remains partial and enforcement challenges persist. For the rest of the continent, minimal policy adoption combined with economic fragility leaves populations exposed to increasingly severe weather events. In addition, energy transition readiness remains weak, with most countries depending heavily on fossil fuels and external finance to support renewable deployment. This situation highlights the urgent need for international support, including climate finance, technology transfer, and capacity-building measures, to prevent widening inequality in climate resilience between Africa and the rest of the world.

Latin America and the Caribbean occupy a middle position in the comparative framework. Countries in this region exhibit moderate CRI scores, though exposure varies widely depending on geography, economic structure, and governance. Brazil, despite its size and emissions profile, lacks a comprehensive national carbon pricing framework, representing a missed opportunity for coordinated transition. While some progress has been made in renewable energy deployment across parts of the region, the overall readiness remains fragmented and inconsistent. Policy adoption is limited, and the absence of harmonized frameworks across countries weakens the region's collective ability to build resilience. This situation illustrates how even middle-income regions with significant renewable potential may fall short if policy adoption is uneven or poorly integrated.

Overall, this comparative analysis shows that resilience is not determined by a single factor such as income but rather by a combination of economic resources, governance structures, institutional coherence, and policy integration. Europe demonstrates how advanced infrastructure and coordinated policy mechanisms reinforce resilience. Asia illustrates both the potential benefits of institutional capacity and the risks of governance gaps. Africa highlights the urgent vulnerabilities associated with weak institutions and limited policy adoption, while Latin America and the Caribbean show how moderate exposure can still translate into fragility if policy frameworks remain fragmented. These findings underscore the importance of tailored, region-specific strategies: high-income regions must expand their role in financing and supporting global resilience, while vulnerable regions require not only financial assistance but also institutional and technological support to ensure their transition pathways are sustainable and inclusive.

5.3 Policy Implications

The comparative analysis of regional climate vulnerability, policy adoption, and transition readiness provides several important lessons for policymakers, international institutions, and stakeholders in both the public and private sectors. These insights emphasize that climate action must be adaptive, inclusive, and sensitive to structural inequalities across countries and regions.

Table 8. Key Policy Implications for Climate Resilience and Transition

Policy Dimension	Implication	Underlying Rationale
Targeting Vulnerable Countries	Prioritize low- and lower-middle-income nations with high CRI scores through finance and support.	Limited fiscal capacity and institutional weakness magnify climate risk exposure.
Carbon Pricing & Adaptation Link	Expand carbon pricing while reinvesting revenues in resilience-building measures.	Aligns mitigation with adaptation co-benefits and creates sustainable funding mechanisms.
Energy Transition Readiness	Promote renewable investment, grid modernization, and decentralized solutions.	Enhances resilience and reduces dependency on fossil imports in emerging economies.
Population-Centered Measures	Develop urban resilience strategies and early-warning systems for high-density, high-risk nations.	Protects vulnerable populations and reduces disaster-related economic and social losses.
Economic Capacity & Climate Action	Strengthen public–private partnerships for large-scale adaptation investments.	Mobilizes domestic and international capital to bridge resource gaps.

The evidence underscores that vulnerable countries—particularly those in South Asia, sub-Saharan Africa, and parts of Latin America—require targeted, sustained support to narrow the adaptation gap. Financial transfers, technical assistance, and infrastructure development are not merely humanitarian imperatives; they are global economic priorities, as instability in vulnerable nations can reverberate through trade, migration, and security systems worldwide. Multilateral institutions such as the World Bank, IMF, and Green Climate Fund have a critical role in operationalizing these transfers by ensuring funds are accessible, transparent, and linked to resilience-building outcomes.

Equally important is the strategic integration of carbon pricing with adaptation measures. While carbon pricing has traditionally been seen as a mitigation tool, it can also generate predictable revenue streams that can be reinvested in infrastructure hardening, disaster preparedness, and social protection systems. Policymakers must design carbon pricing frameworks that not only incentivize emissions reductions but also expand coverage across multiple sectors, strengthen enforcement, and explicitly earmark revenues for resilience-building initiatives. This dual-purpose approach creates a powerful alignment between climate and development objectives.

Another critical policy implication lies in energy transition readiness. Emerging economies face unique challenges, such as growing populations, rapid urbanization, and constrained fiscal capacity, yet these conditions also create opportunities for leapfrogging into sustainable energy systems. By investing in renewable deployment, grid modernization, and decentralized solutions such as solar micro-grids, countries can simultaneously enhance resilience and reduce dependence on volatile fossil fuel imports. Policymakers must also incentivize technological innovation and promote knowledge-sharing across borders to ensure that transitions are efficient, equitable, and scalable.

Population dynamics add another layer of urgency. Countries with large populations—particularly those in Asia and Africa—must prioritize interventions that directly safeguard urban areas where the human and economic costs of climate disasters are most concentrated. Early warning systems, climate-resilient housing, and disaster preparedness education can significantly reduce vulnerability. Crucially, population-focused measures must integrate social and economic planning with environmental policy so that resilience strategies address not only infrastructure but also livelihoods and social equity.

Finally, the relationship between economic capacity and climate action remains central. Higher GDP per capita correlates with stronger resilience, but this does not mean that low-income countries should be left behind. Instead, innovative financing models, including blended finance and public–private partnerships, can mobilize both domestic and international resources to close the resilience gap. Examples include investments in flood defences, sustainable agriculture, and health systems capable of withstanding climate-induced stress.

5.4 Implications for Financial Markets

Climate risks translate directly into financial risks, and the findings highlight several implications for investors, lenders, and global markets.

Table 9. Climate Risk and Financial Market Implications

Market Dimension	Implication	Underlying Rationale
Investor Risk Assessment	High CRI countries carry greater sovereign debt, insurance, and investment risk.	Physical and transition risks raise default and loss probabilities.
Policy-Driven Market Signals	Strong climate policies reduce risk premiums and attract foreign investment.	Carbon pricing and renewables signal commitment to stability and transition.
Integration of Risk Metrics	Use CRI scores with socioeconomic data for portfolio diversification.	Provides a nuanced view of resilience across regions and income levels.

For investors, high CRI scores serve as red flags, indicating elevated physical and transition risks that may affect sovereign creditworthiness, insurance premiums, and broader market performance. Countries unable to demonstrate resilience may face rising borrowing costs and reduced capital inflows, reinforcing cycles of vulnerability.

Conversely, robust policy adoption, especially carbon pricing and renewable energy policies, sends powerful signals to global markets about a country's long-term stability and transition readiness. Such signals shape capital allocation, lower risk premiums, and can catalyse new investments in clean energy and infrastructure. Conversely, nations lacking comprehensive policies risk market penalties, including capital flight and restricted access to foreign investment.

Financial institutions are increasingly adopting integrated approaches that combine CRI scores with GDP per capita, governance quality, and population exposure to assess climate resilience. This multidimensional framework allows investors to identify both risks and opportunities, supporting smarter portfolio diversification and risk-adjusted returns.

Taken together, these findings show that climate vulnerability, economic capacity, and policy adoption do not merely shape environmental outcomes—they are central to determining financial market stability. By aligning climate strategies with financial risk assessment, both policymakers and investors can create positive feedback loops that reduce vulnerability while promoting sustainable growth.

5.5 Integrative Insights on Climate Vulnerability, Policy Adoption, and Financial Resilience

This section synthesizes the key findings of the study, highlighting the interconnected roles of economic capacity, climate vulnerability, policy mechanisms, and energy transition readiness in shaping national resilience and financial market outcomes. Economic capacity, as measured by GDP per capita, directly influences a country's ability to allocate resources toward climate adaptation and mitigation measures. Nations with higher economic capacity are better positioned to implement robust carbon pricing frameworks, invest in renewable energy infrastructure, and support comprehensive adaptation strategies, thereby lowering their overall climate vulnerability. In contrast, lower-income countries, often facing high population density and limited institutional capacity, experience heightened exposure to climate risks, which is reflected in elevated CRI scores.

Policy adoption emerges as a critical mitigating mechanism. The presence of carbon pricing instruments, such as carbon taxes and ETS frameworks, not only incentivizes emissions reduction but also generates revenues that can be reinvested in resilience-building measures. Effective energy transition policies, including the expansion of renewable energy deployment and improvements in energy efficiency, further reinforce national capacity to withstand climate shocks. By integrating these variables—economic capacity, policy adoption, and energy transition—into a cohesive analytical framework, the study demonstrates how countries can strategically reduce vulnerability while simultaneously promoting sustainable development.

Population dynamics and regional context further modulate these interactions. Highly populated nations, particularly those in South and Southeast Asia or sub-Saharan Africa, require targeted adaptation interventions, including climate-resilient infrastructure, early warning systems, and urban planning measures. Regional disparities in governance quality, fiscal capacity, and technological readiness also shape the effectiveness of policy adoption and energy transition strategies. For example, high-income European nations benefit from coordinated carbon pricing policies and well-established renewable energy systems, whereas low-income African nations remain reliant on external financial and technical assistance.

The interplay of these factors has direct implications for financial markets. Countries with strong economic capacity, comprehensive policy adoption, and high energy transition readiness tend to exhibit lower financial risk, fostering stability in investment, credit, and capital flows. Conversely, nations with limited policy instruments and high vulnerability present elevated physical and transition risks, necessitating careful integration of climate risk metrics into investor decision-making, sovereign debt evaluation, and portfolio management.

Figure 3. Conceptual Framework Linking Climate Vulnerability, Economic Capacity, Policy Adoption, and Financial Market Outcomes

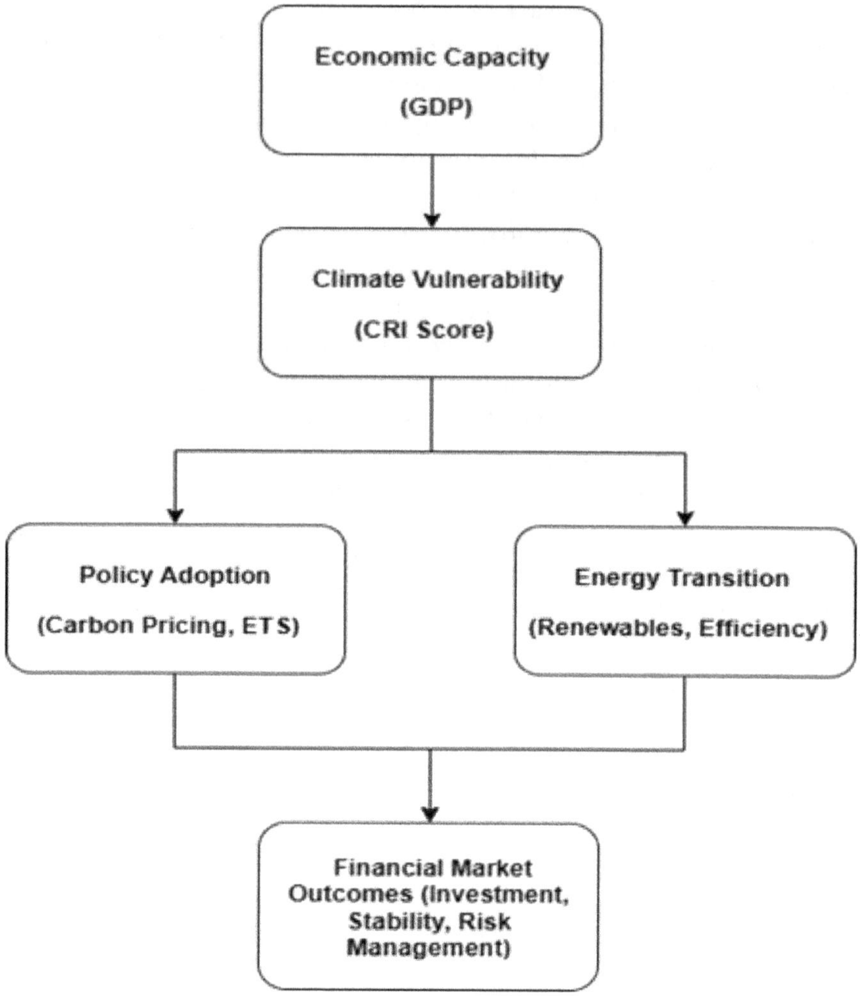

As illustrated in Figure 3, economic capacity, policy adoption, and energy transition jointly mediate the impact of climate vulnerability on financial markets. This framework demonstrates the interconnections between national economic capacity (GDP per capita), climate vulnerability (CRI scores), carbon pricing and ETS adoption, and energy transition readiness (renewable energy deployment and energy efficiency). Together, these factors determine the extent to which countries can manage climate risks and maintain financial stability.

In essence, this integrative analysis highlights that climate resilience is not determined by any single factor but by the combined effect of economic strength, institutional mechanisms, policy implementation, and energy system readiness. Low-income, high-population countries remain the most exposed, emphasizing the need for targeted international support, adaptive policy frameworks, and investment strategies that align economic, environmental, and financial objectives. This section provides a comprehensive synthesis of the study's findings, offering actionable insights for policymakers, investors, and researchers seeking to enhance global climate resilience while ensuring market stability.

6. RECOMMENDATIONS AND FUTURE RESEARCH

The findings of this study underscore the urgent need for targeted and strategic interventions to enhance climate resilience worldwide. Analysis of cross-country data reveals distinct patterns in vulnerability, policy adoption, energy transition readiness, and population exposure, highlighting where interventions can be most effective.

Firstly, vulnerable nations, particularly those classified as low-income and lower-middle-income with high Climate Risk Index (CRI) scores, require immediate support. These countries face structural constraints, including limited financial resources, weak governance, and inadequate adaptation infrastructure, leaving populations exposed to severe climate hazards. International cooperation, climate finance, technology transfer, and capacity-building initiatives are critical to strengthen resilience. Enhancing institutional capacity and improving infrastructure, including disaster preparedness systems, will mitigate potential human and economic losses in these high-risk regions.

Secondly, the study emphasizes the expansion and strengthening of carbon pricing mechanisms. Carbon taxes and ETS have demonstrated a measurable impact on reducing climate vulnerability, particularly when combined with wide coverage and strong enforcement. Policymakers should broaden the scope of these instruments, ensure robust monitoring, and reinvest revenues into both mitigation and adaptation measures. This dual approach not only reduces vulnerability but also signals long-term policy commitment to investors, reinforcing economic stability and resilience.

Thirdly, accelerating the energy transition is essential. Countries with higher renewable energy shares and lower energy intensity exhibit reduced vulnerability, indicating that renewable deployment, energy efficiency improvements, and technological innovation are directly linked to resilience. Coordinating energy transition strategies with carbon pricing policies enhances both adaptation and mitigation outcomes, demonstrating national commitment to sustainable development. Investments in decentralized renewable solutions, smart grids, and energy-efficient technologies can reduce exposure to climate hazards while fostering long-term sustainability.

Fourthly, population-centric adaptation strategies are vital. Large and densely populated nations experience amplified climate risks, necessitating focused urban planning, early warning systems, climate-resilient infrastructure, and integration of socioeconomic planning with environmental policy. These strategies ensure that human lives and economic assets are protected, particularly in regions with high exposure to extreme weather events, such as floods, heatwaves, and storms.

Fifthly, integrating climate risk into financial decision-making is imperative. Financial institutions, investors, and insurers should utilize CRI scores, policy adoption metrics, and energy transition indicators to inform risk assessment and capital allocation. Aligning investment decisions with climate-resilient development objectives can catalyze adaptation and mitigation efforts while reducing systemic risk exposure in global financial markets.

For future research, this study highlights several promising directions. Longitudinal datasets are needed to track the temporal evolution of climate vulnerability, policy effectiveness, and energy transition progress. Sector-specific studies can reveal how climate risks and policy instruments impact industries, agriculture, and urban systems. Further, examining synergies between adaptation, mitigation, and socioeconomic development can inform more integrated policy frameworks, particularly in high-risk regions. Finally, investigating the role of governance quality, fiscal capacity, and technological readiness in mediating vulnerability will provide actionable insights for both national and international stakeholders.

7. CONCLUSION

This study provides a comprehensive examination of the interconnected drivers of climate vulnerability, including economic capacity, policy adoption, energy transition readiness, and population exposure across countries in 2022. The evidence underscores that income level and economic capacity are primary determinants of resilience, with high-income countries demonstrating lower CRI scores due to stronger institutions, robust infrastructure, and effective policy instruments. Conversely, low- and lower-middle-income countries remain highly vulnerable, constrained by limited financial resources, weaker governance, and insufficient adaptation measures.

Policy adoption, particularly carbon pricing through taxes and ETS, emerges as a critical factor in reducing vulnerability. Countries with robust policy frameworks and broad emissions coverage exhibit enhanced resilience, demonstrating the value of integrated mitigation and adaptation strategies. Population size further magnifies exposure, emphasizing the need for population-focused interventions such as urban planning, early warning systems, and resilient infrastructure.

Energy transition readiness significantly contributes to climate resilience. Nations with higher renewable energy shares and lower energy intensity are better positioned to withstand climate shocks. Coordinated approaches between energy transition policies and carbon pricing amplify resilience while signalling commitment to sustainable development.

Furthermore, the study highlights the interplay between climate risk and financial markets. By integrating CRI scores with socioeconomic and policy indicators, financial institutions can better assess exposure, manage risk, and allocate capital strategically, supporting both adaptation and mitigation objectives.

In essence, the research demonstrates that holistic, multi-dimensional approaches are necessary to effectively reduce climate vulnerability. Effective strategies combine strong policy adoption, energy transition readiness, institutional capacity, and financial planning, particularly for low-income and highly exposed countries. By linking climate vulnerability with economic capacity, policy mechanisms, and population exposure, this study provides actionable insights for policymakers, investors, and global stakeholders, contributing to more resilient and sustainable development pathways.

Ultimately, achieving climate-resilient development requires coordinated global efforts, where policy, finance, and technology converge to reduce risk, enhance adaptation, and ensure sustainable futures for vulnerable populations worldwide. The recommendations and research directions outlined here offer a structured roadmap to guide interventions and investments that strengthen resilience across diverse contexts.

REFERENCES

Aldy, J. E., & Stavins, R. N. (2012). The promise and problems of pricing carbon: Theory and experience. *Journal of Environment & Development, 21*(2), 152–180. DOI: 10.1177/1070496512442508

Campiglio, E., Dafermos, Y., Monnin, P., Ryan-Collins, J., Schotten, G., & Tanaka, M. (2018). Climate change challenges for central banks and financial regulators. *Nature Climate Change, 8*(6), 462–468. DOI: 10.1038/s41558-018-0175-0

Eckstein, D., Künzel, V., & Schäfer, L. (2022). *Global Climate Risk Index 2022: Who suffers most from extreme weather events? Weather-related loss events in 2020 and 2001–2020.* Germanwatch. https://germanwatch.org/en/cri

Hallegatte, S., Bangalore, M., & Vogt-Schilb, A. (2016). Assessing socioeconomic resilience to floods in 90 countries. *World Bank Policy Research Working Paper*, (7663).

Ivanovski, K., & Marinucci, N. (2021). Policy uncertainty and renewable energy: Exploring the implications for global energy transitions, energy security, and environmental risk management. *Energy Research & Social Science, 82*, 102415. DOI: 10.1016/j.erss.2021.102415

Pfenninger, S., Hawkes, A., & Keirstead, J. (2014). Energy systems modeling for twenty-first century energy challenges. *Renewable & Sustainable Energy Reviews, 33*, 74–86. DOI: 10.1016/j.rser.2014.02.003

Swart, R. (2019). Assessing physical climate risks for investments: A risky promise. *Climate Services, 14*, 15–18. DOI: 10.1016/j.cliser.2019.04.001

Chapter 11
Role of Financial Markets in Climate Change Adaptation and Mitigation

P. Selvakumar
https://orcid.org/0000-0002-3650-4548
Department of Science and Humanities, Nehru Institute of Technology, Coimbatore, India

Prabakaran
Sri Balaji University, Pune, India

T. C. Manjunath
https://orcid.org/0000-0003-2545-9160
Rajarajeswari College of Engineering, India

Riyazuddin
Maharishi University of Information Technology, Noida, India

M. Gayathri
https://orcid.org/0009-0004-7501-7824
School of Computing, SRM Institute of Science and Technology, Kattankulathur, India

A. R. Kanagaraj
https://orcid.org/0000-0002-6502-0817
Dr. N.G.P. Arts and Science College, India

ABSTRACT

Financial markets act as the primary intermediaries between capital suppliers and users, thereby playing a decisive role in mobilizing, allocating, and channeling investments towards low-carbon, climate-resilient solutions. As the world faces mounting challenges from rising global temperatures, more frequent extreme weather events, biodiversity loss, and the socioeconomic disruptions that follow, the ability of financial markets to redirect capital flows from environmentally wealth funds, private equity, venture capital, and capital markets, holds the keys to accelerating the transition from fossil fuel dependency to clean energy systems, from linear re-

DOI: 10.4018/979-8-3373-6766-8.ch011

silience. One of the most significant contributions of financial markets lies in their capacity to price climate risk. When risks are priced accurately, investors can better differentiate between climate-resilient and climate-vulnerable assets, leading to capital reallocation that strengthens sustainable businesses and reduces exposure to high-carbon sectors.

THE ROLE OF FINANCIAL MARKETS IN ADDRESSING CLIMATE CHANGE

Financial markets act as the primary intermediaries between capital suppliers and users, thereby playing a decisive role in mobilizing, allocating, and channeling investments towards low-carbon, climate-resilient solutions. As the world faces mounting challenges from rising global temperatures, more frequent extreme weather events, biodiversity loss, and the socioeconomic disruptions that follow, the ability of financial markets to redirect capital flows from environmentally wealth funds, private equity, venture capital, and capital markets, holds the keys to accelerating the transition from fossil fuel dependency to clean energy systems, from linear resilience. One of the most significant contributions of financial markets lies in their capacity to price climate risk. By incorporating (Alan et al., 2009)Organizations like the Task Force on Climate-related Financial Disclosures (TCFD) and the International Sustainability Standards Board (ISSB) have created frameworks that encourage companies and investors to assess and disclose material climate risks. When risks are priced accurately, investors can better differentiate between climate-resilient and climate-vulnerable assets, leading to capital reallocation that strengthens sustainable businesses and reduces exposure to high-carbon sectors. This pricing mechanism also has policy implications, as regulators and central banks increasingly recognize climate change as a systemic financial risk, -warning system, guiding both investors and policymakers in managing the transition to a net-zero economy.finance mechanisms provide targeted capital for projects that mitigate climate change or enhance adaptation. The green bond market, once niche, has grown into a trillion-dollar asset class, enabling governments, municipalities, and corporations to fund renewable energy, energy efficiency, and sustainable infrastructure projects. Similarly, carbon markets—both compliance-based and voluntary—create price signals that incentivize emission reductions while unlocking capital for carbon sequestration projects such as afforestation, reforestation, and soil carbon initiatives. Exchange-traded funds (ETFs) and climate-focused indices offer retail and institutional investors alike the ability to align portfolios with sustainability preferences, further mainstreaming climate-conscious investment strategies. These tools not only diversify financial products but also democratize access to sustainable investments, enabling wider participation in the global climate agenda.large institutional investors into climate innovation signals confidence in these technologies, helping to de-risk investments for smaller players. The catalytic effect of capital markets in scaling solutions from laboratory to global deployment cannot be overstated, as many of the technologies essential for achieving deep decarbonization are still at early stages of commercialization.

Beyond innovation and risk pricing, financial markets also act as accountability mechanisms by holding corporations responsible for their climate impact. Shareholder activism, proxy voting, and stewardship practices have become powerfu-facing higher costs of capital, or experiencing stock devaluation. Thus, financial markets embed climate considerations into the very fabric of corporate valuation and competitiveness.trigger systemic financial crises, disrupt global supply chains, and create widespread social inequality. (Alemzero et al., 2021)By aligning monetary policy tools, such as collateral frameworks and refinancing operations, with climate goals, central banks can further guide financial flows toward sustainable sectors. Additionally, sovereign green bonds and climate-aligned sovereign debt instruments provide governments with pathways to finance national climate commitments, thereby linking public finance with market mechanisms.can undermine the credibility of climate finance. Furthermore, the unequal distribution of capital flows highlights the risk of leaving developing countries behind, despite their high vulnerability to climate impacts. Financial markets must therefore evolve to bridge the gap between developed and emerging economies, ensuring equitable access to climate financinequalities.ensuring ecological preservation, recognizing that environmental sustainability underpins economic resilience. By fostering collaboration between public and private actors, aligning market incentives .

CLIMATE CHANGE IMPACTS ON FINANCIAL MARKETS AND STABILITY

Financial markets, by their very nature, are highly sensitive to and sovereigns, while simultaneously introducing volatility across equity, bond, and commodity markets. Insurance companies, for instance, face mounting claims from climate-induced disasters, forcing them to raise premiums or withdraw coverage from high-risk regions, thereby reducing the insurability of assets and heightening systemic financial vulnerabilities.(Barbi & da Costa Ferreira, 2017) At the same time, investors are increasingly aware that industries dependent on fossil fuels and resource-intensive practices face long-term devaluation—creating stranded assets that erode portfolio performance. This growing recognition that climate risks can no longer be considered externalities has elevated climate change from a peripheral environmental concern to a core financial stability issue, one that central banks, regulators, and market participants must confront head-on.prolonged droughts disrupt agricultural yields, leading to volatility in food commodity markets, while heatwaves strain energy systems, driving price fluctuations in electricity and fossil fuels. These physical shocks not only cause localized disruptions but can cascade into global financial markets through interconnected trade networks, investment flows, and supply chains. The unpredictability and increasing intensity of these events make it difficult for investors to assess risk accurately, thereby contributing to heightened volatility, liquidity crunches, and mispricing of assets.

In addition, sudden climate-related disasters can destabilize sovereign creditworthiness, particularly in developing countries heavily reliant on agriculture or natural resources, leading to higher borrowing costs and sovereign debt crises. energy efficiency regulations—companies in carbon-intensive sectors face rising compliance costs, declining profitability, and in some cases, existential threats to their business models. Fossil fuel companies, for example, face stranded assets as reserves that were once considered valuable are rendered unburnable by climate commitments. This phenomenon reverberates across financial markets, affecting not only the equity valuations of energy firms but also the debt held by banks, pension funds, and institutional investors. Transition risks also create competitive imbalances as companies slow to adopt sustainable practices lose market share to innovators in renewable energy, electric mobility, and circular economy solutions. (Charan et al., 2025a)These shifts, while necessary for long-term climate stability, generate short- to medium-term turbulence across sectors and financial instruments, testing the resilience of markets and the adaptability of investors.and even governments may face lawsuits for failing to adequately disclose or manage climate-related risks. Legal actions against fossil fuel firms, corporate boards, and even municipalities have increased, alleging negligence, misinformation, or inadequate risk mitigation strategies. Such legal battles can result in financial penalties, reputational damage, and heightened regulatory scrutiny, all of which affect stock prices, investor confidence, and credit ratings. Liability risks also expand to financial institutions themselves, as stakeholders demand greater accountability for financing activities that exacerbate climate change.

This legal dimension amplifies the uncertainty surrounding climate impacts, as investors must now consider not only physical and transitional disruptions but also potential litigation costs embedded in asset valuations.borders. For example, a climate-induced crop failure in one part of the world can spike global food prices, leading to inflationary pressures, currency fluctuations, and social unrest in multiple regions. Similarly, widespread devaluation of carbon-intensive assets could trigger a financial contagion, akin to the subprime mortgage crisis, if not managed proactively. To mitigate these risks, regulators are embedding climate considerations into stress tests, scenario analyses, and disclosure frameworks. (Charan et al., 2025b)

The Network for Greening the Financial System (NGFS), a coalition of central banks and supervisors, has emphasized that climate risks are "macro-critical" and must be integrated into financial oversight to prevent systemic crises. By pushing for mandatory climate risk disclosures and green taxonomy frameworks, regulators aim to enhance transparency, improve risk pricing, and steer capital flows toward sustainable activities, thereby stabilizing financial markets in the long termClimate change also affects investor sentiment and behavior, which can amplify market volatility. Growing awareness of climate risks has led to the rise of environmental, social, arelated industries, suppliers, and communities. Investor sentiment is also influenced by activist campaigns, shareholder resolutions, and public pressure, all of which can shift market dynamics unpredictably. Financial markets must therefore navigate a complex landscape where climate-driven investor expectations intersect with regulatory requirements and technological disruptions.face heightened risks of sovereign debt distress as climate events erode economic productivity and fiscal capacity

GREEN BONDS AND CLIMATE FINANCE INSTRUMENTS

Among the many strategies to align financial flows with environmental goals, green bonds stand out as one of the most innovative and rapidly expanding markets in sustainable finance, providing governments, corporations, and financial institutions. (Chen et al., 2020)This rapid growth reflects both investor demand for environmentally responsible investments and the increasing recognition that climate risks are material financial risks requiring proactive management. Unlike conventional bonds, green bonds incorporate transparency and accountability mechanisms that ensure proceeds are dedicated to green projects, thereby reinforcing investor confidence and promoting trust in sustainability-linked financial marketOne of the defining features of green bonds is their ability to lower the cost of capital for climate-positive projects, especially in areas where commercial financing might otherwise be limited due to perceived risks, long payback periods, or lower short-term returns. By tapping into the deep pools of global fixed-income markets, issuers can secure financing at favorable rates while signaling their commitment to sustainability. This creates a virtuous decarbonization initiatives, thereby enhancing their competitiveness in a marketplace where sustainability is rapidly becoming a differentiating factor.Beyond green bonds, a diverse array of climate finance instruments has evolved to address the varying risk-return profiles of different investors and the capital sustainability targets, such as carbon intensity reduction or renewable energy adoption, the financial terms of the bond adjust, often through coupon step-ups. This mechanism directly links financial costs to environmental performance, incentivizing corporate accountability. Similarly, blended finance mechanisms combine concessional public finance with private capital, de-risking investments in developing countries and ensuring that climate finance reaches regions most vulnerable to climate change yet often overlooked by commercial investors. Instruments like catastrophe bonds and climate risk insurance schemes also play a complementary role by transferring the financial risks of climate-related disasters to global capital markets, thus enhancing resilience in high-risk regions.particularly attractive because they combine financial returns with positive environmental outcomes. Retail investors, too, are increasingly gaining access to climate finance products through mutual funds, exchange-traded funds (ETFs), and green savings bonds, which democratize participation in sustainable finance. (Demski et al., 2017)This expansion of the investor base enhances liquidity, reduces funding costs, and integrates sustainability considerations into mainstream investment strategies. Moreovesustainability into financial products, these instruments align market incentives with the broader societal goal of climate resilience

.Nevertheless, despite their rapid growth, challenges persist in the green bond and climate finance market. Concerns about greenwashing remain significant, particularly when issuers fail to provide adequate transparency or when proceeds fund projects with marginal environmental benefits. The lack of global harmonization in standards and definitions also creates confusion and hinders comparability, making it difficult for investors to evaluate the true impact of their investments. Additionally, the green bond market remains concentrated in developed economies, with emerging markets and developing countries capturing only challenges requires stronger regulatory oversight, capacity-building efforts, and risk-sharing mechanisms that can unlock greater flows of private capital toward underserved markets.In conclusion, green bonds and climate finance instruments are indispensable components of the global effort to combat climate change, providing scalable pathways to mobilize the trillions of dollars required for the transition to a sustainable, low-carbon economy. (Doelman et al., 2020)They function not only as financing mechanisms but also as signaling tools, fostering accountability, innovation, and trust between issuers and investors. By lowering the cost of capital for green projects, enhancing transparency through robust standards, diversifying financial products, and broadening investor participation, these instruments bridge the gap between environmental imperatives and financial markets. While challenges remain—such as greenwashing, regional imbalances, and standardization gaps—the momentum of green finance is undeniable, and its continued growth will play a decisive role in achieving global climate goals. The evolution of green bonds and climate finance instruments thus reflects a fundamental transformation in global finance: one where environmental sustainability is no longer a niche consideration but a central driver of financial stability, market innovation, and economic prosperity in the climate-conscious era.

SUSTAINABILITY-LINKED FINANCIAL PRODUCTS FOR CLIMATE GOALS

Sustainability-linked financial products for climate goals represent a transformative innovation in global finance, providing a proceeds for designated environmentally friendly activities, sustainability-linked products—such as sustainability-linked bonds (SLBs), sustainability-linked loans (SLLs), and performance-based derivatives—tie the financial and structural characteristics of the instrument directly to the issuer's achievement of predetermined sustainability or climate-related targets. This shift from "use of proceeds" models to "performance-based" structures allows issuers greater flexibility in capital deployment while holding them accountable for holistic improvements in their environmental performance. For example, an issuer may commit to reducing greenhouse gas emissions intensity, increasing renewable energy consumption, improving water efficiency, or achieving specific science-based decarbonization pathways. If these targets are not met, the financial terms of the product, such as coupon payments, interest rates, or repayment conditions, adjust unfavorably, creating tangible financial consequences for non-compliance. Conversely, meeting or exceeding sustainability targets can reduce financing costs, rewarding issuers for their environmental performance. This intrinsic linkage between sustainability outcomes and financial costs makes sustainability-linked products a critical instrument for embedding climate accountability into corporate finance and investment practices, thereby aligning capital flows with long-term climate goalmaterial financial risk requiring proactive management. Since the launch of the first SLB in 2019, issuance volumes have grown into hundreds of billions of dollars, with companies across sectors—, calibration of key performance indicators (KPIs), financial characteristics linked to target achievement, reporting obligations, and verification by independent third parties. (Flori et al., 2021)By setting clear, science-based, and ambitious targets, issuers can demonstrate genuine commitment to climate goals while giving investors the assurance that their capital is incentivizing meaningful change. External verification and impact reporting further strengthen accountability, ensuring that outcomes are measurable, transparent, and comparable across issuers and sectors.

This structured approach helps address the persistent risk of greenwashing, which undermines trust in ESG finance, while simultaneously raising the bar for corporate climate ambition.green finance instruments. Whereas green bonds and similar products restrict proceeds to specific projects—such as wind farms, solar plants, or energy efficiency retrofits—sustainability-linked products allow issuers to apply proceeds across general corporate purposes, as long as they remain accountable for sustainability performance at the organizational level. This flexibility makes them particularly attractive to companies in hard-have a large pool of immediately financeable green projects but still need to transition toward decarbonization. By focusing on overall corporate performance rather than project-level expenditures, sustainability-linked products push entire business models From an investor perspective, sustainability-linked products offer both risk management and opportunity creation. They allow investors to integrate climate risk considerations into portfolios more effectively by ensuring that issuers bear financial consequences for failing to meet climate targets. At the same time, they provide exposure to companies that instruments are particularly valuable because they align with both fiduciary duty and societal responsibility. Furthermore, by linking financing costs to environmental performance, sustainability-linked products create stronger market incentives for climate action than voluntary ESG disclosures or non-financial reporting frameworks. Investors also benefit from increased transparency, as issuers are required to publish regular updates on their progress toward KPIs, allowing for informed decision-making and stewardship practices.sovereign and development finance, where they are being adapted to support national climate goals and sustainable development objectives.

Several countries have begun exploring or issuing sustainability-linked sovereign bonds, linking debt service conditions to progress on emissions reductions, renewable energy deployment, or biodiversity conservation. (Giuzio et al., 2019) This innovation is particularly relevant for developing and emerging economies, which face significant climate vulnerabilities and financing gaps but often lack access to affordable capital. By embedding sustainability-linked mechanisms, sovereign issuers can attract global capital while signaling strong policy alignment with climate objectives. Multilateral development banks and development finance institutions are also integrating sustainability-linked structures into concessional and blended linked instruments but also address global equity concerns by directing finance toward regions that need it most.Despite their potential, sustainability-linked financial products face several challenges that must be addressed to fully realize their role in advancing climate goals. A key concern is the credibility of KPIs and SPTs: if targets are set too conservatively or lack scientific rigor, the instruments risk becoming symbolic rather than transformative. Ensuring that targets are ambitious, measurable, and aligned with the Paris Agreement is critical to avoid accusations of greenwashing. Another challenge lies in the lack of global standardization, as varying definitions, methodologies, and verification practices across jurisdictions create inconsistencies that undermine comparability. Furthermore, the penalty mechanisms for failing to meet sustainability targets are often modest, raising questions about whether they create sufficient incentives for meaningful change. Expanding the range and severity of financial consequences, such as coupon step-ups or credit downgrades, could enhance accountability. (Guru Basava Aradhya et al., 2025) Additionally, small and medium-sized enterprises (SMEs) may struggle to adopt sustainability-linked products due to limited resources for monitoring, reporting, and verification, creating a risk that these instruments remain concentrated among large, multinational corporations. Addressing these issues will require stronger regulatory oversight, harmonized standards, and capacity-building initiatives to broaden accessibility and impact.

CARBON MARKETS AND TRADING FOR CLIMATE MITIGATION

These markets function under two main frameworks: compliance markets, established by governments or international agreements to ensure legal adherence to emissions caps, and voluntary carbon markets, where companies and individuals purchase credits to offset emissions as part of corporate responsibility or net-zero commitments. In compliance systems such as the European Union Emissions Trading System (EU ETS), entities are allocated or must purchase allowances to cover their emissions; those that emit less can sell surplus allowances, while those that exceed their cap must purchase additional units, creating a dynamic incentive structure for efficiency and innovation. In voluntary systems, entities invest in carbon offset projects such as reforestation, renewable energy, soil carbon sequestration, or methane capture, thereby generating tradable credits that can be sold to organizations seeking to neutralize their carbon footprint.(Harmanpreet et al., 2025)By leveraging the efficiency of markets, carbon trading aims to reduce overall emissions at the lowest possible cost, driving capital flows toward low-carbon technologies and climate mitigation projects across the globeThe growth of carbon markets has been remarkable, with compliance markets expanding across multiple regions, including the EU, China, South Korea, California, and other jurisdictions, and voluntary markets scaling as corporations pledge net-zero goals.

The EU ETS, launched in 2005, remains the largest and most mature compliance carbon market, covering sectors such as power generation, heavy industry, and aviation within the European Economic Area. Its success lies in its ability to create a robust price signal for carbon, influencing investment decisions in renewable energy, energy efficiency, and industrial decarbonization. China's national carbon market, launched in 2021 and currently covering the power sector, is poised to become the world's largest by volume as it expands to include other energy-intensive industries. Meanwhile, the voluntary carbon market, central role of carbon markets as a bridge between public policy mandates and private sector climate action.Carbon trading contributes to climate mitigation by embedding economic incentives for emission reductions directly into market operations. When carbon has a price, businesses face a clear financial motivation to cut emissions through efficiency improvements, process innovations, or shifts toward renewable energy sources. For instance, a steel manufacturer facing high carbon costs under an emissions trading system may invest in carbon capture technologies or alternative production methods to remain competitive. Similarly, power utilities might accelerate the retirement of coal plants in favor of renewables or natural gas, as carbon pricing shifts the relative cost advantage toward low-carbon generation. Carbon markets also encourage companies to integrate long-term climate strategies into their financial planning, as the prospect of rising carbon prices increases the cost of inaction. instrument, as it enables ambitious emissions reductions while minimizing economic disruption.

Despite their potential, carbon markets face significant challenges related to credibility, transparency, and equity.(Huang et al., 2021) One persistent concern is greenwashing, where companies may purchase low-cost offsets without meaningfully reducing their own emissions, allowances have sometimes weakened the incentive effect, as seen in the early years of the EU ETS. Furthermore, the uneven distribution of carbon finance highlights equity concerns, as developing countries often struggle to attract investment despite being highly vulnerable to climate impacts. Addressing these challenges requires strong governance, standardized methodologies, and rigorous monitoring, reporting, and verification (MRV) systems to ensure that carbon credits represent real, measurable, and the economy.recognizing that carbon pricing is not only an environmental policy but also a systemic economic variable with far-reaching implications for financial stability.to drive the deep structural changes needed to achieve climate goals.(Raman et al., 2025) To succeed, carbon markets must be embedded in broader policy frameworks, complemented by regulations, subsidies for clean technologies, and investments in adaptation, ensuring that they serve as catalysts rather than substitutes for systemic climate action.robust governance and technological innovation, holds the potential to make carbon trading a cornerstone of global climate action. As countries, companies, and investors converge on the urgent need for decarbonization, carbon markets will not only shape the path toward net-zero emissions but also redefine the relationship between environmental sustainability and economic prosperity in the 21st century.

INVESTMENT STRATEGIES FOR CLIMATE CHANGE MITIGATION

The urgency of climate change has transformed investment from a purely financial exercise into a powerful tool for systemic impact, where institutional investors, asset managers, corporations, governments, and individuals alike can channel capital toward decarbonization and adaptation. Mitigation-oriented investment strategies encompass a spectrum of approaches, ranging from integrating environmental, social, and governance (ESG) considerations into portfolios, to targeted green investments in renewable energy, to the use of innovative financing instruments such as green bonds, sustainability-linked loans, and blended finance structurestargets and net-zero investment frameworks to align portfolios with global climate pathways. (Inoune et al., 2006)This process often involves assessing the climate risk exposure of existing investments, using metrics such as carbon intensity, stranded asset risk, and alignment with a 1.5°C trajectory. By integrating climate scenario analysis and stress testing, investors can better understand how regulatory changes, carbon pricing, and technological shifts could affect asset values, enabling them transparency in emissions reporting, and integrate climate considerations into governance. Collaborative than simply abandoning them. This approach underscores the recognition that systemic decarbonization requires transforming existing industries, not just financing new green ventures.investment strategies. Green bonds, which fund projects such as renewable energy, clean penalties for underperformance. Carbon markets and offset credits also play a role in investment strategies, particularly for investors seeking exposure to projects that directly reduce or remove emissions, such as afforestation or direct air capture. Blended finance, which combines public and private capital to de-risk investments in emerging markets, is increasingly used to scale climate finance to regions most in need of energy transitions. These instruments broaden the toolkit available to investors, allowing climate mitigation strategies to be tailored across risk profiles, geographies, and project types.

Figure 1. Analytical framework to evaluate the level of integration of climate adaptation and mitigation

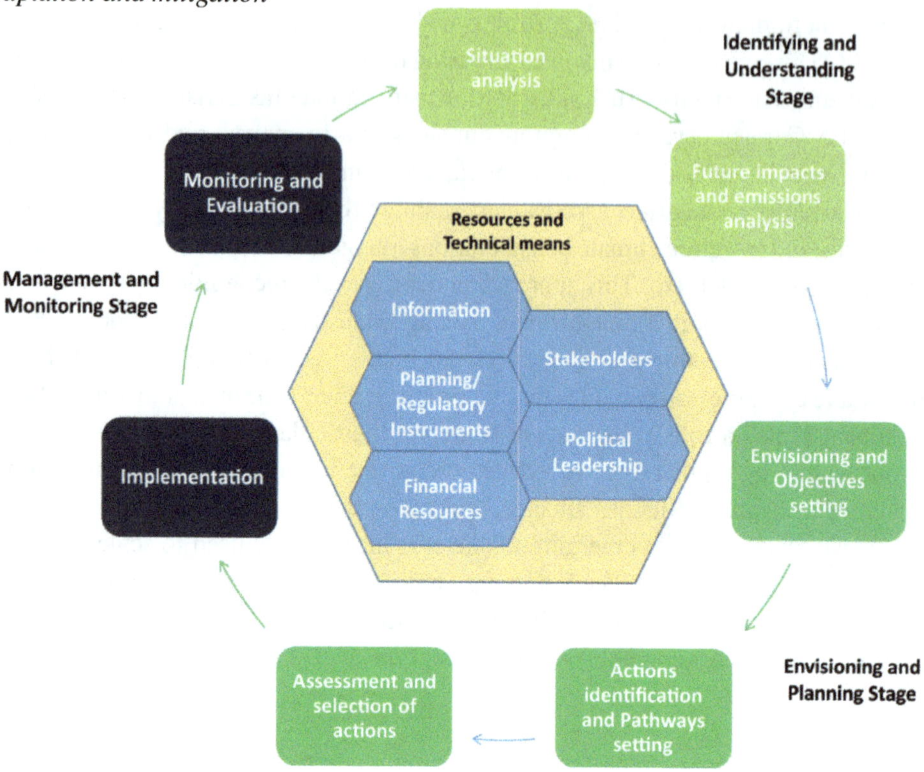

Liability risks relate to legal actions against companies or investors failing to manage climate risks responsibly. By adopting strategies such as climate stress testing, scenario planning, and resilience investing, investors can safeguard portfolios while contributing to mitigation. For example, infrastructure investors increasingly prioritize assets designed to withstand extreme weather, while real estate funds assess exposure to sea-level rise and flood risk. Integrating these risk factors into decision-making not only reduces vulnerability but also channels capital into adaptive and resilient projects that mitigate future emissions indirectly by reducing systemic vulnerabilities.The success of investment strategies for climate change mitigation also depends on policy alignment and international collaboration. Clear policy signals such as carbon pricing, renewable energy subsidies, and emissions regulations provide the regulatory aligning trillions in assets with net-zero pathways. These alliances also highlight the need for harmonized disclosure standards, robust climate data, and transparent reporting mechanisms to build trust and accountability in climate investment strategies.(Iqbal & Bilal, 2021a)Looking ahead, the evolution of investment strategies will be shaped by technological innovation, evolving consumer demand, and regulatory frameworks. Breakthroughs in green hydrogen, battery storage, artificial intelligence for climate modeling, and carbon removal technologies will create new opportunities for investors to contribute to mitigationdivestment, reinvestment, stewardship, innovation, and risk management, reflecting a holistic approach that balances returns with responsibility.investors not only safeguard portfolios but also accelerate global decarbonization. These strategies are no longer niche but mainstream, with institutional investors, corporations, and individuals recognizing their role in shaping climate outcomes. While challenges remain in ensuring transparency, scalability, and equitable access to climate finance, the trajectory is clear: investment strategies will be indispensable in bridging the multi-trillion-dollar gap required for climate mitigation. As finance continues to align with climate goals, it has the power not only to mitigate climate change but also to redefine economic growth and resilience in the 21st century.

RISK MANAGEMENT FOR CLIMATE-RELATED FINANCIAL RISKS

Transition risks emerge from the global shift toward a low-carbon economy, encompassing policy changes such as carbon taxes, stricter emissions regulations, and fossil fuel phase-outs, alongside technological disruptions from renewable energy, electric vehicles, and energy efficiency innovations. Liability risks, meanwhile, arise from litigation or reputational damage faced by companies and financial institutions accused of contributing to climate change or failing to disclose and manage their climate exposures adequately. Together, these risks pose complex, interdependent challenges that require financial institutions, investors, corporations, and regulators to adopt comprehensive risk management frameworks to safeguard financial stability while supporting climate mitigation and adaptation.emissions—into their risk assessments, institutions can move beyondhow carbon-intensive assets may lose value under stricter regulation, or how renewable Mitigation and adaptation strategies form the next stage of risk management. Mitigation strategies often involve reducing exposure to high-risk assets, such as divesting from coal or oil, limiting lending to carbon-intensive sectors, or applying enhanced due diligence for vulnerable projects. Adaptation strategies focus on building resilience within portfolios and business operations—for instance, by financing climate-resilient infrastructure, supporting supply chain diversification, or investing in insurance products that transfer climate risks. Many financial institutions are also deploying climate risk hedging instruments, such as catastrophe bonds, weather derivatives, and insurance-linked securities, which provide financial protection against extreme weather events. These instruments are gaining importance as traditional insurance becomes less viable in regions facing repeated climate disasters.(Iqbal & Bilal, 2021b) By combining mitigation, adaptation, and risk transfer mechanisms, institutions create a diversified defense against climate-related shocks.

Risk management also requires data and technology innovation, as traditional financial models are often inadequate for capturing the non-linear and long-term nature of climate impacts. Advances in climate science, satellite monitoring, geospatial analytics, and artificial intelligence are helping institutions map climate vulnerabilities at asset and portfolio levels with increasing precision. For example, AI-driven platforms can predict flood risk for real estate investments, while geospatial analytics can assess exposure of supply chains to drought-prone regions. Integrating such tools into financial decision-making allows more accurate pricing of climate risks and better allocation of capital. At the same time, partnerships between financial institutions, governments, and academia are expanding the availability of open-source climate data and methodologies, fostering greater consistency in risk assessment.(Sivanand et al., 2025)Governance and capacity building are equally important to effective climate risk management. Boards of directors, risk committees, and senior executives must have the expertise to oversee climate-related risks, requiring training, climate literacy programs, and integration of climate expertise into leadership structures.(Iqbal et al., 2021a) Financial regulators are also playing a proactive role, issuing supervisory guidance on climate risks and integrating climate considerations into prudential frameworks.

Importantly, climate-related risk management must address the issue of systemic risk and interconnectedness. Climate shocks can cascade through markets, supply chains, and economies, amplifying impacts across sectors. For example, a climate disaster damaging agricultural yields may disrupt food supply chains, raise commodity prices, increase inflationary pressures, and ultimately affect consumer spending and corporate profitability. Similarly, a disorderly energy transition could trigger sharp declines in fossil fuel asset valuations, stranding trillions of dollars and causing losses for banks, pension funds, and investors globally. Recognizing these systemic dimensions, financial institutions are increasingly participating in collaborative initiatives, such as GFANZ and the UNEP Finance Initiative, to develop common risk management practices, mpacts of climate transition are essential for robust risk management. Furthermore, the concept of a just transition—which ensures that climate policies and risk management strategies protect vulnerable communities and workers—must be embedded into financial decision-making to avoid exacerbating inequality. As climate-related risks continue to grow in frequency and magnitude, the financial sector's ability to anticipate, assess, and mitigate them will determine not only portfolio resilience but also broader economic and social stability.In conclusion, risk management for climate-related financial risks represents a paradigm shift in finance, moving beyond short-term profit maximization toward safeguarding long-term stability in the face of unprecedented systemic challenges. (Kim et al., 2020)By identifying, assessing, integrating, and mitigating risks, while embracing disclosure, technology, governance, and systemic collaboration, financial institutions can build resilience and enable a smoother transition to a sustainable economy. Climate risk management is no longer optional; it is a fiduciary and regulatory imperative that defines the future of finance. The effectiveness with which institutions manage these risks will shape their competitiveness, resilience, and contribution to global climate goals, making climate risk management a defining feature of financial systems in the 21st century.

INTEGRATING ESG CRITERIA IN FINANCIAL DECISION-MAKING FOR CLIMATE

At its core, ESG integration in finance represents a systematic approach to embedding sustainability considerations into investment analysis, portfolio management, lending practices, and corporate strategy, with a particular focus on mitigating climate-related risks and accelerating the low-carbon transition. By integrating environmental criteria, financiers can redirect capital away from carbon-intensive activities toward greener alternatives, incentivizing innovation in renewable energy, sustainable agriculture, circular economy practices, and climate-resilient infrastructure. For example, ESG integration encourages investors to favor companies with credible net-zero strategies, science-based emissions reduction targets, and transparent climate disclosures. correlates with inadequate risk management, greenwashing, or failure to adapt to regulatory and market shifts. Conversely, companies with strong governance structures demonstrate higher resilience and are better positioned to attract long-term investors seeking alignment with climate goals.assets to best-in-class selection, ESG integration in fundamental analysis, thematic investing, and impact investing. Negative screening excludes investments in sectors such as coal, tar sands, or deforestation-linked activities. (Kopp et al., 2017)Best-in-class approaches reward companies demonstrating superior ESG performance within their sector, while thematic investing directs capital toward climate-aligned themes like renewable energy, clean transportation, or green buildings. Impact investing goes further by intentionally targeting measurable positive environmental and social outcomes alongside financial returns, thereby channeling funds into climate solutions that deliver tangible results. Increasingly, ESG integration also relies on quantitative tools such as ESG scores, ratings, and climate value-at-risk (VaR) assessments, though investors must critically evaluate data quality and consistency, as ESG metrics often vary across providers.

Technology and data innovation are accelerating ESG integration by providing more granular, reliable, and forward-ltraceability, and climate scenario modeling enhance the accuracy of environmental assessments, enabling financial institutions to identify risks and opportunities with greater precision. (Lasco et al., 2014)These tools also support accountability by verifying corporate sustainability claims and reducing reliance on inconsistent ESG ratings. By harnessing advanced analytics, financial institutions can integrate ESG factors more systematically and align portfolios with climate transition pathways.A key element of ESG integration in financial decision-making is its role in driving the reallocation of capital toward climate-positive outcomes. As investors increasingly demand ESG-aligned portfolios, companies face pressure to decarbonize, disclose transparently, and innovate sustainably to attract capital. This reallocation of capital not only reduces systemiwhile those with poor ESG performance face higher capital costs or exclusion from investment universes. This creates a powerful incentive structure that aligns financial performance with climate responsibility.Despite its growing momentum, ESG integration faces challenges and criticisms that must be addressed for it to effectively contribute to climate action. Inconsistent ESG data, varying methodologies among rating agencies, risks of greenwashing, and the difficulty metrics, stronger regulatory oversight, and independent verification of ESG claims are essential. Furthermore, financial institutions must ensure that ESG integration goes beyond box-ticking and is embedded into core strategy, governance, and culture.Looking ahead, the integration of ESG criteria into financial decision-making for climate will continue to evolve, becoming more sophisticated, data-driven, and globally harmonized. (Trinh et al., 2020)The convergence of disclosure standards, the rise of climate stress testing, and increasing stakeholder pressure from regulators, investors, and civil society will deepen the integration of ESG into financial systems. Moreover, as climate change intensifies, ESG integration will increasingly focus on transition planning, resilience, and just transition considerations, ensuring that climate finance supports equitable and sustainable outcomes. Ultimately, ESG integration is not merely about avoiding risks; it is about seizing the opportunity to reorient global finance toward long-term value creation that benefits investors, companies, communities, and the planet.

REGULATORY FRAMEWORKS FOR CLIMATE FINANCE IN FINANCIAL MARKETS

Regulatory frameworks for climate finance in financial markets represent one of the most critical enablers for aligning global capital flows with the urgent goals climate mitigation and adaptation. As climate change intensifies systemic risks to economies and societies, regulators across the world are increasingly recognizing the need to embed sustainability considerations into the very fabric of financial governance. (Mi et al., 2019)These frameworks are designed not only to standardize disclosure and reporting on climate-related risks but also to guide investment flows toward green and sustainable activities, mitigate greenwashing, ensure financial stability, and enhance accountability across the financial system. At their core, regulatory frameworks for climate finance create the rules, incentives, and guardrails that determine how capital markets respond to the climate challenge—whether by reorienting private investment away from high-carbon sectors, mobilizing funds for renewable energy and infrastructure, or ensuring that climate-related risks are transparently assessed and priced into financial decisions. (Mishra et al., 2025)However, the design, implementation, and enforcement of these frameworks are complex, evolving processes that vary significantly across jurisdictions, often reflecting differences in economic priorities, political will, and institutional capacity.sustainability reporting was voluntary, leading to inconsistencies, selective disclosures, and difficulties for investors in evaluating climate risks and opportunities. To address this, regulators have introduced binding frameworks such as the EU Sustainable Finance Disclosure Regulation (SFDR), which requires asset managers and financial market participants to disclose how sustainability risks are integrated into their investment decisions and products. Complementing this is the EU Taxonomy Regulation, a classification system that defines what constitutes a sustainable economic activity, providing clarity to investors and preventing greenwashing.

Similarly, the Corporate Sustainability Reporting Directive (CSRD) in Europe expands mandatory climate-related reporting to a wider set of companies, requiring granular information on sustainability performance, including emissions data and transition plans. Outside the EU, the U.S. Securities and Exchange Commission (SEC) has proposed climate disclosure rules mandating credible role in the low-carbon transition.jurisdictions such as the EU require funds marketed as "sustainable" or "ESG" to disclose detailed information about their investment strategies and sustainability objectives, reducing the risk of misleading claims. (Nirupama et al., 2025)These product-level regulations ensure that climate finance instruments deliver measurable environmental outcomes and maintain investor trust, thereby scaling up flows of capital to climate-positive activities.Despite significant progress, regulatory frameworks for climate finance face ongoing challenges and limitations. A major issue is the fragmentation of standards across jurisdictions, which creates confusion and inefficiencies for global investors managing cross-border portfolios. While harmonization efforts are underway, significant differences remain between the EU, U.S., and Asian approaches to disclosure, taxonomy design, and product labeling. Another challenge is the evolving nature of climate science and economic modeling, which introduces uncertainty into regulatory criteria and stress testing methodologies. Regulators must strike a balance between providing clarity and allowing flexibility as new information and technologies emerge. Moreover, regulatory requirements can impose substantial compliance costs on smaller firms or financial institutions in emerging markets, potentially creating barriers to entry and reducing inclusivity in sustainable finance. (Neuhoff et al., 2009)There is also the risk that stringent regulations in advanced economies could inadvertently reduce capital flows to developing countries, which are most in need of climate finance for adaptation and mitigation but face higher perceived risks.

Furthermore, regulatory frameworks must navigate political and economic resistance, particularly from sectors with vested interests in fossil fuels and carbon-intensive industries. The design of taxonomies, for instance, has sparked controversy over whether to include activities such as natural gas or nuclear power as "green." (Ogunbode et al., 2019)These debates reflect broader tensions between scientific imperatives for climate neutrality and political compromises to ensure energy security or industrial competitiveness. Such conflicts can undermine the credibility of regulatory frameworks if they allow loopholes that dilute their climate integrity. Additionally, enforcement remains a challenge, as regulators must ensure that companies and financial institutions not only disclose climate-related information but also act on it in a meaningful way. Without strong monitoring and penalties for non-compliance, regulatory frameworks risk becoming procedural exercises rather than transformative drivers of climate action. mandatory disclosures, prudential oversight, sustainable taxonomies, product-level standards, and international coordination, these frameworks provide the transparency, accountability, and credibility needed to scale up climate finance. Yet their effectiveness depends on overcoming fragmentation, ensuring robust enforcement, maintaining scientific integrity, and addressing equity concerns to ensure that climate finance reaches vulnerable regions and communities. As the urgency of climate action intensifies, regulatory frameworks must continue to evolve in scope, ambition, and global coherence, transforming financial markets into powerful enablers rather than obstacles in the fight against climate change.

IMPACT OF FINANCIAL MARKETS ON CLIMATE CHANGE OUTCOMES

In practice, financial markets impact climate change outcomes through mechanisms such as pricing of climate-related risks, creation and expansion of green financial products, integration of environmental, social, and governance (ESG) criteria into investment strategies, and enabling or constraining financing for low-carbon innovation.(Park & Gupta, 2012)The extent of this impact, however, depends heavily on transparency, regulatory alignment, investor priorities, and the ability of financial systems to balance short-term returns with long-term sustainability.One of the clearest ways financial markets affect climate outcomes is through the allocation of capital between high-carbon and low-carbon industries. When investors channel funds into renewable energy, sustainable infrastructure, electric mobility, or energy efficiency projects, they create conditions for rapid technological diffusion, cost reductions, and scalability of climate-friendly solutions. For example, falling costs of wind and solar energy were facilitated not only by technological innovation but also by massive flows of private equity, venture capital, and project finance into the sector. As such, the reallocation of global capital toward sustainable industries is not just desirable but essential for aligning financial flows with the Paris Agreement's goal of limiting warming to well below 2°C.infrastructure from extreme weather, and transition risks, such as policy changes, stranded assets, or shifting consumer preferences. By incorporating these risks into valuations, credit ratings, and insurance premiums, financial markets can signal to companies the cost of inaction and reward those that proactively manage their climate exposures. For example, companies failing to disclose emissions or transition plans may face higher borrowing costs or lower stock valuations, while firms that adopt credible net-zero strategies may enjoy investor preference and cheaper access to capital.(Phan et al., 2021).

This risk-pricing mechanism incentivizes corporate behavior that aligns with climate objectives and penalizes unsustainable practices. However, the degree to which markets internalize climate risks remains inconsistent, as many risks are still mispriced or underestimated due to inadequate data, short-termism, or political resistance. Strengthening climate risk disclosure frameworks and enhancing investor awareness is therefore critical to ensuring that financial markets reflect the true costs of climate inaction. These instruments provide investors with structured pathways to support the low-carbon transition while meeting their financial objectives. For instance, the issuance of green bonds by governments and corporations has enabled financing of renewable energy plants, energy-efficient housing, and sustainable transportation networks. Sustainability-linked loans incentivize companies to meet emission-reduction targets by tying interest rates to sustainability performance. In addition, emerging voluntary and compliance carbon markets allow companies and investors to offset emissions while supporting climate mitigation projects such as afforestation or clean cookstove distribution.(Windmeijer, 2005) By mainstreaming such instruments, financial markets create direct pathways nds allow investors to assume climate risk exposure in exchange for high returns, providing funding for post-disaster recovery. Similarly, adaptation finance directed through capital markets supports developing economies in addressing vulnerabilities, bridging the significant adaptation finance gap. The extent to which financial markets mobilize adaptation capital will determine not only economic resilience but also global equity in climate outcomes.Yet the relationship between financial markets and climate outcomes is not purely positive. Systemic challenges and contradictions often undermine the effectiveness of financial markets in driving climate progress. Short-term profit motives may incentivize continued investment in fossil fuels, even as long-term risks grow. Greenwashing—where financial products are marketed as sustainable without substantive impact—erodes investor trust and diverts capital from genuine climate solutions. Market fragmentation, with varying standards and definitions of "green" across jurisdictions, hampers comparability and creates inefficiencies. Additionally, global imbalances mean that much of climate finance raised through markets is concentrated in advanced economies, leaving developing countries underfunded despite facing the greatest climate risks. These challenges highlight that while financial markets are powerful drivers of climate outcomes, they require strong regulatory frameworks, global coordination, and alignment with public policy to achieve their full potential.

ALIGNING FINANCIAL FLOWS WITH CLIMATE GOALS

This alignment requires a fundamental transformation in how capital is allocated, shifting from high-carbon activities and unsustainable development pathways toward green, sustainable, and resilient investments. The task is immense, as trillions of dollars are currently locked in fossil fuel infrastructure, unsustainable land use, and industries with high environmental footprints, which if left unaddressed will lock societies into carbon-intensive trajectories for decades. Despite the growing momentum of green finance, fossil fuel industries continue to attract substantial financial support through bank lending, insurance underwriting, and government subsidies. As a result, the persistence of carbon-heavy investments undermines climate action and risks creating stranded assets—investments that lose value as regulations tighten and clean technologies become more competitive. (Yin, 2019)Financial flows must therefore be deliberately reoriented to prevent new investments in coal, oil, and gas infrastructure and to accelerate the decommissioning of existing high-emission assets. .Aligning financial flows also depends heavily on data, metrics, and taxonomies that provide clarity on what constitutes "green" or "climate-aligned" investment. Without standardized definitions, there is a risk of greenwashing—where funds are labeled as sustainable regulators distinguish between activities that genuinely advance climate goals and those that do not. Similarly, climate-related disclosure frameworks like the Task Force on Climate-Related Financial Disclosures (TCFD) and emerging climate stress tests provide transparency, enabling investors to understand climate risks and opportunities in portfolios. Reliable data and transparent disclosure are thus prerequisites for aligning financial flows, as they build trust, reduce uncertainty, and guide informed decision-making.Another key dimension is integrating climate risk into financial decision-making.

Climate risks—both physical (from extreme weather events, sea-level rise, and shifting climate patterns) and transition-related (arising from policy, technology, and market changes)—pose systemic threats to financial stability. If not properly priced, these risks can lead to market volatility, credit defaults, and cascading economic losses. By embedding climate risk assessments into investment strategies, credit ratings, and lending practices, financial institutions can protect portfolios while aligning with long-term climate goals. Central banks and financial regulators are also stepping in, with initiatives such as the Network for Greening the Financial System (NGFS), which promotes the integration of climate risks into supervision and monetary policy. By requiring financial actors to account for climate risks, regulators ensure that financial flows are not only mobilized for green investments but also safeguarded from unsustainable ones.At the same time, aligning financial flows requires long-term horizon planning. Climate change is a multi-decade challenge, and aligning finance must extend beyond short-term returns to consider intergenerational impacts. Pension funds, sovereign wealth funds, and long-term institutional investors are particularly well-placed to play a leading role given their ability to allocate capital over decades.(Zhao et al., 2021) By embedding climate considerations into their fiduciary responsibilities, these investors can steer trillions of dollars toward sustainable pathways. This long-term alignment also helps prevent sudden shocks from stranded assets, as portfolios gradually transition toward climate-consistent holdings rather than reacting abruptly to regulatory or market shifts.However, aligning financial flows faces significant challenges and barriers. Political resistance, entrenched fossil fuel interests, short-term profit imperatives, and uneven regulatory frameworks often slow progress. Greenwashing remains a persistent concern, with some financial products marketed as sustainable without delivering real impact. Developing economies also struggle with currency volatility, weak financial infrastructure, and limited investor confidence, which raise the cost of capital for climate projects. Overcoming these barriers requires stronger international cooperation, robust regulatory enforcement, and continuous innovation in financial instruments and mechanisms.In conclusion, aligning financial flows with climate goals is not merely a financial exercise but a structural transformation of the global economy. It requires redirecting capital away from harmful activities, scaling up green investments, embedding climate risks into financial decisions, and ensuring that all flows—public and private, domestic and international—contribute to sustainable, low-emission, and climate-resilient development. The process demands transparency, accountability, equity, and long-term vision, supported by strong regulatory frameworks, innovative financial instruments, and global cooperation. The extent to which financial flows are aligned with climate goals in the coming decade will determine not only the success of the Paris Agreement but also the stability, resilience, and sustainability of the global financial system and society at large. Aligning finance is therefore both a moral imperative and an economic necessity, ensuring that global capital works as a driver of climate solutions rather than a perpetuator of climate risks.

EMERGING TRENDS IN CLIMATE FINANCE AND FINANCIAL MARKETS

Emerging trends in climate finance and financial markets reflect the rapidly evolving recognition that addressing climate change is not only an environmental imperative but also an economic and financial necessity. The financial sector is undergoing a structural transformation as capital allocation, investment strategies, and risk management practices increasingly integrate climate-related considerations. Over the past decade, climate finance has expanded from being a niche area of sustainable finance into a central pillar of global financial markets, driven by international agreements such as the Paris Agreement, growing investor demand for responsible investment, regulatory initiatives, and the clear materialization of physical and transition risks. Several powerful trends are shaping the future of climate finance, including the mainstreaming of green financial instruments, the rise of sustainability-linked products, the expansion of carbon markets, the integration of climate risk into financial stability frameworks, the digitalization of climate finance through fintech and blockchain solutions, and the growing emphasis on equity and climate justice in capital flows. These trends illustrate that climate finance is no longer optional or peripheral but is becoming embedded in the global financial architecture, reshaping how capital is raised, invested, and monitored to accelerate the transition toward a low-carbon, resilient global economOne of the most visible trends is the mainstreaming of green bonds and sustainable debt instruments.(Warren et al., 2013).A fourth trend is the emergence of sustainability-linked and performance-based financial products that go beyond use-of-proceeds financing. Unlike green bonds, which finance specific projects, sustainability-linked instruments tie financing terms—such as interest rates—to issuers' achievement of pre-defined sustainability targets, such as reducing emissions or increasing renewable energy share.

This approach embeds accountability into financing structures and incentivizes systemic change across entire organizations rather than just isolated projects. It also broadens the scope of climate finance to sectors and companies that may not have large green capital expenditure pipelines but are committed to transformation.The equity and climate justice dimension of climate finance is also gaining prominence as a key trend. Developed economies are under pressure to fulfill their commitments to mobilize $100 billion annually in climate finance for developing economies and to support adaptation in vulnerable countries.(Viscusi & Zeckhauser, 2006) Beyond sheer financial volume, there is a push to lower the cost of capital for developing countries through blended finance, concessional loans, and innovative guarantees. Mechanisms such as climate risk insurance, resilience bonds, and local currency financing are being explored to make climate finance more inclusive and equitable. This reflects an emerging recognition that without addressing global disparities, climate finance will be insufficient to deliver a just and effective transition. Finally, there is a strong trend toward global regulatory harmonization and standardization. With numerous overlapping frameworks and disclosure standards creating confusion, efforts are underway to streamline and harmonize rules. The ISSB is creating global standards for sustainability reporting, while taxonomies such as the EU Green Taxonomy and China's Green Bond Catalogue are increasingly being referenced as models for classification. This convergence is critical to prevent fragmentation, reduce greenwashing, and create a level playing field that allows capital to flow more efficiently across borders toward climate goals.In conclusion, the emerging trends in climate finance and financial markets reveal a sector undergoing profound transformation, with climate considerations moving from the margins to the mainstream of global finance. (VijayaVenkataRaman et al., 2012)The expansion of green instruments, the rise of performance-based finance, the integration of climate risk into regulation, the digitalization of finance, the growing focus on biodiversity and equity, the mobilization of private capital, and the harmonization of global standards are collectively reshaping the financial landscape.

CONCLUSION

Harnessing financial markets for climate change solutions represents one of the most powerful pathways to align global economic activity with the urgent goals of climate mitigation and adaptation. Financial markets are not only mechanisms for channeling capital but also key arenas where incentives, risks, and returns are shaped in ways that can determine whether economies continue on high-emission pathways or accelerate toward sustainability. By mobilizing vast pools of private and institutional capital, fostering innovation in climate finance instruments, integrating environmental, social, and governance (ESG) principles, and building transparent accountability frameworks, financial markets can provide the scale and speed of financing needed to meet the objectives of the Paris Agreement and net-zero targets. However, realizing this potential requires a deliberate strategy of harnessing financial markets as engines of transformation—redirecting flows from carbon-intensive sectors toward green and resilient investments, supporting climate-vulnerable regions, and embedding climate considerations into the very structure of financial decision-making.A core way to harness financial markets for climate solutions is through the mobilization of capital at scale. Climate-related financing needs are estimated in the trillions of dollars annually, far exceeding the capacity of public finance alone. Financial markets, with their capacity to aggregate savings from households, pension funds, sovereign wealth funds, and private investors, can bridge this gap if directed appropriately. Instruments such as green bonds, sustainability-linked loans, and transition finance products are already demonstrating the ability of markets to raise significant capital for renewable energy, sustainable infrastructure, and low-carbon technologies. By channeling capital through market-based mechanisms, projects such as solar farms, wind parks, electric vehicle infrastructure, and climate-resilient agriculture become bankable and investable, creating a multiplier effect that accelerates decarbonization.

Furthermore, harnessing financial markets requires embedding long-term sustainability into investment strategies. Traditional finance often prioritizes short-term returns, which can be misaligned with climate goals that require decades of commitment. By adopting strategies such as impact investing, sustainable index funds, and climate-aligned benchmarks, markets can shift the focus toward long-term value creation. Asset managers and investors are beginning to recognize that ignoring climate risks can lead to stranded assets, regulatory penalties, and reputational damage. Conversely, early movers in clean technologies, circular economy models, and nature-based solutions are well-positioned to deliver superior returns over time. Thus, harnessing financial markets is as much about changing investment mindsets as it is about introducing new products.Finally, the effectiveness of harnessing financial markets depends on building trust and transparency. Greenwashing—where companies exaggerate or misrepresent their sustainability efforts—remains a significant challenge. Transparent reporting, rigorous verification, and third-party certification are critical to ensuring that climate finance genuinely delivers climate outcomes. By embedding accountability mechanisms, markets can build confidence among investors, governments, and citizens, creating a virtuous cycle where trust leads to increased investment, which in turn accelerates climate action.In conclusion, financial markets hold transformative potential in addressing climate change if harnessed strategically. They can mobilize trillions in capital, integrate climate risks into pricing, drive innovation in financial instruments, expand inclusivity, foster public–private partnerships, and embed sustainability into long-term investment strategies. The challenge lies not in the capacity of financial markets—they are already vast and sophisticated—but in aligning their mechanisms with the urgency of climate goals. By leveraging technology, strengthening regulatory frameworks, and promoting transparency, financial markets can become engines of climate solutions rather than drivers of environmental degradation. In this way, they can serve as both the foundation and the accelerator of a global transition toward a sustainable, resilient, and low-carbon future.

REFERENCES

Alan, S., Attanasio, O., & Browning, M. (2009). Estimating Euler equations with noisy data: Two exact GMM estimators. *Journal of Applied Econometrics*, *24*(2), 309–324. DOI: 10.1002/jae.1037

Alemzero, D. A., Iqbal, N., Iqbal, S., Mohsin, M., Chukwuma, N. J., & Shah, B. A. (2021). Assessing the perceived impact of exploration and production of hydrocarbons on households perspective of environmental regulation in Ghana. *Environmental Science and Pollution Research International*, *28*(5), 5359–5371. DOI: 10.1007/s11356-020-10880-3 PMID: 32960443

Barbi, F., & da Costa Ferreira, L. (2017). Governing climate change risks: Subnational climate policies in Brazil. *Chinese Political Science Review*, *2*(2), 237–252. DOI: 10.1007/s41111-017-0061-3

Charan, J. K., Chandra, S., Vaissnave, V., Ragupathi, T., Satyanarayana, P., Selvakumar, P., & Manjunath, T. C. (2025). Impact of Generative AI on Industries and Organizations. In Zhao, J., Kumar, V., Katina, P., & Richards, J. (Eds.), *Humans and Generative AI Tools for Collaborative Intelligence* (pp. 509–528). IGI Global Scientific Publishing., DOI: 10.4018/979-8-3693-8332-2.ch022

Charan, J. K., Meera, K. L., Sudheer, P., Gopalakrishnan, G., Sharma, M., Selvakumar, P., & Manjunath, T. C. (2025). Applications of Generative AI and Human-AI Collaboration. In Zhao, J., Kumar, V., Katina, P., & Richards, J. (Eds.), *Humans and Generative AI Tools for Collaborative Intelligence* (pp. 423–440). IGI Global Scientific Publishing., DOI: 10.4018/979-8-3693-8332-2.ch018

Chen, A. A., Stephens, A. J., Koon, R. K., Ashtine, M., & Koon, K. M. K. (2020). Pathways to climate change mitigation and stable energy by 100% renewable for a small island: Jamaica as an example. *Renewable & Sustainable Energy Reviews*, *121*, 109671. DOI: 10.1016/j.rser.2019.109671

Demski, C., Capstick, S., Pidgeon, N., Sposato, R. G., & Spence, A. (2017). Experience of extreme weather affects climate change mitigation and adaptation responses. *Climatic Change*, *140*(2), 149–164. DOI: 10.1007/s10584-016-1837-4 PMID: 32355377

Doelman, J. C., Stehfest, E., van Vuuren, D. P., Tabeau, A., Hof, A. F., Braakhekke, M. C., Gernaat, D. E. H. J., van den Berg, M., van Zeist, W.-J., Daioglou, V., van Meijl, H., & Lucas, P. L. (2020). Afforestation for climate change mitigation: Potentials, risks and trade-offs. *Global Change Biology*, *26*(3), 1576–1591. DOI: 10.1111/gcb.14887 PMID: 31655005

Flori, A., Pammolli, F., & Spelta, A. (2021). Commodity prices co-movements and financial stability: A multidimensional visibility nexus with climate conditions. *Journal of Financial Stability*, *54*, 100876. DOI: 10.1016/j.jfs.2021.100876

Giuzio M, Krušec D, Levels A, Melo AS, Mikkonen K, Radulova P (2019) Climate change and financial stability. Financ Stabil Rev 1

Guru Basava Aradhya, S., Chithambar Gupta, V., Selvakumar, P., Raman, M. S., Singh, J., Manjunath, T. C., & Sharma, M. (2025). Health Crises and Digitalization: Digital Health. In Yıldırım, D., Yıldırım, S., & Kandpal, V. (Eds.), *Sustainable Digitalization Strategies in Business and Healthcare* (pp. 171–198). IGI Global Scientific Publishing., DOI: 10.4018/979-8-3373-5087-5.ch007

Harmanpreet, S. P., Chandra, S., Kalshetti, P., Satyanarayana, P., Sharma, M., & Manjunath, T. C. (2025). Digital Transformation in the Public Sector Entities. In A. Santos Ferreira & C. Lourenço dos Santos (Eds.), Enhancing Public Sector Accountability and Services Through Digital Innovation (pp. 353-380). IGI Global Scientific Publishing. DOI: 10.4018/979-8-3693-9251-5.ch011

Huang, J., Wang, X., Liu, H., & Iqbal, S. (2021). Financial consideration of energy and environmental nexus with energy poverty: Promoting financial development in G7 economies. *Frontiers in Energy Research*, *9*, 666. DOI: 10.3389/fenrg.2021.777796

Inoune, M., Iwasaki, M., Otani, T., Sasazuki, S., Noda, M., & Tsugane, S.Japan Public Health Center–Based Prospective Study Group. (2006). Diabetes mellitus and the risk of cancer: Results from a large-scale population-based cohort study in Japan. *Archives of Internal Medicine*, *166*(17), 1871–1877. DOI: 10.1001/archinte.166.17.1871 PMID: 17000944

Iqbal, S., & Bilal, A. R. (2021). Investment performance: Emotional beasts are dragging into the darkness of the castle. *Global Business Review*, •••, 09721509211044309.

Iqbal S, Bilal AR (2021) Energy financing in COVID-19: how public supports can benefit?. China Finance Rev Int

Iqbal, S., Bilal, A. R., Nurunnabi, M., Iqbal, W., Alfakhri, Y., & Iqbal, N. (2021a). It is time to control the worst: Testing COVID-19 outbreak, energy consumption and CO 2 emission. *Environmental Science and Pollution Research International*, *28*(15), 19008–19020. DOI: 10.1007/s11356-020-11462-z PMID: 33184786

Kim, H., Batten, J. A., & Ryu, D. (2020). Financial crisis, bank diversification, and financial stability: OECD countries. *International Review of Economics & Finance*, *65*, 94–104. DOI: 10.1016/j.iref.2019.08.009

Kopp, E., Kaffenberger, L., & Wilson, C. (2017). *Cyber risk, market failures, and financial stability*. International Monetary Fund. DOI: 10.5089/9781484313787.001

Lasco, R. D., Delfino, R. J. P., & Espaldon, M. L. O. (2014). Agroforestry systems: Helping smallholders adapt to climate risks while mitigating climate change. *Wiley Interdisciplinary Reviews: Climate Change*, *5*(6), 825–833. DOI: 10.1002/wcc.301

Mi, Z., Guan, D., Liu, Z., Liu, J., Viguié, V., Fromer, N., & Wang, Y. (2019). Cities: The core of climate change mitigation. *Journal of Cleaner Production*, *207*, 582–589. DOI: 10.1016/j.jclepro.2018.10.034

Mishra, B. R., Selvakumar, P., Kaur, R., Kalshetti, P., Mishra, K. K., Sharma, M., & Manjunath, T. C. (2025). Innovation and Digital Transformation in the Public Sector Entities. In Santos Ferreira, A., & Lourenço dos Santos, C. (Eds.), *Enhacing Public Sector Accountability and Services Through Digital Innovation* (pp. 409–436). IGI Global Scientific Publishing., DOI: 10.4018/979-8-3693-9251-5.ch013

Neuhoff, K., Fankhauser, S., Guerin, E., Hourcade, J. C., Jackson, H., Rajan, R., & Ward, J. (2009). *Structuring international financial support to support domestic climate change mitigation in developing countries*. Climate Strategies.

Nirupama, E., Seenivasan, S., Bhardwaj, K., Karunakaran, N. B., Basava Aradhya, S. G., Selvakumar, P., & Manjunath, T. C. (2025). Transforming Digital Supply Chains With Gamification and Metaverse Integration. In Malik, G., Singh, D., & Bansal, R. (Eds.), *Addressing Practical Problems Through the Metaverse and Game-Inspired Mechanics* (pp. 123–152). IGI Global Scientific Publishing., DOI: 10.4018/979-8-3693-8886-0.ch006

Ogunbode, C. A., Böhm, G., Capstick, S. B., Demski, C., Spence, A., & Tausch, N. (2019). The resilience paradox: Flooding experience, coping and climate change mitigation intentions. *Climate Policy*, *19*(6), 703–715. DOI: 10.1080/14693062.2018.1560242

Park, S., & Gupta, S. (2012). Comparison of SML and GMM estimators for the random coefficient logit model using aggregate data. *Empirical Economics*, *43*(3), 1353–1372. DOI: 10.1007/s00181-011-0519-3

Phan, D. H. B., Iyke, B. N., Sharma, S. S., & Affandi, Y. (2021). Economic policy uncertainty and financial stability–is there a relation? *Economic Modelling*, *94*, 1018–1029. DOI: 10.1016/j.econmod.2020.02.042

Raman, M. S., Koppa, K. B., Chithambar Gupta, V., Bhat, V. A., Guru Basava Aradhya, S., & Selvakumar, P. (2025). Digital Transformation and Sustainability Artificial Intelligence and Sustainable Development. In Yıldırım, D., Yıldırım, S., & Kandpal, V. (Eds.), *Effects of Digitalization and Circular Economy on Sustainable Policy and Climate Change Prevention* (pp. 69–96). IGI Global Scientific Publishing., DOI: 10.4018/979-8-3693-9909-5.ch003

Sivanand, R., Selvakumar, P., & Sharma, N. (2025). Nanotechnology Applications in Cancer Drug Delivery Systems: A Review. *Journal of Environmental Nanotechnology*, *14*(2), 458–470. DOI: 10.13074/jent.2025.06.2511221

Trinh, V. Q., Elnahass, M., Salama, A., & Izzeldin, M. (2020). Board busyness, performance and financial stability: Does bank type matter? *European Journal of Finance*, *26*(7–8), 774–801. DOI: 10.1080/1351847X.2019.1636842

VijayaVenkataRaman, S., Iniyan, S., & Goic, R.VijayaVenkataRaman S. (2012). A review of climate change, mitigation and adaptation. *Renewable & Sustainable Energy Reviews*, *16*(1), 878–897. DOI: 10.1016/j.rser.2011.09.009

Viscusi, W. K., & Zeckhauser, R. J. (2006). The perception and valuation of the risks of climate change: A rational and behavioral blend. *Climatic Change*, *77*(1), 151–177. DOI: 10.1007/s10584-006-9075-9

Warren, R., VanDerWal, J., Price, J., Welbergen, J. A., Atkinson, I., Ramirez-Villegas, J., Osborn, T. J., Jarvis, A., Shoo, L. P., Williams, S. E., & Lowe, J. (2013). Quantifying the benefit of early climate change mitigation in avoiding biodiversity loss. *Nature Climate Change*, *3*(7), 678–682. DOI: 10.1038/nclimate1887

Windmeijer, F. (2005). A finite sample correction for the variance of linear efficient two-step GMM estimators. *Journal of Econometrics*, *126*(1), 25–51. DOI: 10.1016/j.jeconom.2004.02.005

Yin, H. (2019). Bank globalization and financial stability: International evidence. *Research in International Business and Finance*, *49*, 207–224. DOI: 10.1016/j.ribaf.2019.03.009

Zhao, P., Pan, J., Yue, Q., & Zhang, J. (2021). Pricing of financial derivatives based on the Tsallis statistical theory. *Chaos, Solitons, and Fractals*, *142*, 110463. DOI: 10.1016/j.chaos.2020.110463

Chapter 12
The Impact of Green and Renewable Energy Policy on Climate Risk Management

Lamia Mohamed Jamel
https://orcid.org/0000-0002-3244-068X
College of Business Administration, Taibah University, Saudi Arabia

ABSTRACT

The issue of energy transition occupies a central place in contemporary debates on climate governance and sustainable development. Faced with the intensification of extreme climate events and the rise in greenhouse gas (GHG) emissions, green and renewable energy policies appear to be essential levers for reducing climate risks and ensuring the systemic resilience of modern societies. This paper analyzes in an integrated manner the impact of these policies on climate risk management, through a conceptual and theoretical approach based on the paradigms of sustainable development, energy transition, and polycentric climate governance. By mobilizing a multidisciplinary analytical framework, the study highlights the close relationship between the decarbonization of the energy system, economic diversification, and the reduction of environmental vulnerability.

1. INTRODUCTION

Climate change poses one of the most pressing threats to the ecological, economic, and social stability of the planet today. According to the sixth report of the Intergovernmental Panel on Climate Change (IPCC, 2023), the global average temperature

DOI: 10.4018/979-8-3373-6766-8.ch012

has already increased by more than 1.1°C compared to the pre-industrial era, leading to an unprecedented intensification of extreme weather events. These changes are resulting in more frequent droughts, floods, heatwaves, and tropical storms, lastingly affecting ecosystems, agricultural systems, and human infrastructure (UNEP, 2023).

Faced with this situation, the international community has gradually recognized the need to transform energy models dominated by fossil fuels, the main cause of greenhouse gas (GHG) emissions. Public policies on green and renewable energy are therefore essential instruments for mitigating and adapting to climate change (IEA, 2022). They aim to promote clean technologies such as solar, wind, green hydrogen, biomass and geothermal energy, while reducing dependence on hydrocarbons and diversifying energy supply sources (REN21, 2024).

From this perspective, the energy transition is not limited to a technological shift; it represents a systemic transformation affecting the economic, social, and political spheres. According to Sovacool, Hook, Martiskainen, and Baker (2021), the "just transition" to a low-carbon economy requires a profound reconfiguration of institutional structures and governance mechanisms to ensure a fair distribution of the benefits and costs of this transition. This process is based on an integrated approach that combines sustainable development, technological innovation, and climate justice.

Green energy policies also contribute to proactive climate risk management by strengthening societies' resilience to environmental disruptions. Investment in renewable energy promotes the decentralization of energy networks, increasing their adaptive capacity in the event of natural disasters (World Bank, 2023). Furthermore, energy infrastructure based on local and renewable resources reduces economic vulnerability linked to the volatility of international fossil fuel markets (IEA, 2022).

Globally, initiatives such as the European Green Deal (European Union, 2019), the Saudi Green Initiative (2023), and the China-India Renewable Energy Master Plan (REN21, 2024) illustrate the growing importance of green energy policies in state strategic planning. These programs aim not only to reduce GHG emissions but also to strengthen energy security, promote innovation, and support the Sustainable Development Goals (United Nations, 2022).

However, despite significant progress, the effective implementation of these policies still faces several challenges: insufficient financing, persistent dependence on hydrocarbons, slow institutional reforms and unequal access to clean technologies (Meadowcroft, 2022). These obstacles compromise the ability of States to achieve their climate and energy objectives, and to ensure sustainable management of climate change risks.

Thus, the central question guiding this study is: to what extent do green and renewable energy policies influence climate risk management? To answer this question, it is essential to analyze the conceptual foundations of these policies, their socio-economic and environmental effects, as well as the governance mechanisms that determine their long-term effectiveness.

2. CONCEPTUAL AND THEORETICAL FRAMEWORK

The conceptual and theoretical framework of this study is structured around the foundations of sustainable development, energy transition, and climate governance. These approaches offer an integrated understanding of the dynamics linking green energy policies to climate risk management.

2.1. Sustainable development as a founding paradigm

The concept of sustainable development, as defined by the Brundtland Report (1987), is based on the ability to "meet the needs of the present without compromising the ability of future generations to meet their own needs." This paradigm strikes a balance between the economic, social, and environmental dimensions of development (Barbier, 2020). In the energy context, it involves the search for low-carbon solutions that promote economic growth while protecting natural resources.

Green energy policies are fully in line with this paradigm, as they aim to reduce dependence on fossil fuels and promote a circular and low-carbon economy (United Nations, 2022). They thus contribute to the achievement of the Sustainable Development Goals (SDGs), in particular SDG 7 (clean and affordable energy) and SDG 13 (fight against climate change).

2.2. The theory of energy transition

Energy transition theory is based on the idea that energy systems are constantly evolving as a result of technological advances, public policies, and socio-economic changes (Geels, 2002; Sovacool, 2021). According to this approach, the energy transition is a multidimensional process that combines technological, institutional, economic, and cultural transformations.

Geels and Schot (2010) conceptualized this transition through the Multi-Level Perspective (MLP) model, distinguishing three levels:

- The socio-technical landscape (global context, e.g. climate, global economy),
- The socio-technical regime (established energy systems),
- And innovation niches (emerging technologies such as solar or green hydrogen).

Green energy policy acts as a catalytic mechanism in this process, facilitating the transition of niche innovations to the mainstream, thereby accelerating the decarbonization of economic systems (Sovacool et al., 2021).

As part of climate risk management, the energy transition not only helps reduce greenhouse gas emissions, but also strengthens the resilience of infrastructure and communities to climate shocks (IEA, 2022).

2.3. Climate governance and systemic resilience

Climate governance constitutes another essential conceptual pillar. Ostrom (2010) proposed a polycentric approach to environmental governance, emphasizing that the fight against climate change cannot rely solely on states, but requires the cooperation of multiple actors—international institutions, local governments, businesses, and citizens.

This polycentric governance favors the emergence of adaptive energy policies, capable of responding to local specificities while being part of global commitments (Meadowcroft, 2022). In this context, climate risk management depends on the ability of actors to integrate the principles of systemic resilience, that is, the ability of a system to absorb disturbances, adapt and transform itself in response to climate shocks (Folke, 2021).

Green energy policies, when supported by inclusive and transparent governance mechanisms, contribute directly to this resilience by promoting energy diversification, citizen participation and intersectoral cooperation (UNEP, 2023).

2.4. The integrated energy-climate-risk approach

Finally, an integrated approach between energy, climate, and risk is emerging in recent literature. This approach highlights the dynamic interactions between energy policies and climate risk management (World Bank, 2023). It emphasizes that energy sustainability cannot be achieved without proactive management of risks related to extreme events, energy security, and economic stability.

This systemic vision is based on three principles:

1. The interdependence between energy and climate systems.
2. Adaptation as an essential complement to mitigation.
3. Coherence of public policies, ensuring coordination between energy, climate and development objectives (IPCC, 2023).

Thus, the conceptual framework of this research is based on the idea that green energy policies are not only a lever for reducing emissions, but also a strategic tool for managing climate risks, contributing to the construction of a resilient and sustainable economy.

3. GREEN ENERGY POLICIES AND CLIMATE RISK REDUCTION

Green and renewable energy policies occupy a central place in contemporary strategies to combat climate change. They constitute both mitigation instruments, aimed at reducing greenhouse gas (GHG) emissions, and adaptation levers, intended to strengthen the resilience of human and natural systems to climate impacts. In a context of recurring environmental and energy crises, their role in climate risk management is now indisputable (IPCC, 2023; IEA, 2022).

3.1. The role of energy policies in climate mitigation

Mitigation refers to all actions aimed at limiting the extent of climate change by reducing anthropogenic GHG emissions. Green energy policies are fully in line with this logic by promoting the development of low-carbon technologies, energy substitution and energy efficiency (UNEP, 2023).

According to the International Energy Agency (2022), the transition to renewable energy could reduce global energy-related emissions by nearly 70% by 2050, provided public policies are ambitiously implemented. Incentive mechanisms, such as renewable energy subsidies, carbon markets, or emissions taxes, have proven effective in encouraging private investment in the green sector (World Bank, 2023).

For example, the European Union, through the Green Deal launched in 2019, has set a target of carbon neutrality by 2050, accompanied by an investment plan of more than €1 trillion to support clean technologies and energy efficiency (European Commission, 2021). Similarly, China, the world's largest emitter, has adopted a policy of accelerated expansion of solar and wind energy, contributing to a substantial reduction in the carbon intensity of its economy (REN21, 2024).

Thus, green energy policy presents itself as an instrument of climate governance, making it possible to operationalize the commitments made within the framework of the Paris Agreement (2015) and the Sustainable Development Goals (SDGs), in particular SDG 13 relating to the fight against climate change (United Nations, 2022).

3.2. Adaptation and resilience to climate risks

Beyond reducing emissions, green energy policies contribute to proactively managing climate risks by strengthening the resilience of energy, economic, and social systems. According to Folke (2021), resilience is defined as the capacity of a system to absorb shocks, reorganize itself, and evolve towards a new equilibrium without compromising its essential functions.

Integrating renewable energy into national energy systems promotes this resilience through source diversification and grid decentralization. By replacing centralized and vulnerable infrastructure with local microgrids, countries can better cope with supply disruptions related to natural disasters (IEA, 2022). For example, community solar projects in sub-Saharan Africa and the Middle East enable rural areas to achieve greater energy security while reducing their exposure to geopolitical fluctuations in fossil fuel markets (World Bank, 2023).

In the Gulf countries, particularly Saudi Arabia, energy policy is part of an integrated vision of sustainability. The Saudi Green Initiative (2023) plans to produce 50% of electricity from renewable sources by 2030, while launching reforestation and carbon capture programs to offset residual emissions. These initiatives simultaneously contribute to mitigation and adaptation, strengthening the country's capacity to manage climate risks related to desertification, extreme temperatures, and water scarcity (Al-Saidi & Elagib, 2022).

3.3. Multi-level governance of energy policies

Effective implementation of green energy policies relies on multi-level governance, involving international institutions, national governments, local authorities and private actors (Ostrom, 2010; Meadowcroft, 2022). This polycentric governance promotes coordination between different decision-making levels and allows policies to be adapted to specific socio-economic contexts.

National energy policies are often based on internationally defined strategic frameworks, such as the Paris Agreement or Nationally Determined Contributions (NDCs). At the local level, cities and regions play a crucial role in the adoption of innovative solutions, such as sustainable mobility, positive energy buildings, and community outreach programs (UN-Habitat, 2022).

This bottom-up approach complements traditional top-down strategies, ensuring the political and institutional coherence necessary for integrated climate risk management (Bulkeley & Betsill, 2021).

3.4. Limits and challenges of green energy policies

Despite their undeniable benefits, green energy policies face several structural challenges. The first is economic: the high initial costs of renewable infrastructure and dependence on public financing hamper their massive deployment, particularly in developing countries (World Bank, 2023). The second challenge is institutional:

intersectoral coordination and policy stability often remain insufficient to ensure the continuity of investments (Meadowcroft, 2022).

On the social level, the energy transition can generate inequalities, particularly in regions dependent on fossil fuels, where job repurposing and energy justice are becoming crucial issues (Sovacool et al., 2021). Finally, on the technological level, the limitations linked to energy storage and the intermittency of renewable sources require an intensification of research and development (IEA, 2022).

3.5. Towards sustainable energy-climate-development integration

To maximize their impact on climate risk management, green energy policies must be integrated into a holistic vision of sustainable development. This integration requires the creation of coherent policy frameworks linking energy, climate, and socio-economic objectives.

As the United Nations Environment Programme (2023) highlights, only an integrated and inclusive approach, linking mitigation and adaptation, will ensure the long-term sustainability of energy systems and strengthen the capacity of societies to face future climate challenges.

4. SOCIO-ECONOMIC AND ENVIRONMENTAL IMPACTS OF GREEN ENERGY POLICIES

Green and renewable energy policies have multidimensional effects that go beyond the strict framework of energy production. Their implementation profoundly influences economic dynamics, social conditions, and environmental quality. Analyzing these impacts helps us understand how the transition to sustainable energy systems contributes both to reducing climate vulnerabilities and promoting inclusive and resilient development.

4.1. Economic impacts: green growth and competitiveness

Energy transition policies promote the emergence of a green economy, defined by the OECD (2021) as a model of sustainable growth aimed at improving human well-being while reducing environmental risks. This dynamic relies on investment in renewable technologies, energy efficiency, and low-carbon innovation.

According to the World Bank (2023), global investment in renewable energy will exceed $500 billion in 2022, representing nearly 70% of new installed electricity capacity. These policies are creating new markets, stimulating entrepreneurship, and strengthening industrial competitiveness in sectors such as solar photovoltaic, wind, hydrogen, and energy storage (REN21, 2024).

The economic benefits are also visible at the national level: energy diversification reduces dependence on hydrocarbon imports, stabilizes trade balances, and mitigates the effects of oil price fluctuations (IEA, 2022). In the case of the European Union, the energy transition contributes to energy sovereignty and the relocation of certain strategic industries (European Commission, 2021).

However, these transformations require significant upfront public investment and long-term planning. Developing countries, faced with fiscal and institutional constraints, often face barriers to mobilizing sufficient financing, hence the need for international partnerships and equitable climate financing mechanisms (UNEP, 2023).

4.2. Impacts on employment and social transformation

The energy transition is leading to a restructuring of the labor market, with rapid growth in green jobs. According to the International Labor Organization (ILO, 2022), renewable energy generated more than 13 million jobs worldwide in 2021, a figure that is constantly growing. New opportunities include construction, maintenance, research, management, and energy services.

These policies also contribute to a structural transformation of skills and vocational training. They require the development of educational programs focused on clean technologies, environmental management, and sustainable engineering (ILO, 2022).

However, the transition does not benefit all social groups equally. In some regions dependent on fossil fuel industries, it is causing job losses and socio-economic tensions (Sovacool et al., 2021). Hence the importance of the concept of "just transition," which aims to ensure that the economic and environmental benefits of the transition are shared equitably, while supporting affected workers and communities (Newell & Mulvaney, 2023).

4.3. Environmental impacts: air quality and resource conservation

The environmental benefits of green energy policies are among the most visible and measurable. Gradually replacing fossil fuels with renewable sources significantly reduces emissions of CO_2, methane, and nitrogen oxides—the main causes of global warming and air pollution (IPCC, 2023).

According to the World Health Organization (WHO, 2021), exposure to air pollution causes approximately seven million premature deaths each year. The deployment of renewable energy and clean technologies therefore contributes to a direct improvement in public health by reducing the concentration of air pollutants in urban areas.

Furthermore, green energy policies promote better management of natural resources. Solar and wind technologies, although intensive in critical metals, consume much less water and space than fossil fuels to produce the same amount of energy (IRENA, 2022). This represents a major asset in arid regions such as the Middle East, where water and food security are closely linked to energy sustainability (Al-Saidi & Elagib, 2022).

4.4. Social impacts and energy equity

Green energy policies can also help reduce energy inequality and strengthen social justice. Access to clean and affordable energy improves living conditions, promotes education, supports health, and stimulates local development, particularly in rural areas (United Nations, 2022).

The democratization of energy, thanks to solar micro-grids, local cooperatives and decentralized production systems, allows citizens to become actors in the transition, thus strengthening participatory governance and social cohesion (Ostrom, 2010; UN-Habitat, 2022).

However, the social benefits of the energy transition depend on governments' ability to ensure equitable inclusion. Without appropriate redistributive policies and institutional frameworks, the transition risks reproducing or exacerbating existing inequalities between developed and developing countries (Meadowcroft, 2022; Sovacool et al., 2021).

4.5. Sustainability issues and indirect effects

Despite their benefits, green energy policies have some indirect effects that raise sustainability issues. The massive development of green technologies requires the exploitation of rare metals (lithium, cobalt, nickel), the extraction of which can generate significant environmental and social impacts (World Bank, 2023).

Furthermore, the construction of large renewable infrastructure—dams, wind farms, solar fields—can have impacts on biodiversity and local ecosystems (UNEP, 2023). These issues call for rigorous ecological planning and a life cycle approach to energy technologies (IRENA, 2022).

Thus, the real challenge of green energy policies lies in their ability to reconcile environmental efficiency, economic viability and social justice, essential conditions for sustainable and equitable management of climate risks.

5. GOVERNANCE AND CHALLENGES OF IMPLEMENTING GREEN ENERGY POLICIES

The success of green energy policies depends largely on the quality of governance and institutional capacity to plan, coordinate, and monitor the energy transition. In a context of systemic complexity and interdependence between public, private, and community actors, renewable energy governance is positioned as a central lever for integrated climate risk management (Sovacool et al., 2021).

Energy governance is not limited to the regulation of the energy sector, but encompasses all political, institutional and financial frameworks that condition the transformation of economic systems towards sustainability (Meadowcroft, 2022). It is based on coherence between national policies, local strategies and international commitments such as the Paris Agreement or the Sustainable Development Goals (SDGs).

5.1. Multi-level governance and institutional coordination

Green energy policies require multi-level governance, involving international institutions, national governments, local communities, and non-state actors. This approach allows for effective coordination between global decarbonization priorities and regional specificities in terms of resources, infrastructure, and institutional capacities (Bulkeley et al., 2021).

International institutions such as the International Energy Agency (IEA), the International Renewable Energy Agency (IRENA) and the United Nations Environment Programme (UNEP) play a coordinating and standardizing role, establishing common reference frameworks, indicators and technical standards.

At the national level, effective governance relies on clear institutional roles and intersectoral coherence. Overlapping responsibilities between energy, environment, and economic development ministries can hamper the implementation of integrated policies (OECD, 2022). The development of dedicated energy transition agencies, such as the Saudi Energy Efficiency Center or the Energy Transition Commission in the United Kingdom, illustrates an attempt at institutional streamlining and improved strategic planning (Saudi Green Initiative, 2023).

5.2. Participatory governance and stakeholder inclusion

The participation of citizens, businesses, and civil society is a pillar of sustainable energy governance. Participatory governance promotes the transparency, legitimacy, and social acceptability of green policies (Ostrom, 2010). It also helps strengthen collective responsibility in the face of climate issues.

Models of energy cooperatives or community projects are emerging in several European and African countries, promoting decentralized production and local reappropriation of energy resources (REN21, 2024). These approaches contribute to the democratization of energy and the reduction of inequalities of access (Newell & Mulvaney, 2023).

However, stakeholder inclusion remains a challenge, particularly in developing countries where governance is often centralized and technical and administrative capacity is limited (UNDP, 2022). The absence of strong participatory mechanisms can lead to public distrust and slow the implementation of renewable energy projects.

5.3. Financial and technological challenges

The energy transition relies on massive investments in infrastructure, research, and innovation. According to the World Bank (2023), countries must triple their investments in clean energy by 2030 to achieve climate targets of carbon neutrality.

However, financing the transition remains a major challenge, particularly for emerging economies and least developed countries. Fiscal constraints, the high initial cost of green technologies and weak local financial markets limit the ability to mobilize the necessary resources (UNEP, 2023).

Climate finance mechanisms such as the Green Climate Fund, green bonds, and public-private partnerships (PPPs) represent essential levers to overcome these obstacles (IEA, 2022). However, their effectiveness depends on transparency, financial governance, and the credibility of regulatory frameworks (OECD, 2022).

On the technological front, the challenges concern the stability of the electricity grid, the management of energy storage, and the availability of critical materials needed to manufacture batteries and solar panels (IRENA, 2022). Public policies must therefore integrate a dimension of innovation and circularity, promoting recycling and the eco-design of green technologies.

5.4. Political and institutional challenges

Energy governance often faces political resistance, particularly in countries dependent on hydrocarbons. Economic interests linked to fossil fuels can influence public decision-making and delay structural reforms (Sovacool & Griffiths, 2020).

Furthermore, energy transitions require legal reforms to adapt regulatory frameworks to new technologies and business models. The absence of clear legislation on renewable energy, tax incentives, or feed-in tariffs can hamper private investment (UNDP, 2022).

Political and institutional stability also plays a key role. In contexts where governance is fragmented or marked by corruption, the implementation of green policies can be compromised (Meadowcroft, 2022). The creation of independent regulatory authorities and the promotion of transparency in resource management are key elements of good energy governance.

5.5. Challenges related to climate justice and global equity

The implementation of green energy policies must also be approached from the perspective of climate justice. Countries in the Global South, although less responsible for historical emissions, suffer disproportionately from the impacts of climate change (IPCC, 2023).

The issue of global energy equity is acute: developed countries have far greater financial and technological capacity, while developing countries lack the means to achieve a rapid transition (UNFCCC, 2023). International cooperation, shared green technology, and equitable financing are therefore essential conditions for the collective success of the transition (World Bank, 2023).

5.6. Towards adaptive and resilient energy governance

Faced with these challenges, the future of green energy policies rests on adaptive governance, capable of adjusting strategies based on feedback, technological developments and climate constraints (Bulkeley et al., 2021).

Such governance must integrate long-term planning, institutional resilience and enhanced international cooperation to coordinate climate change mitigation and adaptation efforts.

The emergence of hybrid governance frameworks—combining state regulation, citizen participation, and private innovation—constitutes a promising development towards a just, inclusive, and sustainable energy transition (Meadowcroft, 2022; Sovacool et al., 2021).

6. DISCUSSION

The implementation of green and renewable energy policies is currently one of the most powerful strategic levers for managing climate risks globally. Nevertheless, the transition to a low-carbon economy remains a complex, multidimensional process that is unevenly distributed across geographic, economic, and political contexts. This discussion examines the key lessons learned from the preceding analysis, emphasizing the tensions between environmental ambition, socioeconomic feasibility, and institutional coherence, as well as the conditions for successful sustainable and equitable energy governance.

6.1. Consistency between energy transition and climate risk management

Green energy policies contribute directly to reducing climate risks by limiting greenhouse gas (GHG) emissions and promoting the adaptation of energy systems to environmental hazards (IPCC, 2023). By diversifying energy sources, states strengthen their systemic resilience, thereby reducing dependence on fossil fuels that are vulnerable to geopolitical and climate disruptions (IEA, 2022).

However, coherence between energy and climate strategies sometimes remains insufficient. In several countries, transition policies are still sectoral, disconnected from national adaptation strategies, or hampered by conflicts of economic interest (Meadowcroft, 2022). The lack of integration between energy, water, transport, and land-use policies limits the scope of environmental benefits (Bulkeley et al., 2021).

Thus, climate risk management requires a systemic and integrated approach, linking energy planning, climate governance and territorial development. The most effective policies are those that align carbon neutrality objectives with adaptation mechanisms focused on the sustainability of infrastructure and the protection of vulnerable populations (UNEP, 2023).

6.2. The central role of governance and international cooperation

The energy transition cannot be envisaged without strong, inclusive, and multi-level governance. As Sovacool et al. (2021) point out, the success of green energy policies depends less on technological advances than on the institutional capacity to coordinate, regulate, and incentivize.

Countries that have adopted clear and participatory governance frameworks—such as the European Union, Japan, or the United Arab Emirates—have made rapid progress in decarbonizing the energy mix. Conversely, contexts characterized by excessive centralization or institutional instability show significant delays (OECD, 2022).

International cooperation plays a key role in this process. Developing countries, facing financial and technical constraints, rely heavily on multilateral assistance and the transfer of green technologies (UNFCCC, 2023). However, the uneven distribution of climate finance and the slow pace of commitments by industrialized countries raise questions of climate justice and global equity (Newell & Mulvaney, 2023).

In this sense, global energy governance must evolve towards a polycentric and collaborative model, promoting coordination between international institutions, private actors and local communities, while guaranteeing transparency and accountability (Ostrom, 2010).

6.3. The socio-economic paradoxes of the energy transition

Although the transition to green energy generates new jobs, stimulates innovation, and improves air quality, it also produces paradoxical effects. On the one hand, it strengthens economic competitiveness and paves the way for sustainable growth (OECD, 2021); on the other, it can accentuate social and territorial inequalities, particularly in regions dependent on fossil fuel industries (Sovacool & Griffiths, 2020).

These imbalances highlight the need for a "just transition," which integrates social dimensions into climate and energy policies. The transition must be accompanied by compensatory measures such as professional retraining, training, and support for affected communities (ILO, 2022).

Furthermore, green energy policies must be designed with local specificities in mind. A uniform approach, based solely on global technological standards, risks neglecting the needs of the most vulnerable populations, particularly in rural or arid areas (Al-Saidi & Elagib, 2022). The territorialization of energy policies therefore appears to be an essential condition for ensuring their social acceptability and environmental effectiveness.

6.4. Environmental dilemmas: between technological progress and new risks

Environmentally, the energy transition is not without its ecological dilemmas. While renewable technologies enable a massive reduction in GHG emissions, their intensive deployment generates increased demand for rare metals (lithium, cobalt, nickel) and puts pressure on ecosystems (World Bank, 2023).

This paradox highlights the need for a circular economy in the energy sector, based on the recycling, reuse, and substitution of critical materials (IRENA, 2022). Moreover, green infrastructure, such as dams and wind farms, can disrupt local biodiversity, calling for more rigorous environmental planning (UNEP, 2023).

Thus, the sustainability of the energy transition will depend on the ability of decision-makers to reconcile technological innovation and ecological preservation, by integrating environmental impact assessments from the design phase of projects.

6.5. Conditions for the success of a sustainable energy transition

The success of a sustainable energy transition depends on a set of structural and strategic conditions. First, political stability and institutional transparency are essential to attract private investment and ensure the continuity of public policies (Meadowcroft, 2022). Second, citizen participation and inclusive governance strengthen the legitimacy and effectiveness of the measures adopted (Ostrom, 2010; UNDP, 2022).

Furthermore, energy policies must be accompanied by flexible legal frameworks that allow for experimentation, innovation and adaptation to rapid changes in the technological and economic context (Bulkeley et al., 2021).

Finally, the energy transition must be thought of as an evolutionary process, part of adaptive governance (IPCC, 2023). This requires continuous policy evaluation, knowledge sharing, and intersectoral cooperation. Building a truly sustainable global energy system will depend on the ability of states to integrate social, economic, and environmental dimensions into a coherent and inclusive model (Newell & Mulvaney, 2023).

7. CONCLUSION AND RECOMMENDATIONS

7.1. General conclusion

The transition to green and renewable energy is now a central focus of global climate governance and a fundamental tool for reducing climate risks. Analysis of energy policies implemented around the world highlights a dual dynamic: on the one hand, a growing movement toward decarbonizing energy systems and, on the other, structural disparities in the capacity of states to implement these policies sustainably and equitably.

The results of this work show that green energy policies have multi-scale positive effects. They contribute to reducing greenhouse gas (GHG) emissions, improving air quality, economic diversification and creating green jobs (IRENA, 2022; ILO, 2022). They also contribute to strengthening climate resilience by reducing dependence on hydrocarbons and improving the energy security of States (IEA, 2022).

However, the implementation of these policies remains fraught with structural and institutional challenges. Financial constraints, slow regulatory reforms, lack of intersectoral coordination, and resistance from traditional fossil fuel sector stakeholders are all major obstacles (OECD, 2022; Sovacool & Griffiths, 2020).

The success of the energy transition therefore depends not only on technological progress, but above all on the quality of governance, international cooperation and social justice that accompany this transformation (Newell & Mulvaney, 2023; Meadowcroft, 2022). Ultimately, climate risk management cannot be dissociated from the way in which societies structure their energy institutions and guide their development choices.

7.2. Strategic recommendations

In light of the above findings, several strategic directions can be formulated to strengthen the effectiveness and equity of green energy policies in managing climate risks.

1. Strengthen integrated and multi-level governance

The first recommendation is to establish coherent, inclusive, and coordinated energy governance between the international, national, and local levels. Such governance must be based on transparency, citizen participation, and institutional accountability (Ostrom, 2010). States should strengthen the capacities of energy planning institutions and promote intersectoral synergies between energy, transport, water, and urban planning (Bulkeley et al., 2021).

2. Accelerate technological transition and innovation research

Governments must invest in research and development (R&D) to improve storage, green hydrogen, and carbon capture technologies (IRENA, 2022). Open innovation and public-private collaboration are essential to reduce the costs of renewable technologies and ensure their accessibility in developing countries (World Bank, 2023).

Furthermore, the energy transition must be accompanied by a circular economy approach, promoting the recycling of critical materials and the reduction of the ecological footprint of green infrastructure (UNEP, 2023).

3. Establish inclusive and sustainable financial mechanisms

Mobilizing green finance is a central pillar of a successful energy transition. States should develop instruments such as green bonds, energy transition funds, and public-private partnerships (PPPs) to attract private investment (IEA, 2022).

International financial institutions, including the World Bank and the Green Climate Fund, must also strengthen their support to countries in the South, in order to reduce structural inequalities and ensure global energy justice (UNFCCC, 2023).

4. Promote a just and socially inclusive transition

Any sustainable energy strategy must integrate the principle of just transition, guaranteeing the protection of workers, vocational training and the economic reconversion of regions dependent on fossil industries (ILO, 2022; Sovacool et al., 2021).

Public policies must also aim to reduce energy poverty by facilitating universal access to clean, affordable and reliable energy, in line with Sustainable Development Goal 7 (United Nations, 2022).

5. Integrate energy planning with climate resilience

Energy policies must be part of an integrated climate risk management approach, combining mitigation and adaptation. This involves taking into account natural hazards (droughts, heatwaves, storms) in energy infrastructure planning (IPCC, 2023).

States should develop national energy resilience plans, aligned with long-term climate strategies, to ensure continuity of supply in the event of an environmental crisis (UNEP, 2023).

6. Strengthen international cooperation and technology transfer

The energy transition cannot be successful without strengthened international cooperation. Developed countries must honor their financial commitments under the Paris Agreement and promote technology transfer to low-income countries (UNFCCC, 2023).

The establishment of regional energy innovation platforms would accelerate the sharing of good practices, harmonize regulations and strengthen collective resilience in the face of climate risks (IEA, 2022; REN21, 2024).

7.3. Research perspectives

Finally, it is essential to continue academic research on the interactions between energy policies, social justice and climate governance. Future work should focus on:

- Adaptive governance models to integrate climate variability and technological evolution.
- Performance indicators for assessing the sustainability and equity of energy policies.
- The impact of geopolitical dynamics and supply chains on global energy security.

- The role of digitalization and artificial intelligence in smart grid management and climate risk prediction.

These guidelines will help deepen understanding of the role of green energy policies in building a development model that is resilient, equitable and climate-sustainable.

References

Al-Saidi, M., & Elagib, N. A. (2022). Ecological modernization and renewable energy transitions in the Gulf Cooperation Council countries. *Renewable & Sustainable Energy Reviews*, *160*, 112278.

Barbier, E. B. (2020). Greening the post-pandemic recovery: A sustainable growth strategy. *Environmental and Resource Economics*, *76*(4), 685–703. DOI: 10.1007/s10640-020-00437-w PMID: 32836827

Brundtland, G. H. (1987). *Our Common Future*. Oxford University Press.

Bulkeley, H., & Betsill, M. (2021). Cities and climate change governance. Routledge Handbook of Environmental Governance, 231–245.

Bulkeley, H., Broto, V., & Edwards, G. (2021). *Governing climate change: The new politics of urban sustainability*. Routledge.

European Commission. (2021). *The European Green Deal: 2021 Progress Report*. European Union.

Folke, C. (2021). Resilience (Republished). *Ecology and Society*, *26*(4), 11–25.

Geels, F. W. (2002). Technological transitions as evolutionary reconfiguration processes. *Research Policy*, *31*(8–9), 1257–1274. DOI: 10.1016/S0048-7333(02)00062-8

Geels, F. W., & Schot, J. (2010). The dynamics of transitions: A socio-technical perspective. In *Transitions to Sustainable Development* (pp. 11–104). Routledge.

Intergovernmental Panel on Climate Change (IPCC). (2023). *Sixth Assessment Report: Climate Change 2023*. United Nations.

Intergovernmental Panel on Climate Change (IPCC). (2023). *Sixth Assessment Report: Climate Change 2023*. United Nations.

International Energy Agency (IEA). (2022). *World Energy Investment Report 2022*. IEA Publications.

International Energy Agency (IEA). (2022). *World Energy Outlook 2022*. IEA Publications.

International Labor Organization (ILO). (2022). *World Employment and Social Outlook: The Role of Green Jobs*. ILO.

International Renewable Energy Agency (IRENA). (2022). *Innovation Outlook: Renewable Power-to-Hydrogen*. IRENA.

International Renewable Energy Agency (IRENA). (2022). *World Energy Transitions Outlook 2022*. IRENA.

Meadowcroft, J. (2022). Governance, transitions, and the green state. *Environmental Politics*, *31*(4), 575–594.

Newell, P., & Mulvaney, D. (2023). The political economy of just transitions: Toward inclusive and sustainable energy transitions. *Energy Research & Social Science*, *99*, 103081.

Organisation for Economic Co-operation and Development (OECD). (2021). *Green Growth and Sustainable Development Report 2021*. OECD.

Organisation for Economic Co-operation and Development (OECD). (2022). *Policy Coherence for Sustainable Energy Transition*. OECD.

Ostrom, E. (2010). Polycentric systems for coping with collective action and global environmental change. *Global Environmental Change*, *20*(4), 550–557. DOI: 10.1016/j.gloenvcha.2010.07.004

REN21. (2024). Renewables 2024 Global Status Report. Paris: REN21 Secretariat.

Saudi Green Initiative. (2023). *Annual Report 2023*. Ministry of Energy.

Sovacool, B. K., & Griffiths, S. (2020). The cultural barriers to a low-carbon future: A review of energy justice. *Energy Policy*, *138*, 111227.

Sovacool, B. K., Hook, A., Martiskainen, M., & Baker, L. (2021). The justice and equity implications of the clean energy transition. *Nature Energy*, *6*(7), 728–739.

UN-Habitat. (2022). *Climate Action in Cities: Integrated Approaches for Urban Resilience*. UN-Habitat.

United Nations. (2022). *Sustainable Development Goals Progress Report 2022*. UN Publications.

United Nations Development Program (UNDP). (2022). *Energy Governance for Sustainable Development*. UNDP.

United Nations Environment Program (UNEP). (2023). *Adaptation Gap Report 2023*. UNEP.

United Nations Framework Convention on Climate Change (UNFCCC). (2023). *Climate Finance Delivery Plan 2023–2026*. UNFCCC Secretariat.

World Bank. (2023). *Energy Sector Directions Paper: Powering a Green Future*. World Bank.

World Bank. (2023). *Financing the Energy Transition: Global Outlook 2023*. World Bank.

World Health Organization (WHO). (2021). *Air Pollution and Public Health Report 2021*. WHO.

Chapter 13

A Systematic Review of Climate Risk Integration in Banking Risk Management

Jiafan Li
https://orcid.org/0000-0001-6280-8193
University of Huddersfield, UK

Erhan Kilincarslan
https://orcid.org/0000-0001-8007-8905
University of Huddersfield, UK

ABSTRACT

This chapter presents a systematic literature review of 54 peer-reviewed studies on how commercial banks integrate climate-related risks into their risk management frameworks. Drawing upon institutional, agency, stakeholder, and resource-based theories, the analysis identifies three interrelated dimensions of bank responses: climate adaptation, loan pricing, and climate risk disclosure. Findings reveal that banks, particularly in China and the European Union, increasingly incorporate environmental considerations into strategic and operational decisions through green credit, ESG engagement, digital innovation, and board diversity. However, persistent challenges, such as symbolic compliance, brown lending, and uneven regulatory enforcement, limit substantive progress. The review highlights how institutional pressures, managerial incentives, stakeholder expectations, and organisational capabilities jointly shape banks' climate behaviours.

DOI: 10.4018/979-8-3373-6766-8.ch013

1. INTRODUCTION

Risk management in banking refers to the systematic process of assessing how an organisation's strategies influence its exposure to financial, operational, and strategic risks (Srivastav and Hagendorff, 2015). Effective risk management strengthens banks' resilience against adverse shocks, safeguarding solvency, liquidity and public confidence. Historical crises, such as the collapse of Lehman Brothers in 2008 and the 2023 liquidity crisis at Credit Suisse, underscore how even well-capitalised banks remain vulnerable to systemic contagion (Ratnovski, 2013). To mitigate these risks, banks operate under stringent regulatory oversight. The Basel Framework, developed by the Basel Committee on Banking Supervision, represents the cornerstone of global banking regulation, setting principles for credit, market, operational, and liquidity risk. Importantly, a bank's risk management function must remain independent from revenue-generating activities to preserve objectivity and ensure compliance (Lim et al., 2017).

Extensive research has examined different aspects of banking risk management. Aebi et al. (2012) and Srivastav and Hagendorff (2015) emphasised the role of corporate governance, while Kunz and Heitz (2021) reviewed how risk culture is embedded within management control systems. More recently, scholars have linked risk management to environmental strategy, showing that banks increasingly recognise both risks and opportunities associated with climate change (Cucinelli et al., 2024). However, the literature remains fragmented in explaining how climate-related risks, an emerging category of systematic risk, are incorporated into traditional risk management frameworks.

The World Economic Forum (2025) identifies climate change as one of the most critical threats to global stability in the coming decade. Climate risks manifest through three interrelated dimensions: (1) physical risks from events such as floods, hurricanes, and wildfires that damage assets and impair collateral values; (2) transition risks arising from policy, regulatory, or technological shifts towards a low-carbon economy; and (3) liability risks linked to litigation, stranded assets, and reputational losses from failing to meet climate commitments. These risks have profound implications for banks' credit exposures, market stability, and long-term growth prospects.

Recent scholarship increasingly addresses this issue through the systematic literature review (SLR) method. Breitenstein et al. (2021) reviewed environmental hazards and financial sector risk management; Nguyen et al. (2023) investigated how Vietnamese banks advance green financing; de Bandt et al. (2025) analysed the impact of climate risks on credit exposures; and Vyshnevskyi and Sohn (2025) explored central banks' roles in supporting climate action. Despite these advances, the literature remains fragmented: some focus on central banks, others on commercial banks, and many take a broad financial sector perspective without distinguishing institutional contexts.

What remains underexplored is a systematic understanding of how banks integrate climate-related risks into their existing risk management frameworks. Such integration extends beyond regulatory compliance to encompass credit allocation, portfolio management, and internal governance structures. This challenge has become more salient following the Paris Agreement, under the United Nations Framework Convention on Climate Change (UNFCCC), which mobilised 195 Parties toward global decarbonisation.[1] Complementary initiatives, such as the United Nations Environmental Programme Finance Initiative (UNEP FI) and the Equator Principles (EPs), have promoted common standards encouraging banks to embed climate-related risks into their strategic and operational risk management practices.[2]

To address this gap, our study conducts a systematic review of literature on climate risk integration in banking risk management. Following Snyder (2019), we applied eight keyword combinations in Scopus, resulting in a final sample of 54 peer-reviewed articles. These studies are synthesised under three key thematic areas: 1) Climate Adaptation, (2) Loan Pricing, and (3) Climate Risk Disclosure. Findings reveal that green credit has become a central mechanism for managing climate-related risks, enhancing banks' long-term profitability, sustainability, and resilience. The Paris Agreement marks a turning point, encouraging banks to adopt green initiatives, improve ESG performance, deploy digital technologies, and diversify boards to strengthen governance. The Paris Agreement has also prompted banks to recalibrate loan pricing and volumes for carbon-intensive (brown) firms. With the introduction of mandatory disclosure requirements, banks now combine regulatory compliance with voluntary transparency initiatives to meet stakeholder expectations. Moreover, firms providing higher quality climate disclosures are increasingly rewarded with improved loan conditions and credit access, enhancing overall asset quality. However, challenges persist, including greenwashing, uneven regulatory enforcement, continued brown lending, and the unintended risks of rapid digital transformation.

This study makes several key contributions. First, it provides an academic contribution by presenting the first SLR dedicated to climate risk integration in banking risk management, synthesising global evidence on how banks adapt to climate-related challenges. Second, it contributes to policy debates by identifying gaps in supervisory frameworks related to digital transformation, asset quality, and legal enforcement, providing actionable insights for regulators such as the Basel Committee, NGFS, UNFCCC and national authorities. Third, it offers practical guidance for banking professionals by clarifying best practices and barriers to embedding climate risk in governance and risk management systems, while anticipating potential regulatory backlash amid geopolitical uncertainty. Finally, it sets out a future research agenda to deepen understanding of banks' substantive versus symbolic adaptation behaviours and to inform the development of more effective regulatory and institutional responses.

The remainder of this chapter is structured as follows. Section 2 presents the theoretical framework. Section 3 outlines the research methodology. Section 4 discusses the main findings. Section 5 proposes a future research agenda, and Section 6 concludes with the key insights and implications of the study.

2. THEORETICAL FRAMEWORK

This study draws upon four complementary theoretical lenses – the institutional theory, agency theory, stakeholder theory and resource-based view, to explain why and how banks respond to climate-related risks, either proactively or passively, and why their actions may be substantive or symbolic in nature. Together, these frameworks provide a multidimensional understanding of the strategic, organisational, and behavioural mechanisms shaping banks' responses to climate pressures.

Institutional theory suggests that banks operate within a dense network of formal regulations, social expectations, and normative pressures that collectively influence their behaviour (Birindelli et al., 2022; Doan et al., 2025). As articulated by DiMaggio and Powell (1983), three mechanisms of institutional isomorphism (i.e., coercive, mimetic and normative pressures) help explain how banks adapt to evolving climate-related demands. Coercive pressures arise from legal and regulatory requirements that compel banks' compliance with national and supranational climate policies. Normative pressures stem from social norms and expectations for sustainable development, encouraging banks to adopt green initiatives to enhance legitimacy and attract socially responsible investment. Mimetic pressures drive banks operating in less developed institutional environments to emulate leading peers that have successfully embedded climate risk into management practices. Collectively, these forces motive banks to integrate environmental initiatives into their strategic and operational decisions to maintain institutional legitimacy and stakeholder trust (Islam et al., 2025). Yet, these same forces can also produce symbolic compliance, where banks appear environmentally engaged without implementing meaningful change. Such behaviour, often termed "greenwashing", may undermine financial stability and disadvantage institutions that genuinely commit to green transformation (Delmas and Montes-Sancho, 2010). Hence, institutional theory not only explains conformity but also underscores the need to differentiate between substantive and symbolic risk management behaviours within the banking sector.

Agency theory provides a complementary micro-level perspective by focusing on conflicts between shareholders and management. Investments in climate adaptation often involve substantial upfront costs and long-term horizons, yielding uncertain or delayed returns. When shareholders prioritise short-term profitability, management may underinvest in sustainability initiatives, increasing vulnerability to future climate shocks (Doan et al., 2025). This misalignment can result in risk strategies that privilege immediate financial performance over resilience and long-term value creation.

At the same time, stakeholder theory expands this logic by recognising that banks have fiduciary and ethical responsibilities to multiple constituencies, e.g., shareholders, depositors, employees, regulators, and society at large (Friedman and Miles, 2002). From this perspective, environmental initiatives are not mere compliance activities but strategic investments that enhance reputation, trust, and market competitiveness. Transparent and high-quality climate risk disclosure helps reduce information asymmetry between banks and stakeholders, lowering reputational risks and limiting symbolic behaviours such as greenwashing or selective reporting (Friedrich et al., 2023). In this sense, stakeholder engagement and disclosure practices reinforce accountability and can mitigate the agency problem by aligning managerial incentives with long-term sustainability objectives.

Finally, the resource-based view (RBV) offers an internal organisational lens. It posits that a firm's sustained competitive advantage depends on its ability to develop and leverage unique, valuable, and hard-to-imitate resources and capabilities (Kraaijenbrink et al., 2019; Birindelli et al., 2022). Within the banking sector, climate risk management capabilities, such as data analytics for environmental risk assessment, green product innovation, and institutional learning from past climate shocks, can constitute such strategic resources. For instance, Doan et al. (2025) demonstrated that banks with branch networks exposed to previous extreme weather events develop superior risk-assessment competencies and credit resilience. Consequently, the RBV suggests that climate-related risk management strategies should be viewed as a dynamic capability, which is continuously reviewed, updated, and integrated into broader strategic frameworks (Lockett et al., 2009).

Overall, these theoretical perspectives highlight that banks' engagement with climate-related risks is shaped by external institutional pressures, internal governance dynamics, and unique organisational resources. Understanding how these dimensions interact is essential to distinguishing between genuine and superficial adaptation efforts, thereby deepening insight into how climate risk integration can enhance both financial stability and sustainable development.

3. METHODOLOGY

A systematic literature review (SLR) provides a structured and replicable approach to identifying, synthesising, and evaluating existing knowledge on a specific research domain (Snyder, 2019). This study adopts the PRISMA (Preferred Reporting Items for Systematic Reviews and Meta-Analyses) protocol, which is widely recognised for ensuring transparency, reliability, and methodological rigour in literature reviews (Fan et al., 2022). The review process consisted of four sequential stages, summarised below:

Stage 1: Search Strategy

Following Snyder (2019), a snowballing approach was first employed to identify relevant keywords from seminal reviews and empirical studies. Accodrindlgy, eight keyword combinations were formulated to capture the intersection of climate change, banks, and risk management:

- "Climate" AND "bank" AND "risk" OR "risk management"
- "Carbon" AND "bank" AND "risk" OR "risk management"
- "Green" AND "bank" AND "risk" OR "risk management"
- "Environment" AND "bank" AND "risk" OR "risk management"

This iterative process ensured the inclusion of both broad and specific terminologies, maximising coverage and minimising potential omissions.

Stage 2: Database Selection

Scopus was selected as the primary database due to its comprehensive coverage of peer-reviewed research in finance, management, and sustainability disciplines (Mio et al., 2020). The search was restricted to peer-reviewed journal articles published in English between 2000 and 2025, which is a period encompassing the Kyoto Protocol and the evaluation of climate-finance debates. The subject area was further limited to "Economics, Econometrics and Finance" and "Business, Management and Accounting" to ensure thematic relevance and reduce noise from unrelated domains. This approach balances inclusivity with quality control, yielding a robust dataset for systematic synthesis.

Stage 3: Screening and Eligibility

In line with the PRISMA protocol, a multi-stage screening process was conducted to ensure relevance and quality. Figure 1 outlines the steps taken in the SLR, which follows the established research design adopted in climate risk and banking research (e.g., Naili and Lahrichi, 2022; Brüggemann and Lueg, 2025; de Bandt et al., 2025). Using titles, keywords, and abstracts, the initial search identified 2,332 articles, which were reduced to 2,207 after removing duplicates. Three exclusion criteria were then applied: i) exclude studies that not directly examine the relationship between climate risk and risk management in banks; ii) exclude papers outside recognised academic outlets, using the Chartered Association of Business Schools's (CABS) *Academic Journal Quality Guide 2024* to ensure the inclusion of high-quality studies; and iii) prioritise empirical studies and exclude purely qualitative, conceptual, or case-based analyses, discussions or surveys.

Figure 1. PRISMA Chart

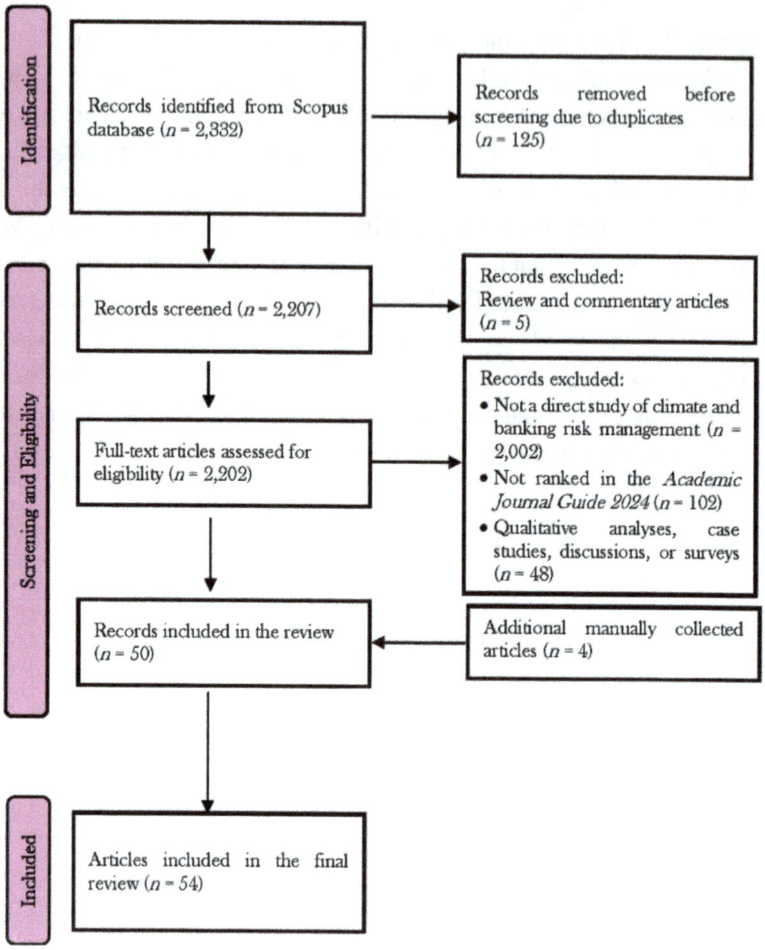

After these filters, 50 relevant studies remained. An additional four articles were identified through manual backward and forward citation searches, resulting in a final sample of 54 articles. This comprehensive process ensures both methodological rigour and conceptual breadth.

We also provide descriptive statistics to offer further insight into the scope and quality of the reviewed studies. Table 1 presents the geographical distribution of the sample, which predominantly comprises global or cross-country analyses (33.33%), followed by studies focusing on China (25.93%), the United States (18.52%), and the European Union (12.96%). Table 2 summarises the distribution of publications by journal and their rankings according to the *Academic Journal Guide (AJG) 2024*, indicating that the review primarily draws on high-quality academic sources.

Table 1. Distribution of articles across countries

Country/Region	**Count**	**Share**
Global/Cross-country	18	33.33%
China	14	25.93%
United States	10	18.52%
European Union	7	12.96%
India, Japan, Taiwan, Turkey, United Kingdom	1	9.26%
Total	54	100.00%

Notes: "Global/Cross-country" refers to studies that analyse multiple countries or adopt a worldwide perspective rather than focusing on a single national context. The individual countries grouped in the last row are each represented by one study in the sample.

Table 2. Publication outlets of reviewed articles

Journal Title	*AJG 2024* **Ranking**	**Count**	**Share**
Finance Research Letters	2	8	14.81%
Business Strategy and the Environment	3	4	7.41%
Journal of Financial Stability	3	3	5.56%
Economics Letters	3	2	3.71%
Global Finance Journal	2	2	3.71%
International Review of Economics and Finance	2	2	3.71%
International Review of Financial Analysis	3	2	3.71%
Journal of Financial Intermediation	4	2	3.71%
Research in International Business and Finance	2	2	3.71%
Review of Accounting and Finance	2	2	3.71%
Agricultural Finance Review	1	1	1.85%
British Journal of Management	4	1	1.85%
Corporate Social Responsibility and Environmental Management	1	1	1.85%
Emerging Markets Review	2	1	1.85%
Energy Economics	3	1	1.85%
European Journal of Finance	3	1	1.85%
Financial Markets, Institutions and Instruments	3	1	1.85%
International Journal of Finance and Economics	3	1	1.85%

continued on following page

Table 2. Continued

Journal Title	*AJG 2024* **Ranking**	**Count**	**Share**
International Journal of Managerial Finance	2	1	1.85%
International Journal of the Economics of Business	2	1	1.85%
Journal of Banking and Finance	3	1	1.85%
Journal of Business Ethics	3	1	1.85%
Journal of Cases on Information Technology	1	1	1.85%
Journal of Cleaner Production	1	1	1.85%
Journal of Corporate Finance	4	1	1.85%
Journal of Financial Research	3	1	1.85%
Journal of Financial Services Research	3	1	1.85%
Journal of International Economics	4	1	1.85%
Journal of International Financial Markets, Institutions and Money	3	1	1.85%
Journal of Real Estate Research	1	1	1.85%
Managerial Finance	1	1	1.85%
Pacific Basin Finance Journal	2	1	1.85%
Review of Finance	4*	1	1.85%
Review of Financial Studies	4*	1	1.85%
Technological Forecasting and Social Change	3	1	1.85%
Total		54	100%

Note: The Chartered Association of Business Schools' *2024 Academic Journal Guide (AJG 2024)* assigns journals a rating on a five-point scale (1, 2, 3, 4, and 4*), where 4* denotes the highest-quality, world-leading journals.

Stage 4: Data Analysis and Thematic Classification

The final dataset was systematically reviewed and coded according to emerging thematic patterns. Through iterative analysis, the studies were classified into three interrelated themes: (1) Climate Adaptation, (2) Loan Pricing, and (3) Climate Risk Disclosure, each representing a critical dimension of banks' responses to climate-related risks. The thematic structure was developed collaboratively by two researchers, and all discrepancies were resolved through consensus discussion. Table 3 shows the distribution of articles across the identified themes.

Table 3. Distribution of articles by themes in banks' climate risk management

Themes	Number of articles
Climate adaptation	30
Loan Pricing	21
Climate risk disclosure	5

Notes: Few articles address more than one theme; therefore, the total number slightly exceeds the sample of 54 studies.

While several studies overlapped across themes, the categorisation highlights both convergence and divergence in how banks integrate climate risks within risk management frameworks. Collectively, these themes provide a comprehensive, multidimensional perspective on how banks operationalise climate risk integration, which is an area often explored separately but without a holistic approach.

4. FINDINGS AND DISCUSSION

4.1 Climate Adaptation

Out of the 54 reviewed studies, 30 examine how banks adapt their risk management frameworks to climate-related risks. Climate adaptation refers to an organisation's capacity to anticipate, adjust to, and mitigate the adverse effects of climate change on its operations and financial performance. Enhanced adaptation is consistently associated with higher profitability, improved resilience, and lower credit risk exposure (Doan et al., 2025).

Evidence indicates that banks in China and the European Union (E.U.) are at the forefront of integrating climate adaptation into risk management practices. Following the Paris Agreement, lower funding costs and reputational benefits have incentivised banks to strengthen their environmental commitment. Prior experience with climate shocks has also enhanced institutional learning and diversification strategies. Additionally, high ESG performance, the adoption of digital technologies, and board diversity have emerged as important drivers of effective climate-related adaptation.

These patterns can be interpreted through the lens of institutional theory, which emphasises how coercive, normative, and mimetic pressures shape organisational behaviour. The expansion of green credit in China and the E.U. reflects strong coercive pressures. For instance, in China, through the Green Credit Guidelines[3] and top-down legal enforcement, and in Europe, through supranational commitments under the Paris Agreement and E.U. Taxonomy regulations. Simultaneously, normative expectations for environmental accountability and sustainable finance encourage banks to embed green lending standards into risk management frameworks to preserve legitimacy and stakeholder trust. Moreover, banks operating in emerging markets often adopt similar practices through mimetic isomorphism, emulating global leaders to signal responsibility and attract international investment. These institutional mechanisms explain why the diffusion of green credit practices is uneven across jurisdictions and why some banks exhibit substantive adaptation while others engage in more symbolic compliance.

Based on the synthesis of the reviewed literature, the theme of climate adaptation is categorised into five interrelated subthemes: (1) green credit, (2) green commitment and prior experience, (3) ESG engagement, (4) digital transformation and technology, and (5) board diversity. Each subtheme captures a distinct but complementary mechanism through which banks strengthen their resilience to climate-related risks.

Table 4. Articles on climate adaptation

Study	Data		Green credit	Green commitment & prior experience	ESG engagement	Digital transformation & technology	Board diversity
	Period	Country/ Region					
Afzal et al. (2024)	2013-2022	E.U.				√	
Baek and Kang (2025)	2005-2022	Global/Cross-country			√		
Birindelli et al. (2022)	2011-2019	Global/Cross-country		√			
Birindelli et al., 2024)	2011-2020	Global/Cross-country		√			
Chen et al. (2025)	2014-2022	China		√			
Cucinelli et al. (2024)	2008-2021	E.U.		√			
de Novellis et al. (2025)	2015-2021	Global/Cross-country					√
Doan et al. (2025)	2002-2019	Global/Cross-country		√			
Erhemjamts et al. (2024)	2003-2018	U.S.			√		
Feng et al. (2024)	2007-2022	China	√				
Feng et al. (2025)	2015-2021	China					

continued on following page

Table 4. Continued

Study	Data		Green credit	Green commitment & prior experience	ESG engagement	Digital transformation & technology	Board diversity
	Period	Country/ Region					
Galletta et al. (2021)	2011-2015	Global/Cross-country					√
Galletta et al. (2023)	2011-2020	Global/Cross-country			√		
Galletta and Mazzù (2022)	2011-2020	Global/Cross-country					
Hassan et al. (2025)	2013-2022	Global/Cross-country					√
Huang et al. (2025)	2012-2022	China	√				
Islam et al. (2025)	2002-2023	Global/Cross-country		√			
Islam and Singh (2025)	1997-2017	U.S.		√			
Liu et al. (2024)	2010-2020	China				√	
Mirza et al. (2025)	2012-2022	E.U.	√				
Oh and Kim (2023)	2012-2019	U.S.		√			
Song et al. (2025)	2008-2021	China	√				
Sun et al. (2025)	2012-2021	China				√	
Wang and Fu (2025)	2017-2023	China				√	
Wen et al. (2025)	2009-2023	China			√		
Wu (2025)	2009-2023	China	√				
Wu and Zhang (2025)	2009-2024	China	√				
Zhang et al. (2025)	2013-2021	China				√	
Zheng et al. (2025)	2013-2023	E.U.	√				
Zhou et al. (2024)	2009-2021	China	√				

(1) Green Credit

Green credit has become a cornerstone of banks' climate adaptation strategies, supporting environmentally friendly industries and projects that generate both economic and social benefits (Feng et al., 2024). It enhances environmental risk management across all phases of the credit cycle, from pre-loan screening and approval to post-loan monitoring (Huang et al., 2025). In China, commercial banks use green credit criteria to assess borrowers' environmental performance during loan approval (Song et al., 2025).

Empirical evidence constantly shows that green credit reduces banks' risk-taking and enhances long-term profitability and liquidity (Feng et al., 2025; Huang et al., 2025; Wu, 2025; Wu and Zhang, 2025). Chinese banks increasingly apply differentiated interest rates and targeted lending mechanisms to incentivise low-carbon transition. Yet, this effect appears heterogenous across bank sizes. Feng et al. (2024) observed that the risk-mitigating impact of green credit is short-lived, particularly among smaller and less profitable banks that either lack of the capacity to manage green portfolios or use green lending strategically to offset other risk exposures. Similarly, Zhou et al. (2024) reported that while green loan reduces insolvency risks for large Chinese banks, they may heighten it for small and medium-sized banks.

Evidence from Europe is more limited but promising. Mirza et al. (2025) found that green and blue credit[4] initiatives increase European banks' profitability and lower loan default risks by leveraging specialised environmental expertise. However, Zheng et al. (2025) showed that many European banks remain hesitant to finance small and medium-sized green enterprises due to perceived uncertainties and higher due-diligence costs. Although green credit can improve margins and sustainability alignment, it does not fully stabilise the sector, highlighting an ongoing tension between profitability and environmental responsibility.

(2) Green Commitment and Prior Experience

Banks' green commitment, their strategic and operational dedication to climate actions, plays a crucial role in systemic risk reduction and long-term sustainability (Degryse et al., 2023; Birindelli et al., 2024). Strong environmental performance enhances banks' legitimacy and signals resilience to stakeholders (Doan et al., 2025).

Empirical findings show that banks with higher green commitment enjoy lower funding costs, as investors and depositors accept reduced returns in exchange for perceived environmentally responsibility (Islam et al., 2025). During extreme weather events, banks integrate CSR and climate risk pricing to adjust for potential losses (Chen et al., 2025). Consequently, medium- to high-commitment banks report lower credit risk and higher loan quality, while even low-commitment banks benefit from operating in environmentally advanced countries (Birindelli et al., 2022).

European banks reinforce climate resilience through training programmes, CSR committees, and sustainability-linked remuneration, which strengthen internal engagement (Cucinelli et al., 2024). Similarly, Chinese banks with prior exposure to flood-prone regions demonstrate enhanced risk management capabilities and maintain stable credit spreads under stress (Chen et al., 2025). In the United States, geographically diversified banks exhibit greater resilience to abnormal temperature shocks, reallocating credit effectively due to superior local information networks (Islam and Singh, 2025). Agricultural banks also adapt strategically by increasing capital buffers and diversifying portfolios to withstand environmental volatility (Oh and Kim, 2023). Nonetheless, such strategies only partially mitigate climate-related vulnerabilities, underscoring the need for integrated, cross-market adaptation frameworks.

(3) ESG Engagement

A growing body of literature highlights ESG engagement as a critical driver of sustainable banking resilience. Higher ESG scores correlate with lower risk-taking, stronger stability, and improved long-term profitability, while simultaneously attracting socially responsible investors (Baek and Kang, 2025).

Banks with robust ESG profiles display superior asset quality and lower operational risks (Galletta et al., 2023). They also benefit from lower funding costs and reputational capital that supports proactive risk management (Baek and Kang, 2025; Wen et al., 2025). Erhemjamts et al. (2024) found that U.S. banks facing high physical climate risk tend to invest more heavily in ESG activities to mitigate adverse financial impacts. In turn, ESG performance reduces the incidence of symbolic compliance; banks with fewer ESG controversies take fewer risks and demonstrate more authentic integration of sustainability into governance (Galletta and Mazzù, 2022; Galletta et al., 2023). Therefore, ESG is not merely a reporting mechanism but a strategic tool for risk mitigation and legitimacy building in an era of climate uncertainty.

(4) Digital Transformation and Technology

Digital transformation enhances banks' resilience and operational agility under climate-related pressures (Sun et al., 2025). Technologies such as AI, big data, blockchain, cloud computing, and Fintech enable more precise and real-time environmental risk assessment (Liu et al., 2024; Sun et al., 2025; Wang and Fu, 2025; Zhang et al., 2025).

In Europe, green technology adoption (GTA) contributes to profitability and risk reduction through green innovation and human resource development (Afzal et al., 2024). In China, digitalisation strengthens supervisory capacity, reduces information asymmetry, and limits greenwashing by improving monitoring efficiency (Liu et al., 2024; Sun et al., 2025). The People's Bank of China's Fintech Development Plan (2022–2025) further institutionalises the integration of financial technology into green finance. Wang and Fu (2025) confirmed that FinTech improves green credit expansion by refining risk identification and loan pricing accuracy.

However, technology's benefits are not evenly distributed. National commercial banks outperform regional institutions due to their superior financial resources and scale (Zhang et al., 2025). Moreover, as Liu et al. (2024) caution, rapid digitalisation introduces new operational and cyber risks, such as data breaches and fraud, which remain understudied in bank risk management. This underscores the dual nature of technological transformation as both an enabler and potential source of new vulnerabilities.

(5) Board Diversity

Board diversity has gained increasing attention as a governance factor that enhances banks' climate resilience. Greater female representation on boards tends to foster stronger environmental oversight and more balanced decision-making (de Novellis et al., 2025). For instance, banks with at least four women directors are more likely to implement comprehensive environmental risk policies (Hassan et al, 2024).

However, the effect appears non-linear. When female board representation exceeds approximately 30%, boards may adopt a more conservative stance, reducing risk appetite, and limiting proactive disclosure, a phenomenon often referred to as *greenhushing*[5] (de Novellis et al., 2025). Conversely, women-led banks demonstrate lower greenhushing tendencies, reflecting their greater responsiveness to reputational and regulatory pressures. Additionally, managerial incentives linked to environmental performance further strengthen governance quality and the implementation of sustainability strategies (Galletta et al., 2021). Board diversity, therefore, operates as a governance-level adaptation mechanism, shaping how banks internalise climate-related risks and balance profit with purpose.

The patterns observed across these subthemes also reflect insights from the RBV, which emphasises that firms achieve sustained competitive advantage through the accumulation and deployment of unique, valuable, and hard-to-imitate capabilities. Banks that invest in digital transformation, cultivate board diversity, and build organisational learning from past climate shocks effectively develop internal competencies that enhance resilience to environmental risks. These resources, ranging from data analytics for environmental assessment to governance structures that support strategic decision-making, represent dynamic capabilities that allow banks to anticipate, absorb, and adapt to climate-related disruptions more effectively than peers. In this sense, climate risk management is not merely a compliance activity but a strategic process of capability-building. The heterogeneity in banks' performance across countries and institutions therefore illustrates how differences in internal resources shape the depth and success of climate adaptation strategies.

4.2 Loan Pricing

A total of 21 studies in our review examine how banks incorporate climate-related risks into loan pricing and credit allocation. Loan pricing reflects banks' assessment of both borrowers' environmental profiles and their exposure to physical or transition risks. The Paris Agreement and the rise of governmental and societal awareness of climate issues have become decisive drivers prompting banks to adjust interest rates, loan volumes, and contract terms for firms and households (Bruno and Lombini, 2023). This shift is particularly visible among institutions exposed to fossil-fuel or carbon-intensive sectors, often referred to as "brown lending".

Empirical evidence shows that, following the Paris Agreement, European banks reallocated credit away from polluting firms, with loan volumes roughly 3 percent lower than for cleaner firms (Reghezza et al., 2022). Similarly, Degryse et al. (2023) observed that banks offer lower interest rates to green firms only after the Paris Agreement's implementation. Loans to these firms are generally shorter in maturity and contain fewer covenants, reflecting banks' preference for flexible, lower-risk exposure, an effect termed "when green meets green". Banks also differentiate between short-term climate shocks (e.g., abnormal temperature) and long-term physical threats (e.g., sea-level rise) when setting prices and maturities.

Based on the synthesis of findings, this theme is organised into two interrelated subthemes: (1) interest rate and volume adjustments, and (2) brown lending dynamics, each illustrating a distinct dimension of how climate risk considerations shape credit decisions.

Table 5. Articles on loan pricing

Study	Data		Interest rate & volume adjustments	Brown lending dynamics
	Period	Country/Region		
Aslan et al. (2022)	2010-2019	Turkey		√
Bell et al. (2023)	2008-2017	U.K.		√
Bruno and Lombini (2023)	2010-2021	Global/Cross-country	√	
Chaudhry and Kumari (2024)	1997-2013	India		√
Conlon et al. (2024)	1980-2020	U.S.	√	
Degryse et al. (2023)	2010-2018	Global/Cross-country	√	
Delis et al. (2024)	2002-2016	Global/Cross-country		√
Demetriades and Politsidis (2025)	2001-2021	Global/Cross-country		√
Duanmu et al. (2022)	2002-2016	U.S.	√	
Ehlers et al. (2022)	2005-2018	Global/Cross-country		√
Gambacorta et al. (2024)	2018-2019	E.U.		√
Ge et al. (2025)	2000-2019	Global/Cross-country	√	
Ho and Wong (2023)	2010-2021	Global/Cross-country		√
Hrazdil et al. (2024)	2007-2016	U.S.	√	
Ivanov et al. (2024)	2011-2023	U.S.	√	
Mueller et al. (2025)	2013-2019	U.S.		√
Mueller and Sfrappini (2025)	2010-2019	Global/Cross-country	√	
Nguyen et al. (2022)	1992-2018	U.S.	√	
Shikimi (2025)	2018-2021	Japan		√
Reghezza et al. (2022)	2014-2018	E.U.	√	
Zhu and Zhao (2022)	2009-2019	China	√	

(1) Interest rate and volume adjustments

Banks increasingly use loan pricing to internalise environmental risk. Institutions tend to charge higher interest rates or restrict credit volumes for borrowers with poor environmental performance, particularly under stronger regulatory regimes.

Even before the Paris Agreement, Hrazdil et al. (2023) found that U.S. banks responded to firms' first negative climate-related news by issuing loans with higher spreads, shorter maturities, stricter covenants, reflecting concern for reputational rather than environmental risk. Under California's 2011 cap-and-trade policy, U.S. banks shortened maturities, raised rates, and tightened lending terms (Ivanov et al., 2023).[6] In China, the 2012 China Green Credit Guidelines similarly led banks to charge higher rates to firms with low innovation or heavy carbon intensity (Zhu and Zhao, 2022).

After the Paris Agreement, banks worldwide intensified this differentiation. They increased interest rates and reduced loan volumes to polluting borrowers, especially in countries with higher climate-policy awareness (Bruno and Lombini, 2023). Banks with weaker credit quality or profitability were particularly likely to decrease their exposure to polluting firms (Reghezza et al., 2022). However, regulatory stringency remains a key determinant: Mueller and Sfrappini (2025) found that banks expand credit to brown borrowers in low-stringency environments, such as the U.S., while cutting exposure in stricter regimes, such as Europe.

Banks also recalibrate pricing in response to climate shocks. U.S. banks tend to decrease lending and increase loan-loss reserves after extreme events (Conlon et al., 2024) and tighten mortgage standards in disaster-affected regions (Duanmu et al., 2021). The time horizon of the threat matters: Ge et al. (2024) showed that borrowers exposed to short-term abnormal-temperature risks face higher interest charges, whereas Nguyen et al. (2022) found that banks raise long-term mortgage rates in areas vulnerable to sea-level rise. Overall, banks' pricing behaviour reflects an evolving differentiation between immediate physical risks and gradual transition risks in credit markets.

(2) Brown lending dynamics

Before the 2015 Paris Agreement, banks showed limited incentive to fund the green transition of fossil-fuel-intensive firms (Delis et al., 2024). For instance, U.K. banks did not price transition risk into mortgages until the Clean Growth Strategy (2018).[7] Since 2015, however, mounting regulatory and reputational pressures have prompted many banks to embed climate risk into loan contracts (Ehlers et al., 2021). Banks in Japan (Shikimi, 2025), Turkey (Aslan et al, 2022) and other emerging markets (Ho and Wong, 2023) have increased rates or reduced credit supply to fossil fuels and other carbon-intensive borrowers. Green-labeled banks impose higher spreads and stricter non-pricing terms, such as shorter maturities or additional collateral, when lending to high-brown firms (Ho and Wong, 2023; Delis et al., 2024; Demetriades and Politsidis, 2025).

Yet, findings are not uniform. Chaudhry and Kumari (2024) reported that Indian banks still view energy consumption positively, granting lower-cost debt to energy-intensive, high-growth firms. Similarly, Ehlers et al. (2021) found that while fossil-fuel industries receive higher loan premiums, there is no systematic difference between banks that are UNEP FI or EPs members and those that are not, implying that "green" affiliation does not always translate into differentiated pricing.

Contradictions also appear in the U.S. context. Demetriades and Politsidis (2025) observed that banks charge roughly 7 percent higher interest rates yet provide 22 percent larger loan volumes to fossil-fuel borrowers, reflecting a trade-off between profitability and risk control. Mueller et al. (2025) further showed that U.S. banks engage in brown-loan securitisation as a form of *regulatory arbitrage*. They offload brown exposures when transition risk rises (e.g., during policy shifts) and re-absorb them when risks subside. This mechanism enables banks to maintain profitable lending while circumventing capital constraints.

In the United States, profit motives remain a central factor behind the persistence of brown lending. Demetriades and Politsidis (2025) found that in syndicated loans, where fossil-fuel and other carbon-intensive firms require large volumes of financing, banks find complete withdrawal economically unfeasible. Continued participation in these syndicated arrangements enables them to preserve fee income, maintain client relationships, and protect market share, even while recognising the environmental risks involved. Similarly, Aslan et al. (2022) observed that Turkish banks adjust credit allocation across provinces based on local pollution intensity but face difficulties in accurately assessing the additional funding costs of lending in more polluted regions. This information asymmetry may result in the mispricing of environmental risk or the neglect of firms with positive sustainability signals. Moreover, Gambacorta et al. (2024) found that several European banks project an overly optimistic environmental tone in their annual reports while simultaneously extending significant credit to brown sectors, which is a form of *environmental window-dressing*[8] that raises questions about the authenticity of their green commitments.

These findings align closely with agency theory, which explains the tension between short-term profit objectives and long-term sustainability goals within financial institutions. Managers, incentivised by performance-based rewards and shareholder expectations for immediate returns, may prioritise profitability over environmental responsibility. The persistence of brown lending, even when regulatory or reputational pressures increase, illustrates this agency conflict. In contexts where executive compensation is tied to quarterly results or loan growth, the incentive to maintain profitable relationships with carbon-intensive clients can outweigh the perceived benefits of risk reduction through green transition. As a result, banks may adopt risk management practices that appear environmentally conscious but continue to expose them to climate transition risk, reflecting symbolic rather than substantive adaptation. Agency-driven short-termism thus offers a behavioural explanation for the inconsistency observed between banks' stated green commitments and their actual lending practices.

4.3 Climate Risk Disclosure

Five studies in our review focus on the mandatory and voluntary disclosure of climate-related risks in the banking sector. Climate risk disclosure serves a core mechanism for enhancing transparency, reducing systemic uncertainty, and signalling sound risk management. Disclosing climate-related exposures, banks communicate to investors, regulators, and other stakeholders that they are proactively managing environmental risks and aligning with sustainable finance principles (Hu and Borjigin, 2025).

Despite this progress, most climate disclosures remain voluntary, with only a few jurisdictions, e.g., New Zealand, the European Union, the United Kingdom, and China, introducing mandatory reporting requirements beginning in fiscal year 2023 or later. These policies aim to harmonise environmental transparency with financial stability goals. Friedrich et al. (2022) found that European increasingly acknowledge climate-related risks and now provide higher-quality disclosure within the Risk Management sections of annual reports.

Empirical evidence also reveals that disclosure frameworks influence banks' lending and pricing behaviours. Demetriades and Politsidis (2025) observed that loan spreads for fossil-fuel borrowers are unaffected by whether those firms operate in countries with general mandatary disclosure regimes. However, when mandatory ESG disclosure is implemented, banks charge such firms higher loan premiums, reflecting perceived transition risks and the cost of compliance. This finding underscores the importance of policy-driven green lending incentives from the supply side to rebalance credit flows toward sustainable activities.

In addition, voluntary disclosure by corporate borrowers has become an increasingly important determinant of loan pricing. Banks tend to reward firms that voluntarily disclose climate-risk information because such transparency reduces information asymmetry and improves risk assessment accuracy (Ge et al., 2024). Consistent this, Lin et al. (2025) found that Taiwanese banks offer lower loan interest rates to firms that disclose their greenhouse-gas emissions or maintain lower overall emission levels. These findings collectively suggest that transparent climate-risk communication functions as both a risk-mitigation tool and a market-based incentive for environmentally responsible behaviour among borrowers and lenders alike.

These findings can also be understood through the lens of stakeholder theory, which emphasises that banks operate within a network of relationships that extend beyond shareholders to include depositors, regulators, investors, and the wider public. Climate risk disclosure represents a key mechanism through which banks maintain legitimacy and accountability to these diverse stakeholders. By voluntarily revealing emissions data and climate-risk exposures, banks signal responsiveness to societal and regulatory expectations, thereby strengthening trust and reducing reputational uncertainty. In turn, this transparency helps to align the interests of management and stakeholders, mitigating agency conflicts and promoting more responsible lending behaviour. The positive relationship between disclosure quality and loan pricing outcomes illustrates how stakeholder-oriented strategies can serve both ethical and strategic purposes, enhancing information efficiency while reinforcing banks' credibility and long-term resilience in the transition toward sustainable finance.

Table 6. Articles on climate risk disclosure

Study	Data		Mandatory disclosure	Voluntary disclosure
	Period	Country/Region		
Demetriades and Politsidis (2025)	2001-2021	Global/Cross-country	√	
Friedrich et al. (2023)	2017-2020	E.U.		√
Ge et al. (2025)	2000-2019	Global/Cross-country		√
Hu and Borjigin (2025)	2015-2024	U.S.		√
Lin et al. (2025)	2013-2021	Taiwan		√

5. GUIDANCE FOR FUTURE RESEARCH

Future research should deepen understanding of the motivations and behavioural mechanisms underlying banks' climate-related actions, such as green credit allocation, brown lending persistence, and climate-related loan pricing, particularly under varying institutional pressures. A key question remains whether banks strengthen their risk management frameworks substantively or symbolically in response to climate threats. Addressing this issue is essential for assessing the credibility of their climate adaptation in banking and for advancing the long-term green transition.

Another promising direction involves exploring the application of digital technologies in risk management beyond the dominant contexts of China and Europe. Although existing studies highlight the benefits of technologies such as AI, big data, blockchain, cloud computing and FinTech for improving green credit assessment and transition risk management, their potential vulnerabilities remain underexplored. Further research should investigate how digitalisation can protect banks from cybersecurity threats, data fraud, and privacy breaches, as well as how regulatory frameworks might strengthen public supervision and data governance. Since large national banks typically have greater resources for technological investment than small or underperforming institutions, further investigation into how different banks can overcome financial and human capital barriers to digital transformation would provide valuable policy insight.

The governance dimension also warrants greater attention. Existing research largely examines board diversity through quantifiable metrics such as gender, ownership, and incentives. Future work should extend this focus to the social and moral aspects of leadership, aligned with the Sustainable Development Goals (SDGs). Ethical leadership and moral reasoning within boardrooms may determine whether banks pursue sustainability as a long-term strategic commitment or merely a reactive measure to regulatory or reputational pressure. Qualitative or mixed-method approaches could help reveal how directors balance competing priorities, such as profitability, CSR, ESG, and stakeholder expectations, and how internal governance dynamics influence the prevalence of greenwashing, greenhushing, or authentic sustainability initiatives.

Further research is needed to understand how banks manage physical climate risks through local operations and branch networks. While studies have shown that U.S. agricultural banks use branch experience to manage exposure to floods, abnormal temperatures, and sea-level rise, little is known about how credit unions, community banks, or regional institutions respond to droughts, storms, and wildfires in other parts of the world. Examining how banks design branch planning strategies, whether centralised or geographically diversified, could shed light on how they maintain credit continuity and customer support during extreme events. This line of inquiry would also illuminate the role of banks in building community resilience and disaster preparedness in regions highly exposed to physical climate risks.

Understanding the incentives and profitability trade-offs underlying banks' climate behaviour remains another critical research area. Evidence suggests that banks engage in green transitions partly because of lower funding costs and reputational gains, yet these motivations often weaken during economic downturns or periods of high interest rates (Islam et al., 2025). Future research should therefore examine the design of diversified incentive structures that sustain banks' environmental commitments even under adverse market conditions. This is particularly important for small and micro banks, institutions with limited capital buffers, or those in developing economies. Researchers should explore questions such as why smaller banks increase green lending despite higher insolvency risks, why banks concerned with ESG reputation continue extending brown credit, and whether higher interest rates on carbon-intensive loans reflect short-term opportunism or long-term transition strategies. A related question is how banks can safely shift from reliance on interest income to more diversified revenue sources without compromising asset quality or increasing credit costs.

The regulatory and institutional environment remains a decisive factor in shaping banks' climate responses. Studies show that Chinese and European banks engage in green credit and digital transformation largely due to strong top-down legal enforcement, whereas elsewhere, coercive regulatory pressure is weaker. Yet even in highly regulated systems, strict enforcement may yield unintended outcomes such as greenwashing, information overload, or reduced profitability from slow-maturing green investments (Demetriades and Politsidis, 2025). Future research should therefore investigate how regulatory design can balance environmental objectives with financial performance, encouraging substantive rather than symbolic adaptation. The U.S. withdrawal from the Paris Agreement in 2025 further highlights how political reversals and policy uncertainty can trigger regulatory arbitrage and capital losses (Mueller et al., 2025). Exploring how consistent institutional environments and cross-border coordination could strengthen banks' long-term climate resilience and support a global transition toward Net Zero remains a vital research priority.

Finally, the geographical focus of existing studies remains heavily concentrated on large commercial banks in the U.S., China, and the E.U., leaving significant gaps in knowledge about banks operating in emerging and developing economies. Countries such as Turkey and India provide contrasting insights: Turkish banks encourage green lending in polluted regions (Aslan et al., 2022), whereas Indian banks appear to reward carbon-intensive firms with lower borrowing costs (Chaudhry and Kumari, 2024). Expanding research into underrepresented regions would contribute to a more balanced understanding of how contextual factors, such as governance quality, legal enforcement, and market maturity, affect banks' approaches to climate risk management.

6. CONCLUSIONS AND IMPLICATIONS

This chapter has provided a comprehensive synthesis of how commercial banks integrate climate-related risks into their risk management frameworks. Drawing on 54 studies across different regions, it examined banks' responses through three interconnected dimensions: climate adaptation, loan pricing, and climate risk disclosure. The evidence shows that climate risk management has evolved from a regulatory obligation into a strategic imperative for financial stability, profitability, and legitimacy. Banks, particularly those in China and the European Union, have taken a leading role in embedding environmental considerations into their governance and lending practices. The Paris Agreement has acted as a powerful catalyst, accelerating the provision of green credit to environmentally responsible firms and motivating institutions to strengthen their green commitments. These actions have brought both reputational and financial benefits, including lower funding costs and enhanced stakeholder confidence. In parallel, higher ESG performance, investment in digital technologies, and greater board diversity have improved banks' capacity to manage and absorb climate shocks, while the expansion of climate risk disclosure has enhanced transparency and reduced information asymmetries between banks, regulators, and investors. Together, these developments illustrate that managing climate-related risks is increasingly viewed not as a compliance burden but as an essential element of long-term competitiveness and resilience.

Nonetheless, the review also highlights persistent challenges. Climate adaptation within the global banking industry remains uneven and incomplete. Many institutions continue to extend credit to carbon-intensive sectors, often charging higher loan premiums but without meaningfully reducing their overall exposure. Others engage in symbolic rather than substantive actions, such as selective reporting or environmental window-dressing, to maintain legitimacy without significant transformation. Such patterns demonstrate that progress is closely tied to the stringency of regulation, the strength of institutional commitment, and the availability of financial and technological resources to support genuine change. The persistence of these inconsistencies raises a broader question of how to sustain authentic commitment to climate transition in a fragmented global context where political and regulatory approaches diverge widely.

Interpreting these patterns through theoretical lenses provides further insight. From an institutional perspective, banks' climate behaviours reflect the combined influence of coercive, normative, and mimetic pressures. Coercive pressures from regulatory bodies and international agreements push banks toward compliance, while normative expectations from society and mimetic imitation of leading peers encourage legitimacy-seeking behaviour. This interplay helps explain both the spread of green credit initiatives and the endurance of symbolic compliance in less mature institutional environments. Agency theory adds a behavioural dimension, highlighting that managerial incentives and short-term shareholder expectations can discourage long-term sustainability investments. The continued funding of high-carbon industries, despite growing climate awareness, exemplifies the tension between immediate profitability and long-term resilience. Stakeholder theory offers a broader view, showing that effective climate risk management depends on transparency and accountability to multiple constituencies, such as investors, regulators, clients, and the public. Banks that provide high-quality climate disclosures are rewarded through lower financing costs and stronger reputational standing, indicating that responsiveness to stakeholder expectations can generate tangible strategic advantages. Complementing these external perspectives, the resource-based view emphasises the internal side of resilience: banks that build unique capabilities, such as advanced data analytics, green innovation capacity, and diverse governance structures, develop dynamic competencies that enhance their ability to anticipate and adapt to climate-related shocks. Taken together, these theoretical perspectives explain why some banks achieve substantive integration of climate risks into their management systems, while others remain locked in symbolic adaptation.

The evidence also yields several broader implications for policy and practice. Policymakers should recognise that mandatory regulations alone cannot ensure authentic environmental integration. Effective climate transition in the financial sector requires a balanced approach that combines regulatory stringency with market-based incentives, normative guidance, and technological support. Coordination among international bodies such as the Basel Committee, the NGFS, and the UNFCCC will be essential to harmonise disclosure standards, reduce regulatory arbitrage, and prevent greenwashing. For banking practitioners, the findings highlight the importance of investing in digital infrastructure for risk analytics, embedding ESG principles throughout governance and credit processes, and fostering diverse and sustainability-conscious boards. These practices not only strengthen risk management but also enhance access to capital, investor trust, and long-term profitability. Ultimately, banks that treat climate adaptation as a strategic investment rather than a compliance cost are more likely to withstand transition shocks, align profitability with sustainability, and contribute meaningfully to global net-zero objectives.

In conclusion, the global banking sector stands at a decisive juncture. The ability of banks to integrate climate considerations into their strategic, operational, and governance frameworks will determine not only their individual resilience but also the stability of the wider financial system. Achieving this requires sustained collaboration among regulators, financial institutions, and civil society to close the gap between symbolic and substantive adaptation. As climate change continues to reshape economic and environmental landscapes, embedding climate risk management within the core logic of banking is not just an ethical necessity but a strategic imperative for building a stable, sustainable, and inclusive global economy.

REFERENCES

Aebi, V., Sabato, G., & Schmid, M. (2012). Risk management, corporate governance, and Bank performance in the financial crisis. *Journal of Banking & Finance, 36*(12), 3213–3226. DOI: 10.1016/j.jbankfin.2011.10.020

Afzal, A., Hasnaoui, J., Firdousi, S., & Noor, R. (2024). Climate change and the European banking sector: The effect of green technology adaptation and human capital. *Review of Accounting and Finance, 23*(3), 394–418. DOI: 10.1108/RAF-10-2023-0341

Aslan, C., Bulut, E., Cepni, O., & Yilmaz, M. (2022). Does climate change affect bank lending behavior? *Economics Letters, 220*, 110859. DOI: 10.1016/j.econlet.2022.110859

Baek, S., & Kang, M. (2025). Does ESG enhance asset quality and funding cost management in banking diversification? *Finance Research Letters, 73*, 106542. DOI: 10.1016/j.frl.2024.106542

Bell, J., Battisti, G., & Guin, B. (2023). The greening of lending: Evidence from banks' pricing of energy efficiency before climate-related regulation. *Economics Letters, 230*, 111212. DOI: 10.1016/j.econlet.2023.111212

Birindelli, G., Bonanno, G., Dell'Atti, S., & Lannuzzi, A. (2022). Climate Change Commitment, credit risk and the country's environmental performance: Empirical evidence from a sample of international banks. *Business Strategy and the Environment, 31*(4), 1641–1655. DOI: 10.1002/bse.2974

Birindelli, G., Dell'Atti, S., Di Tommaso, C., Lannuzzi, A., & Pacelli, V. (2024). The impact of banks' climate engagement on systemic risk. does committing a little or a lot make a difference?'. *Research in International Business and Finance, 70B*, 102392. DOI: 10.1016/j.ribaf.2024.102392

Breitenstein, M., Nguyen, D. K., & Walther, T. (2021). Environmental hazards and risk management in the financial sector: A Systematic Literature Review. *Journal of Economic Surveys, 35*(2), 512–538. DOI: 10.1111/joes.12411

Brüggemann, T., & Lueg, R. (2025). Bank Responses to Physical and Transition Risks in Lending: A Diagnostic Framework from a Systematic Literature Review. *Business Strategy and the Environment, 0*, 1–18. DOI: 10.1002/bse.70176

Bruno, B., & Lombini, S. (2023). Climate transition risk and bank lending. *Journal of Financial Research, 46*(S1), 59–106. DOI: 10.1111/jfir.12360

Chaudhry, N., & Kumari, D. (2024). How do banks price carbon risk? evidence from India. *Pacific-Basin Finance Journal, 84*, 102304. DOI: 10.1016/j.pacfin.2024.102304

Chen, S., Xu, Q., Zeng, J., & Zhao, Y. (2025). Ark amidst floods: Evidence from personal credit at a Chinese bank. *International Review of Financial Analysis, 106*, 104514. DOI: 10.1016/j.irfa.2025.104514

Conlon, T., Ding, R., Huan, X., & Zhang, Z. (2024). Climate risk and financial stability: Evidence from syndicated lending. *European Journal of Finance, 30*(17), 2001–2031. DOI: 10.1080/1351847X.2024.2343111

Cucinelli, D., Nieri, L., & Piserà, S. (2024). Climate and environment risks and opportunities in the banking industry: The role of Risk Management. *International Journal of Finance & Economics, 0*, 1–17. DOI: 10.1002/ijfe.3083

de Bandt, O., Kuntz, L., Pankratz, N., Pegoraro, F., Solheim, H., Sutton, G., Takeyama, A., & Xia, F. D. (2025). The effects of climate change-related risks on banks: A literature review. *Journal of Economic Surveys, 39*(4), 1553–1594. DOI: 10.1111/joes.12665

de Novellis, G., Pedrazzoli, A., Pennetta, D., & Venturelli, V. (2025). Greenwashing, greenhushing, and the path to Green Banking. *Global Finance Journal, 67*, 101147. DOI: 10.1016/j.gfj.2025.101147

Degryse, H., Goncharenko, R., Theunisz, C., & Vadasz, T. (2023). When green meets green. *Journal of Corporate Finance, 78*, 102355. DOI: 10.1016/j.jcorpfin.2023.102355

Delis, M. D., Greiff, K., Losifidi, M., & Ongena, S. (2024). Being stranded with fossil fuel reserves? climate policy risk and the pricing of Bank Loans. *Financial Markets, Institutions &amp. Instruments, 33*(3), 239–265.

Delmas, M. A., & Montes-Sancho, M. J. (2010). Voluntary agreements to improve environmental quality: Symbolic and substantive cooperation. *Strategic Management Journal, 31*(6), 575–601. DOI: 10.1002/smj.826

Demetriades, E., & Politsidis, P. N. (2025). Bank lending to fossil fuel firms. *Journal of Financial Stability, 76*, 101349. DOI: 10.1016/j.jfs.2024.101349

DiMaggio, P. J., & Powell, W. W. (1983). The Iron Cage Revisited: Institutional isomorphism and collective rationality in organisational fields. *American Sociological Review, 48*(2), 147. DOI: 10.2307/2095101

Doan, A., Tran, A., Tran, H., & Nguyen, C. (2025). A global perspective on climate adaptation, ownership, and Bank Performance. *International Journal of the Economics of Business*, •••, 1–43. DOI: 10.1080/13571516.2025.2479754

Duanmu, J., Li, Y., Lin, M., & Tahsin, S. (2022). Natural disaster risk and residential mortgage lending standards. *Journal of Real Estate Research*, *44*(1), 106–130. DOI: 10.1080/08965803.2021.2013613

Ehlers, T., Packer, F., & de Greiff, K. (2022). The Pricing of Carbon Risk in Syndicated loans: Which Risks Are Priced and why? *Journal of Banking & Finance*, *136*, 106180. DOI: 10.1016/j.jbankfin.2021.106180

Erhemjamts, O., Huang, K., & Tehranian, H. (2024). Climate risk, ESG performance, and ESG sentiment in US Commercial Banks. *Global Finance Journal*, *59*, 100924. DOI: 10.1016/j.gfj.2023.100924

Fan, D., Breslin, D., Callahan, J. L., & Iszatt-White, M. (2022). Advancing literature review methodology through rigour, generativity, scope and transparency. *International Journal of Management Reviews*, *24*(2), 171–180. DOI: 10.1111/ijmr.12291

Feng, T., Xu, J., Zhou, Z., & Luo, Y. (2025). How green credit policy affects commercial banks' credit risk? *Journal of Cases on Information Technology*, *26*(1), 1–21. DOI: 10.4018/JCIT.333858

Feng, Y., Pan, Y., Sun, C., & Niu, J. (2024). Assessing the effect of green credit on risk-taking of commercial banks in China: Further analysis on the two-way Granger causality. *Journal of Cleaner Production*, *437*, 140698. DOI: 10.1016/j.jclepro.2024.140698

Friedman, A. L., & Miles, S. (2002). Developing stakeholder theory. *Journal of Management Studies*, *39*(1), 1–21. DOI: 10.1111/1467-6486.00280

Friedrich, T. J., Velte, P., & Wulf, I. (2023). Corporate climate reporting of European banks: Are these institutions compliant with climate issues? *Business Strategy and the Environment*, *32*(6), 2817–2834. DOI: 10.1002/bse.3272

Galletta, S., Goodell, J., Mazzù, S., & Paltrinieri, A. (2023). Bank reputation and operational risk: The IMPACT OF ESG. *Finance Research Letters*, *51*, 103494. DOI: 10.1016/j.frl.2022.103494

Galletta, S., & Mazzù, S. (2022). ESG controversies and Bank Risk Taking. *Business Strategy and the Environment*, *32*(1), 274–288. DOI: 10.1002/bse.3129

Galletta, S., Mazzù, S., & Naciti, V. (2021). Banks' business strategy and Environmental Effectiveness: The monitoring role of the Board of Directors and the managerial incentives. *Business Strategy and the Environment, 30*(5), 2656–2670. DOI: 10.1002/bse.2769

Gambacorta, L., Polizzi, S., Reghezza, A., & Scannella, E. (2024). Do banks practice what they preach? Brown lending and environmental disclosure in the Euro Area. [Preprint]. *Journal of Financial Services Research.* Advance online publication. DOI: 10.1007/s10693-024-00435-9

Ge, W., Qi, Z., Wu, Z., & Yu, L. (2025). Abnormal temperatures, climate risk disclosures and bank loan pricing: International evidence. *British Journal of Management, 36*(2), 726–744. DOI: 10.1111/1467-8551.12867

Hassan, M. K., Hasan, R., Bataineh, H., & Haque, M. (2025). Bank Board Diversity and Environmental Risk Management: Does critical mass matter in project finance decisions? *Managerial Finance, 51*(6), 879–895. DOI: 10.1108/MF-03-2024-0168

Ho, K., & Wong, A. (2023). Effect of climate-related risk on the costs of bank loans: Evidence from syndicated loan markets in emerging economies. *Emerging Markets Review, 55*, 100977. DOI: 10.1016/j.ememar.2022.100977

Hrazdil, K., Anginer, D., Li, J., & Zhang, R. (2024). Climate reputation and bank loan contracting. *Journal of Business Ethics, 192*(4), 875–896. DOI: 10.1007/s10551-023-05517-7

Hu, Z., & Borjigin, S. (2025). Climate Information Disclosure Quality and systemic risk in the U.S. Banking Industry. *Journal of Financial Stability, 79*, 101420. DOI: 10.1016/j.jfs.2025.101420

Huang, F., Zhu, H., & Wu, S. (2025). Green Credit, monetary policy, and bank risk-taking. *Finance Research Letters, 79*, 107289. DOI: 10.1016/j.frl.2025.107289

Islam, E., & Singh, M. (2025). Geographic diversification, climate risk, and bank lending: Evidence from farm loans. *Journal of Financial Intermediation, 63*, 101152. DOI: 10.1016/j.jfi.2025.101152

Islam, M. J., Moreira, F., & Douch, M. (2025). Does banks' environmental engagement impact funding costs? *Journal of International Financial Markets, Institutions and Money, 103*, 102184. DOI: 10.1016/j.intfin.2025.102184

Ivanov, I. T., Kruttli, M. S., & Watugala, S. W. (2024). Banking on Carbon: Corporate Lending and Cap-and-Trade Policy. *Review of Financial Studies, 37*(5), 1640–1684. DOI: 10.1093/rfs/hhad085

Kraaijenbrink, J., Spender, J. C., & Groen, A. J. (2019). The resource-based view: A review and assessment of its critiques. *Journal of Management, 36*(1), 349–372. DOI: 10.1177/0149206309350775

Kunz, J., & Heitz, M. (2021). Banks' risk culture and management control systems: A Systematic Literature Review. *Journal of Management Control, 32*(4), 439–493. DOI: 10.1007/s00187-021-00325-4

Lim, C. Y., Woods, M., Humphrey, C., & Seow, J. (2017). The paradoxes of risk management in the banking sector. *The British Accounting Review, 49*(1), 75–90. DOI: 10.1016/j.bar.2016.09.002

Lin, H., Shen, H., Hsieh, Y., & Hsiao, H. (2025). Corporate carbon emissions and the costs of bank borrowing: Evidence from an emerging market. *Corporate Social Responsibility and Environmental Management, 0*(6), 1–17. DOI: 10.1002/csr.70087

Liu, Z., Li, J., & Sun, H. (2024). Climate transition risk and bank risk-taking: The role of Digital Transformation. *Finance Research Letters, 61*, 105028. DOI: 10.1016/j.frl.2024.105028

Lockett, A., Thompson, S., & Morgenstern, U. (2009). The development of the resource-based view of the firm: A critical appraisal. *International Journal of Management Reviews, 11*(1), 9–28. DOI: 10.1111/j.1468-2370.2008.00252.x

Mio, C., Panfilo, S., & Blundo, B. (2020). Sustainable development goals and the strategic role of Business: A Systematic Literature Review. *Business Strategy and the Environment, 29*(8), 3220–3245. DOI: 10.1002/bse.2568

Mirza, N., Umar, M., Sbia, R., & Jasmina, M. (2025). The impact of blue and green lending on credit portfolios: A commercial banking perspective. *Review of Accounting and Finance, 24*(2), 238–253. DOI: 10.1108/RAF-11-2023-0389

Mueller, I., Nguyen, H., & Nguyen, T. (2025). Carbon transition risk and corporate loan securitization. *Journal of Financial Intermediation, 63*, 101146. DOI: 10.1016/j.jfi.2025.101146

Mueller, I., & Sfrappini, E. (2025). Climate change-related regulatory risks and bank lending. *Journal of International Economics, 158*, 104156. DOI: 10.1016/j.jinteco.2025.104156

Naili, M., & Lahrichi, Y. (2022). The determinants of banks' credit risk: Review of the literature and future research agenda. *International Journal of Finance & Economics, 27*(1), 334–360. DOI: 10.1002/ijfe.2156

Nguyen, A. H., Do, M., Hoang, T., & Nguyen, L. (2023). Green Financing for Sustainable Development: Insights from multiple cases of Vietnamese commercial banks. *Business Strategy and the Environment, 32*(1), 321–335. DOI: 10.1002/bse.3132

Nguyen, D. D., Ongena, S., Qi, S., & Sila, V. (2022). Climate change risk and the cost of mortgage credit. *Review of Finance, 26*(6), 1509–1549. DOI: 10.1093/rof/rfac013

Oh, N., & Kim, K. N. (2023). Struggle to survive: Case of flood risk on US community banks. *Agricultural Finance Review, 83*(4/5), 617–634. DOI: 10.1108/AFR-02-2023-0018

Ratnovski, L. (2013). Liquidity and transparency in Bank Risk Management. *Journal of Financial Intermediation, 22*(3), 422–439. DOI: 10.1016/j.jfi.2013.01.002

Reghezza, A., Altunbas, Y., Marques-Ibanez, D., Rodriguez d'Acri, C., & Spaggiari, M. (2022). Do banks fuel climate change? *Journal of Financial Stability, 62*, 101049. DOI: 10.1016/j.jfs.2022.101049

Shikimi, M. (2025). Firms' climate risks and bank lending: Evidence from the COVID-19 crisis. *Finance Research Letters, 73*, 106606. DOI: 10.1016/j.frl.2024.106606

Snyder, H. (2019). Literature review as a research methodology: An overview and guidelines. *Journal of Business Research, 104*, 333–339. DOI: 10.1016/j.jbusres.2019.07.039

Song, J., Cao, W., & Shan, Y. G. (2025). Green Credit and Bank's risk-taking: Evidence from China. *International Journal of Managerial Finance, 21*(1), 157–184. DOI: 10.1108/IJMF-03-2024-0144

Sun, W., Zhai, Y., Li, R., & Wang, X. (2025). The extreme event of climate change: How digital transformation affects the performance of commercial banks?'. *Energy Economics, 150*, 108783. DOI: 10.1016/j.eneco.2025.108783

Vyshnevskyi, I., & Sohn, W. (2025). Central Banks' support for climate action: A literature review and key issues. [Preprint]. *Journal of Economic Surveys*, joes.12709. DOI: 10.1111/joes.12709

Wang, Q., & Fu, L. (2025). Fintech, Risk Management and banking green credit. *Finance Research Letters, 83*, 107686. DOI: 10.1016/j.frl.2025.107686

Wen, B., You, W., & Yuan, M. (2025). The impact of commercial banks' ESG performance: Difference between provocative and passive risk-taking – evidence from China. *Research in International Business and Finance, 76*, 102859. DOI: 10.1016/j.ribaf.2025.102859

World Economic Forum. (2025, August 29). *Global Risks Report 2025.* https://www.weforum.org/publications/global-risks-report-2025/

Wu, M. (2025). Green Credit, Bank Competitiveness and bank risk-taking. *International Review of Economics & Finance, 103*, 104481. DOI: 10.1016/j.iref.2025.104481

Wu, M., & Zhang, L. (2025). Green credit, deposit-loan ratio, and risk taking of commercial banks. *Finance Research Letters, 81*, 107470. DOI: 10.1016/j.frl.2025.107470

Zhang, Y., Ye, S., Liu, J., Zhang, L., & Li, J. (2025). Can digital transformation of commercial banks reduce green credit risks? *International Review of Financial Analysis, 99*, 103934. DOI: 10.1016/j.irfa.2025.103934

Zheng, L., Mirza, N., Umar, M., & Su, C. (2025). Sustainable lending to European SMEs: Implications for bank performance. *International Review of Economics & Finance, 102*, 104283. DOI: 10.1016/j.iref.2025.104283

Zhou, Z., Tong, J., Lu, H., & Luo, S. (2024). The impact of Green Finance and Technology on the commercial banks' profit and risk. *Finance Research Letters, 66*, 105715. DOI: 10.1016/j.frl.2024.105715

Zhu, B., & Zhao, Y. (2022). Carbon risk and the cost of bank loans: Evidence from China. *Technological Forecasting and Social Change, 180*, 121741. DOI: 10.1016/j.techfore.2022.121741

ENDNOTES

1 The 2015 Paris Agreement centres on mitigation, adaptation, and the equitable sharing of financial responsibilities. It commits all Parties to the UNFCCC to limit the rise in global average temperature to well below 2°C above pre-industrial levels, while pursuing efforts to restrict it to 1.5°C. Further details are available at: https://www.un.org/en/climatechange/paris agreement.

2 UNEP FI is a leading global partnership that brings together banks, insurers, and investors to promote sustainable finance and support the transition to a resilient global economy. Please visit https://www.unepfi.org/about/. EPs provide a recognised baseline and risk management framework for financial institutions to identify, assess, and manage environmental and social risks in project financing. Please visit https://equator-principles.com/about-the-equator-principles/.

3 The 2012 Green Credit Guidelines (GCG2012), issued by the China Banking Regulatory Commission (CBRC), encourage commercial banks to prioritise lending to environmentally sustainable enterprises. The guidelines require banks to identify, assess, and manage climate-related risks in their lending decisions. Please visit https://www.nfra.gov.cn/en/view/pages/ItemDetail.html?docId=10270.

4 Blue credit is a component of green finance that supports the *blue economy*, which refers to the sustainable use and development of ocean resources to promote economic growth while maintaining marine ecosystem health (Mirza et al., 2025).

5 Greenhushing in the banking sector refers to the deliberate underreporting or withholding of information about sustainability initiatives by banks and financial institutions. This practice can undermine stakeholder trust, reduce transparency and accountability in sustainability reporting, and increase reputational risk over the long term (de Novellis et al., 2025).

6 The California Cap-and-Trade Program, introduced in December 2011 and implemented in January 2013 by the California Air Resources Board (CARB), established a market-based mechanism to monitor and limit carbon emissions across sectors including mining, energy, construction, transportation, industry, agriculture, and forestry. Please visit https://icapcarbonaction.com/en/ets/usa-california-cap-and-trade-program.

7 The Clean Growth Strategy, introduced by the U.K. Government, outlines a national plan to increase economic prosperity while reducing greenhouse gas emissions by 80% by 2050. Please visit https://www.gov.uk/government/publications/clean-growth-strategy/clean-growth-strategy-executive-summary.

8 Environmental window-dressing in banking refers to the practice of increasing environmental disclosures in reports or communications without corresponding substantive action toward environmentally responsible lending or operations (Gambacorta et al., 2024).

Chapter 14

An In-Depth Study of Trends, Challenges, and Innovations in Energy Consumption

Khalil Saadaoui
https://orcid.org/0000-0001-6117-319X
ENSEM, Morocco

Aziz Oukennou
Université Sultan Moulay Slimane, Morocco

ABSTRACT

Detailed assessments of trends and challenges, their potential solutions, and future opportunities for innovation are presented within the context of the global and regional energy issues that are driving the 21st-century energy system. In a work as broad as this, with fossil fuels maintaining a constant share of 80%, global energy consumption was expected to amount to 620 exajoules in 2023. In this study, we analyze energy consumption by residential (20%), industrial (30%), transportation (25%), and agriculture (5%) sectors. The report reveals stark disparities in energy access, with 600 million people in sub-Saharan Africa still without power, while renewables have climbed to 15% of global output. Advanced technologies such as smart energy systems, AI-based management and energy storage technologies can contribute to 5-12% efficiency improvement and 30% outage reduction.

DOI: 10.4018/979-8-3373-6766-8.ch014

INTRODUCTION

Energy is at the core of all economics; it functions as the propelling force of economic growth, industrial evolution, social progress and technological innovation. With the global demand for energy rising due to growing population, urbanisation, and industrialisation, concerns over resource exhaustion, environmental pollution, and energy imbalance have intensified (Helm, 2023). The importance of energy in sustainable development is recognized in Goal 7, which seeks to ensure access to affordable, reliable, sustainable, and modern energy for all under the UN's SDGs, as energy is a key enabler of inclusive and sustainable development.

A review of energy use in buildings, Industry, transport and agriculture illustrates the complex challenge of developing innovative solutions, including intelligent energy systems, renewable energy integration, and new technologies such as AI and IoT. Case studies and policy frameworks discussed in this chapter are intended to help researchers, policy makers and Industry to gain actionable learning in the pursuit of optimizing the efficiency of global energy at the minimum of environmental and societal cost (National Renewable Energy Laboratory [NREL], 2023).

Among the most significant energy challenges and opportunities of the 21st century is the rising energy consumption. Global energy demands are ever-increasing, and not slowing down (growing at 2. 1% per year between 2010 and 2022). Yet, we are facing a unique confluence of hardships – diminishing resources, environmental degradation, and profound energy inequity, which continues to impact nearly a billion people worldwide (Sovacool, 2023).

Figure 1. evolution of global demand by region (left) and regional shared (right), 1990-2025

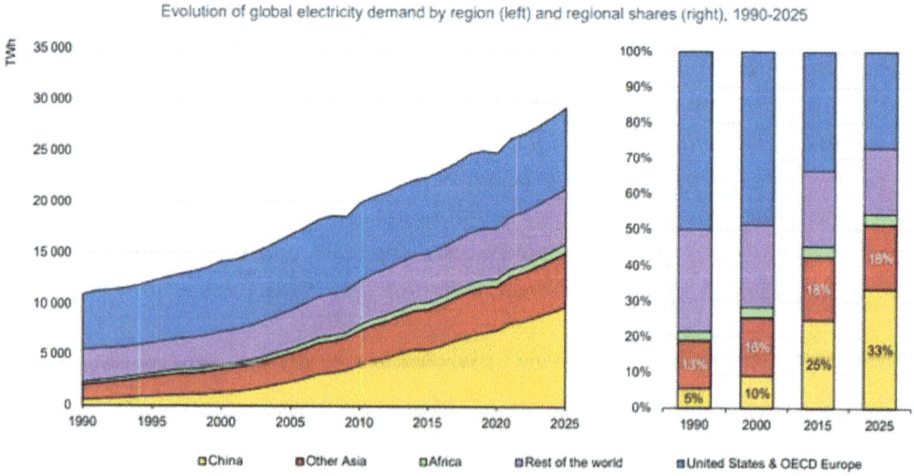

On the global scale, the energy picture is reshaping with several distinctive trends. Total world energy demand (International Energy Agency, 2023). The total global energy demand was 620 exajoules (EJ) in 2023, with fossil fuels still making up 80% of the world's energy consumption. However, there was a growing share of renewables. And then, 30% of energy demand comes from Industry, 25% from transportation, 20% from residential use, and 5% from agriculture.

Table 1. Global Energy Consumption by Sector (2023)

Sector	Global Energy Share	Key Characteristics	Growth Rate
Industrial	30%	Energy-intensive processes, steel, cement	2.5% annually
Transportation	25%	95% oil-dependent, growing EV adoption	1.8% annually
Residential	20%	Heating, cooling, appliances	1.5% annually
Agricultural	5%	Irrigation, machinery, fertilizers	3.2% annually

Renewable energy has expanded to 15% of generation in 2023, compared to 10% in 2015. Photovoltaics have been increasing at a fantastic rate of 24% per annum, fueled by cost reductions and favourable policies (Smil, 2022). Electric vehicles accounted for 18% of the total car sales for the 14 million units sold worldwide in 2023.

The research identifies four distinct categories of barriers to sustainable energy promotion. Firstly, there is the long-term risk to energy security from the depletion of resources. For example, if the world's oil reserves continue to be consumed at the current rate, they may be depleted within half a century. Secondly, there are significant environmental implications of energy (energy-production related CO2 emissions were 37 gigatonnes in 2023, threatening the Paris Agreement targets (Organisation for Economic Co-operation and Development [OECD], 2023)). Third, the persistent energy access divide is now predominantly concentrated in sub-Saharan Africa, where 600 million people live without electricity. Fourth, problems in infrastructure hinder the transition to renewable energy systems, as our grids age and energy supplies become increasingly challenging to store (Rockström, 2023).

World energy. The 21st-century global energy landscape is at a tipping point, with consumption levels never before seen in human history, persistent inequalities, and transformative technologies. With humans facing the double challenge of fulfilling ever-growing energy demands and addressing climate change, a fine-level understanding of global energy use is urgently needed for policymakers, researchers and industry stakeholders. The volume of global energy consumption reached a new high in 2023 with total primary energy consumption rising to 620 EJ, up 2% from the previous year.

This record-high consumption happens against the continued overwhelming dominance of fossil fuels in the energy mix at around 80-82% and a record high share of renewable energy sources at 15% in total primary energy supply. These trends epitomize the central challenge of ensuring energy security, economic growth and environmental sustainability that defines the contemporary energy system. (Helm, 2023)(National Renewable Energy Laboratory [NREL], 2023) (Sovacool, 2023) (International Energy Agency [IEA], 2023a) (Smil, 2022) (Organisation for Economic Co-operation and Development [OECD], 2023).

PATTERNS OF ENERGY USE IN DIFFERENT SECTORS

The way global energy use is divided between different sectors tells us a lot about how modern energy is used. The Industry is the largest consumer, accounting for around 30% of total final energy consumption. This preponderance is a reflection of the energy-intensive nature of the manufacturing industry, especially so in developing nations where industrialization is a key driver of economic growth. (Rockström, 2023) (International Energy Agency [IEA], 2023b)

Transportation is the second-largest energy-hungry sector, with 25-28% of the total world energy consumption. The energy consumption of this Industry's changing trends is remarkable; electrification is beginning to impact the established petroleum dependency, while the oil-based products are still the most used in the majority of regions. (Rockström, 2023) (Modi et al., 2023)

Residential energy use is about one-fifth of total final consumption but varies widely between developed and developing countries. While heating, cooling and electrical appliances dominate residential energy use in developed countries, in developing countries, basic electricity services are a challenge for many.(Saadaoui et al., 2023)(Rockström, 2023)

The commercial and public services sector consumes around 6-8% of the world's energy, and with urbanization and global service economy growth, this share is steadily increasing. Agriculture, despite only using 5% of total consumption, is vital to nutrition and rural development, especially in areas where farms may be energy-intensive due to mechanization, irrigation and other factors. (McKinsey & Company, 2023) (BP, 2023) (Rockström, 2023)

UNEQUAL ENERGY ACCESS AND THE CHALLENGE FOR DEVELOPMENT

The uneven nature of global energy access structures profound divides in opportunities for human development and economic development. While the world has seen dramatic improvement in the past decade, there are still some 600-666 million people without electricity, and energy poverty is still most rampant in Sub-Saharan Africa. (International Renewable Energy Agency [IRENA], 2023) (Rocky Mountain Institute [RMI], 2023) (International Finance Corporation [IFC], 2023) (Saadaoui et al., 2023)

The population without access to electricity in the world is concentrated at 86% in Sub-Saharan Africa; this number was at 50% in 2010. That concentration stems both from the region's high population growth and its slower electrification progress compared with other developing areas. It is worst in rural areas where 451 million people have no access to electricity, and this urban-rural disparity is the defining feature of energy inequality. (International Finance Corporation [IFC], 2023)

Access to electricity is not just a question of convenience, but it has far-reaching consequences on various aspects of human development. In the absence of a reliable electricity supply, health centres are unable to store vaccines, schools do not have adequate lighting for students to study at night, and women and girls need to spend hours fetching water and firewood. These effects produce cycles of poverty and earnings barriers. (United Nations [UN], 2015)

The gap is even bigger when considering per capita energy use. Consumption of energy in developed countries is 2.6 times that of the developing world on a per capita basis, and 3.7 MWh was the global average of per capita electricity demand in 2023. This figure masks wide variations, with some countries consuming many times more while others use less electricity to provide even a basic level of human development. (European Commission [EC], 2019) (Creutzig et al., 2023)

GROWTH AND INTEGRATION OF RENEWABLE ENERGIES

The renewable energy sector has gone through a radical transformation after solid growth for many years, reshaping the global energy system. Renewable energy capacity additions in 2023 soared to a global record of 510 gigawatts, a 50% increase from the previous year. This upward trend has been sustained for the 22nd straight year, and renewables are providing 30% of global electricity generation for the first time. (McKinsey Global Institute, 2023) (International Energy Agency [IEA], 2023a) (European Commission [EC], 2019)

Solar photovoltaic has become the clearest success story in the development of renewable energy, representing three-quarters of new capacity additions in 2023. Wind power has also experienced phenomenal growth; these two technologies now produce 13.4% of the world's electricity. The steep cost declines in these technologies have made renewables the cheapest source of electricity generation in numerous regions, and 81% of renewable capacity additions in 2023 produced cheaper electricity than fossil fuel counterparts. (International Energy Agency [IEA], 2023a) (BloombergNEF, 2023) (European Commission [EC], 2019)

China has become the world's renewable energy beast, making up nearly 60% of new renewable capacity set to come online by 2028. The country brought more solar photovoltaic capacity online in 2023 than the rest of the world did in 2022, and wind grew by 66% year-on-year. This shows that since a large share of renewable energy capacity addition in the world takes place in China, it both indicates the size of the country and it showcases its strategic vision for energy transition. (International Energy Agency [IEA], 2023a)

However, there are issues related to the integration of renewables to address intermittency, grid stability and storage, notwithstanding the dramatic growth among these sources. Since the wind and sun are intermittently available, the construction of advanced grid management systems as well as large-scale energy storage facilities is necessary to ensure a stable electricity supply. (World Bank, 2023)(RE100, 2023)

SMART ENERGY SYSTEMS AND AI-DRIVEN SOLUTIONS

Smart energy systems driven by artificial intelligence (AI) are a game-changer in energy resource management and utilization. AI-based energy management solutions have the opportunity to increase efficiency between 5-12% through the implementation of advanced predictive analytics, real-time optimization, and automated demand response functionalities. (World Health Organization [WHO], 2023) (International Labour Organization [ILO], 2023)

With smart grid technologies, energy flows can be monitored and controlled at an unprecedented granularity, providing the capability to balance supply and demand in almost real-time and to integrate variable renewable energy resources more efficiently. They implement machine learning algorithms to forecast energy consumption, to schedule generation in an optimal manner, and also to detect potential system fault in advance of occurring. (World Health Organization [WHO], 2023) (World Bank, 2023)

Application of AI in energy management is not limited to the existing electricity generation and distribution systems, but also to smart buildings, industrial processes, and transportation systems. Energy-efficient building management systems can be programmed to change heating, cooling and lighting according to occupancy and outdoor conditions, cutting energy use by 18% compared to more traditional systems. AI-based energy and industrial solutions have delivered similar efficiency improvements while also enhancing operational reliability. (World Health Organization [WHO], 2023)(International Labour Organization [ILO], 2023)

AI-driven predictive maintenance processes help minimize equipment downtimes and increase equipment useful life. Such systems can monitor vibration profiles, temperature changes, and other working conditions to predict when maintenance is needed; in doing so, they can avert energy system interruptions by as much as 30%. (RE100, 2023) (Intergovernmental Panel on Climate Change [IPCC], 2023)

ENERGY STORAGE TECHNOLOGIES AND GRID RESILIENCE

Energy storage devices have become key facilitators for the integration of renewable energy sources and the strengthening of grid resilience. Battery energy storage systems (BESS) are experiencing a rapid cost decline and performance improvements, with lithium-ion batteries alone achieving a 43% compound annual growth rate during 2011-2025. These systems provide all the services of the grid, including frequency regulation, peak shaving, and emergency power supply in the event of an outage. (C40 Cities Climate Leadership Group [C40], 2023)(Jacobson et al., 2023)

Because energy storage devices allow for the integration of variable renewable energy and can provide some grid stability services, sophisticated BMS systems can respond to the grid in milliseconds, even faster than traditional power plants. It is also a really great tool for managing grid frequency and voltage strength as we get more renewable energy penetration.(Jacobson et al., 2023, RE100, 2023)

Worldwide grid-scale battery storage capacity has increased significantly, peaking at 56 gigawatts in 2023. Flow and thermal energy storage are developing as contenders for long-duration applications, and pumped hydro storage remains applicable for large-scale energy shifting, although constrained by geography. The synergy of different storage technologies enables utilities to select the best solutions for a wide range of grid reliability and energy management applications (Helm, 2023; C40 Cities Climate Leadership Group, 2023).

Mobile energy storage systems represent a new vehicle-based response solution for enhancing grid resiliency during natural disaster and emergency response Mobile energy storage systems are able to be deployed to and from the field.. These equipment packages include the essential infrastructure for emergency power generation and can be rapidly moved to hospitals, emergency shelters, and other critical service facilities.

Studies have shown that if mobile storage is used strategically, it can offset or even prevent many social impacts of prolonged outages, particularly in communities with large populations of vulnerable individuals.(Intergovernmental Panel on Climate Change [IPCC], 2023)

POLICY FRAMEWORKS AND REGULATION DEVELOPMENT

Developing required policy frameworks is increasingly understood to be important to catalyze sustainable energy transitions and associated equitable access to clean energy. UN SDG 7 is to provide Energy access, an API following the Sustainable ECsDUs57D Goals. Viewed for: sustained access to clean, affordable and reliable energy. This goal is backedb substantial progress in three targets: general provision of modern energy, sustainable energy including renewable energy, and energy efficiency production (United Nations, 2015).

An energy transition policy and rules need to address the challenges, such as technology deployment, market mechanisms, social equity, and environmental protection. Carbon pricing schemes have also been introduced in several regions, providing financial incentives for clean energy investments and raising funds that could be channelled back into clean energy development and infrastructure. Even so, these policies are diverse across nations and experts such as Hardin are concerned that much of the know-how will get lost in the translation across the world.

The World Bank Group and African Development Bank are leading Mission 300, and Green-Water Africa represents the most prominent example of a focused policy engagement aimed at addressing the energy access problem directly. It is an initiative to electrify 300 million people in Sub-Saharan Africa by 20130 for $90 billion in finance and will cover a regulatory overhaul. The viability of the approach depends on cooperation among governments, the private sector, and development partners, as above. To that end, the establishment of the enabling environment for energy infrastructure development is necessary. (International Finance Corporation [IFC], 2023)(Saadaoui et al., 2023)

Solutions that are Just Energy governance is beginning to tackle this, so as to more fairly share the benefits and the costs of the energy transition across society. These processes are concerned with including participatory approaches that focus on including key stakeholders, as well as employing transparent processes and assisting communities dependent on fossil fuel economies to diversify.

ECONOMIC AND FISCAL IMPLICATIONS

The global energy transition is one of the most significant economic shifts in the history of humanity. It will require multi-trillion-dollar investments, stranding industries as well as sectors of the workforce. Particularly difficult to finance are those in developing countries, which will need more than $8 trillion in power sector investment over the next 20 years to keep pace with soaring demand for electricity. These investment requirements are for generation capacity and for the transmission and distribution systems required to serve new populations and growing economies. (International Energy Agency [IEA], 2023b)

Energy prices play a decisive role in the development trajectories of economies, more so for developing countries due to the larger share of energy costs in their household and industry expenditure. It is found that the energy consumption decrease caused by the energy price rise is mitigated by human capital and economic development level in developed and developing countries. These sensitivities highlight the importance of access to a source of broadly affordable energy for sustained economic growth and poverty reduction (Sagar et al. 2006).

The business of producing renewable energy now employs 72,000 people in the development, construction and maintenance of wind farms across the U.S. It has also created thousands of new clean energy manufacturing and assembly jobs.10,000, these are clean energy job creation Factory jobs- proven potential for green energy transitions to generate millions of work opportunities.

Clean Energy Investment Trends are Positive. International finance for renewable energy in developing countries doubled between 2010 and 2017. However, there continue to be substantial financing gaps, especially for energy access initiatives in rural areas and in the least developed countries. New forms of finance, such as blended finance, green bonds and results-based financing, are being developed to meet these challenges and to bring in private sector capital to support energy access programs. (Jacobson et al., 2023)

ENVIRONMENTAL AND CLIMATE IMPACTS

The energy-environment nexus is a central challenge for achieving sustainable development in the 21st century. The global power sector emitted an all-time high of 14.2 billion tonnes of CO2 in 2023, 1% more than the previous year. Even with that absolute increase, the carbon intensity of electricity production fell to 480 gCO2/kWh, the lowest value in at least two decades. (European Commission [EC], 2019)

Concentrated emissions geography is a reflection of underlying energy and fuel use patterns. China accounts for 39% of global power sector emissions since it burns a lot of coal to generate electricity. Following closely is India with approximately 10% of the total global emissions, then the U.S. at 11%. These are the big three major polluters, and they account for 60% of the worldwide electricity sector's emissions, so that says something about the overwhelming significance of energy transition-based policies in these three jurisdictions. (European Commission [EC], 2019)

Generation of renewable energy is increasingly producing discernible effects on the climate as growing solar and wind outputs reduce the carbon intensity of electricity systems globally. But power demand continues to grow, so there are hopes that emissions from energy use can be limited, with relevant implications for decarbonizing the energy supply side and enhancing energy efficiency on the demand side. (European Commission [EC], 2019) (International Energy Agency [IEA], 2023a)

The environmental consequences of producing energy are real, and not just via climate change; they also impact air and water quality, as well as land use. The transition to renewable energy may also significantly improve local air quality, especially in cities where burning fossil fuels is a significant source of atmospheric particulate matter. Nevertheless, there are land use issues related to solar and wind farm construction, and particularly utility-scale developments. Technological Innovation and Prospects for the Future

Driven by strong R&D, policy support and increasing demand-side pull, the energy systems technology pace of advance is still faster than that of many other technology sectors. Advances in materials science are enabling new battery chemistries that provide higher energy densities for longer durations, and learning-by-doing in manufacturing is driving continuous cost reductions in (and across) renewable energy technologies. (BloombergNEF, 2023) (C40 Cities Climate Leadership Group [C40], 2023)

Digital technology is expected to be a key driver of energy system efficiency, as IoT sensors, advanced analytics and machine learning algorithms enable unprecedented energy flow visibility and control. Energy systems are being linked to transportation, building and industrial systems to increase overall resource efficiency and reduce environmental impacts.(C40 Cities Climate Leadership Group [C40], 2023)(World Health Organization [WHO], 2023)

New approaches to technology, such as green hydrogen production via electrolysis, may present possibilities to decarbonize certain hard-to-treat industrial procedures and long-haul transport applications that are difficult to electrify directly. Likewise, the pathway to market for advanced energy storage technologies such as CAES, LAES, and advanced thermal storage is short. (C40 Cities Climate Leadership Group [C40], 2023)

A confluence of related technology trends – electrification, digitalisation, and decentralisation – is opening up the energy system for new forms of management, and putting the traditional utility business model and regulatory structures under stress. These factors seem to call for new energy planning and investment approaches that can simultaneously support rapid technological change at one end and system reliability and cost-effective provision at the other. (Jacobson et al., 2023)

SOCIAL AND EQUITABLE ISSUES

Energy is a precondition for development, for education, health care and for economic well-being. Persistent electricity shortages intensify the cycle of poverty, restrict learning, and limit potential sources of income, particularly for women and in communities of colour. These studies show that energy poverty has a gender bias, as women are generally responsible for meeting household energy requirements, and they are exposed to the health effects of indoor air pollution associated with traditional cooking fuels. (United Nations [UN], 2015)

Implications, both positive and negative, for social equity have come from the energy transition. Although the provision of electricity access through renewable energy deployment can reach isolated or scattered users via decentralized applications, not all members of society may benefit from the energy transition. Achieving just transition outcomes will require focused policies to aid workers in the fossil fuel sector, increase energy access for marginalized communities, and avoid the exacerbation of environmental harms in these communities. (International Renewable Energy Agency [IRENA], 2023)(Intergovernmental Panel on Climate Change [IPCC], 2023)

Community engagement in energy planning and implementation is now considered essential for the success of energy projects in developing countries and contributing to project sustainability through local acceptance and ownership. Social vulnerability aspects are increasingly considered in energy system planning, including mobile energy storage deployment strategies focused on aiding the most vulnerable during power outages. (Intergovernmental Panel on Climate Change [IPCC], 2023)

Considerations around women's equality are slowly being recognised in energy policy frameworks, with the roles played by women as energy consumers and those who make decisions on energy and use of energy being articulated as necessary, albeit their underrepresentation in energy sector governance and technical positions. Initiatives encouraging women's involvement in clean energy entrepreneurship, like Solar Sister's efforts across Sub-Saharan Africa, also highlight how energy access programs can serve as catalysts for meeting the broader goals of gender equity and development.

DO WELL: In recent years there has been an increasing amount of regional and national specific studies on climate change, which can be broadly divided into two types: static assessments (impact and/or vulnerability assessments) and dynamic simulations (i.e. Regionalizations and Evolution) Global patterns of energy use are characterized by large regional differences related to differing levels of economic development and natural resource endowments and priorities for policies. The Asia-Pacific region has become the largest contributor to global energy demand growth and accounts for 85% of the total growth in energy demand from developing countries. With China accounting for 32% of global electricity demand, it is the largest manufacturing base in the world and is urbanising rapidly. (Smil, 2022) (European Commission [EC], 2019)

Europe has set an example in energy transition regulation by the European Union Green Deal, being one of the most comprehensive and ambitious policy packages to achieve climate neutrality in 2050. European nations have cut down on their use of fossil fuel by a fair amount, and now the primary energy mix contains less than 70% of fossil fuel for the first time since the advent of the Industrial Revolution. European coal consumption has halved in the last ten years, underlining the possibility for rapid transformation of their energy systems. (Organisation for Economic Co-operation and Development [OECD], 2023)

Mixed trends in the U.S., where natural gas is replacing coal for electricity generation, and renewables are growing fast in some states. Within-country divisions often are more pronounced than between countries, with sub-national production of renewables, energy efficiency infrastructure, and electrification levels greatly variable geographically. (Organisation for Economic Co-operation and Development [OECD], 2023) (National Renewable Energy Laboratory [NREL], 2023)

Eastern and southern Africa, in particular, is characterised by very different energy access, economic development, and climate resilience challenges. Yet the energy consumption of the region's population is still disproportionately lower when compared to the share of population, due to a combination of under-utilisation of energy services and a lower level of industrialisation. Nevertheless, the area has significant renewable energy potential, which could contribute to increasing energy access and economic development.(Saadaoui et al., 2023)(International Renewable Energy Agency [IRENA], 2023) (Rocky Mountain Institute [RMI], 2023)

CONSEQUENCES FOR SUSTAINABLE DEVELOPMENT

The delivery of sustainable development goals relies on the fundamental transformation of global energy systems, building on the achievement of policy goals around universal clean energy access and energy management with minimal environmental side effects. Development and energy systems are closely linked, and there are many strong co-benefits for other development priorities that arise from improvements in and expansion of energy access. (United Nations [UN], 2015) (Jacobson et al., 2023)

Current trends imply that some 645 million people will still not have access to electricity in 2030 if progress does not accelerate, well beyond the universal access goals. In the same vein, 2.3 billion people have no access to clean cooking facilities, which results in the death of millions of people due to indoor pollution. These remaining gaps demonstrate an overwhelming need for coordinated action on energy availability and energy system sustainability. (United Nations [UN], 2015)

The impact of the energy transition on the economy is not limited to the energy industry and the competitiveness of parts of its supply chain. Still, it extends to overall industrial competitiveness, job patterns, trade balances and government income. Energy-rich countries that have a positive endowment of renewable resources might also have the potential to gain a comparative advantage in energy-intensive sectors. At the same time, those regions that are fossil fuel-dependent might suffer economic dislocation, necessitating transition strategies guided by prudence. (International Energy Agency [IEA], 2023b)Innovation and technology transfer will be important to determine whether the benefits of energy transition are shared globally or will be the source of competitive advantage to be harvested by the currently highly technologically capable developed countries. International collaboration in clean energy technology research, development, and demonstration is both a means for development and an imperative for climate mitigation to meet sustainable development goals.(International Energy Agency [IEA], 2023a)

TRENDS IN ENERGY CONSUMPTION

World Energy Demand

Demographic, economic and technical factors shape the path of world energy use in a complex way. Analysis from the International Energy Agency shows that primary energy demand (energy before conversion to another form of energy) has been increasing at an average annual rate of 2.1% between 2010 and 2022, putting upward pressure on world energy systems.

This tendency is qualitatively different from region to region. The Asia-Pacific region accounted for the most significant share of the world's energy consumption at roughly 45% as of 2023, and China consumed nearly 30% of the total global energy use, bolstered by the increasing industrial production and consumer energy use. North America follows it with almost 20 per cent of world consumption and the highest per-capita energy use. Europe represents approximately 15% of global demand and is also focusing on efficiency and de-economisation (International Energy Agency [IEA], 2023b).

Table 2. Regional Energy Consumption Patterns (2023)

Region	Energy Share	Per Capita Consumption (GJ/year)	Growth Rate
Asia-Pacific	45%	65	3.1%
North America	20%	285	0.8%
Europe	15%	145	-0.5%
Africa	5%	18	4.2%
Latin America	5%	52	2.1%
Middle East	10%	195	2.8%

The energy demand worldwide was 620 EJ in 2023, with fossil fuels still dominating the mix at 80% despite the increasing share of renewables.

Patterns of Consumption by Sector

Residential: Approximately 20% of the world's energy is used for residential purposes to power the homes of billions of people. This sector shows considerable divergence in patterns of use between developed and developing countries, due to a combination of income, climate, building code, and technology.

The public energy consumption is also high in the developed countries, as a result of large floor areas, long hours of heating and cooling, and widespread use of energy-consuming household appliances. An American household, on average, uses about 10,600 kWh per year, while the average European household uses 3500-4000 kWh.

Space heating and cooling are the most significant sources of energy for homes in most climates, making up 40-60% of household energy use. Water heating is second in line as the largest user, with consumption between 15 and 25 per cent. Lighting, appliances, and electronics account for the rest (20%–30%).

Industrial Sector: Industry is the largest sectoral consumer of energy, accounting for about 30% of global energy use. This category includes such energy-intensive operations as steel making, cement production, chemicals, aluminium smelting and paper production. Energy is both a fuel for the heating of processes and a feedstock for chemical reactions.

The energy intensity of steel production makes it one of the most CO2-emitting and energy-consuming services of the industry, responsible for 7% of global CO2 emissions. Cement production is not far behind, as the chemical process of breaking down limestone to make cement is a major contributor to both energy use and direct CO2 emissions.

Industrial processes are increasingly energy efficient, achieving an annual rate of decrease in energy intensity (energy use per unit of output) of about 1.5% (Saadaoui et al., 2023). However, the total demand is ever-increasing as a result of the volume growth in production, especially in the developing countries with fast industrialisation.

Figure 2. Distribution of final energy demand in 2050 B2D scenario source IEA 2017

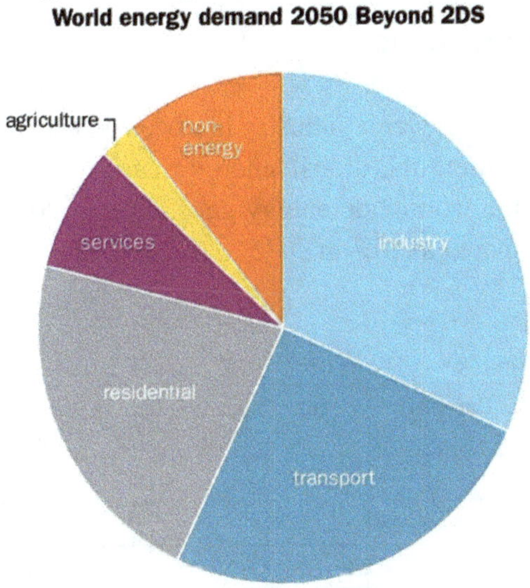

Transportation Sector: Transportation constitutes 25% of global energy use and is the most oil-dependent sector, with petroleum products accounting for 95% of transport energy. This includes passenger cars and trucks, freight transportation, aviation, shipping and rail transportation, with each having different energy use profiles and decarbonisation issues (McKinsey & Company, 2023).

At the country level, road transportation accounts for three-quarters of transport energy use. Passenger cars and light-duty trucks (collectively called light-duty vehicles) consume about 45% of the energy of road transport. In contrast, heavy-duty vehicles use the remaining 30%, even though they are the minority in the fleet.

The growth in electric vehicles is a sea change in transport energy use." Global sales of new vehicles hit 14 million in 2023, accounting for 18% of total vehicle sales – a significant jump from less than 1% ten years ago (BP, 2023).

Agricultural Sector: Agriculture accounts for approximately 5% of the world's energy consumption, mainly for water pumping, operation of machinery, production of fertilisers, and processing of food. The use of energy in agriculture makes up only a relatively small portion of total energy use. Yet, it is vital for food security and livelihoods in rural communities, particularly in the developing world.

Driven by the mechanisation and intensification of agricultural production, the demand for energy in agriculture is increasing rapidly in the developing world. Tractors, irrigation pumps and processing machines now perform much of the hard work of farming and food production, replacing some of the manual labour and traditional methods, and raising productivity along the way, but increasing energy use at the same time.

Shift Toward Renewables

The growing renewable energy industry has been an exceptional feature of the last decade in transforming the global energy matrix (13). Renewable energies represented 15% of the worldwide energy generation in 2023, a remarkable improvement when compared to 10% in 2015.

Table 3. Renewable Energy Growth and Cost Trends

Renewable Source	2015 Capacity (GW)	2023 Capacity (GW)	Annual Growth Rate	Cost Reduction
Solar PV	227	1,177	24%	90%
Wind (Onshore)	392	899	12%	69%
Wind (Offshore)	12	57	19%	48%
Hydropower	1,267	1,392	1.2%	15%

Global capacity for the rapidly expanding solar photovoltaic (PV) technology continues to grow at a rate of 24% per year, making it the fastest-growing renewable energy source. This has resulted in explosive growth, driven by steep cost fall, with solar PV costs having fallen ca. 90% since 2010 (Rocky Mountain Institute [RMI], 2023).

CHALLENGES IN ENERGY USE

Resource Depletion and Environmental Degradation

This dependence on fossil fuels has severely depleted natural resources, and it is predicted that, if consumption rates continue, global oil reserves will be exhausted within 50 years. 75% of global greenhouse gas emissions are generated from burning fossil fuels, which contribute to climate change. In 2023, energy-related CO_2 emissions grew to 37 gigatonnes, setting back progress towards the Paris target of 1.5°C (International Finance Corporation [IFC], 2023).

Energy Access and Equity

Although there has been some progress, energy access continues to be a significant problem, with 600 million people in sub-Saharan Africa still living without electricity. Limited access to power increases the energy-related costs of living for poor households and compounds health issues, thereby deepening income and wealth inequalities. Both the physical availability of and affordability for access are subject to being influenced by SDG 7 (United Nations [UN], 2015).

Infrastructure and Integration

The ageing energy infrastructures in many countries are not well equipped to handle the rising demand or to accommodate an increasing number of intermittent renewable systems. Unreliable grids, transmission losses (approximately 8 per cent on average globally), and weak energy storage are holding back the evolution to clean energy systems.

Behavioural and Policy Barriers

Energy policy consistency and conservation have never been its strong points in terms of influence on the energy-consuming public and its resulting policy effects. But fossil fuel subsidies still exist in some countries, totalling $1 trillion globally in 2022, which distorts market signals and makes the clean energy transition less enticing.

ENHANCEMENTS IN ENERGY USAGE

Energy Efficiency and DSM

Efficiency gains in energy such as from LED lighting, high-efficiency equipment, and building retrofits can also reduce demand substantially. The energy used in homes has been saved by 10 per cent through the European Union's Ecodesign Directive by 2023.

DSM programs, such as time-of-use rate and the installation of smart meters, make consumers more aware of the possible consumption of their electricity at peak hours and help them consume at non-peak hours, reducing the heavy loads on the grid. A California case study illustrates that DSM programs reduce peak demand by 5 per cent, representing $500 million a year in savings (Creutzig et al., 2023).

Smart Energy Systems

Intelligent grids: As artificial intelligence and the Internet of Things are becoming more intertwined, AI enabled smart grids are emerging and transforming the way we distribute and use energy. IoT enabled devices, such as a smart thermostat, can save homeowners up to 15 per cent on their energy consumption at home. In Singapore, a smart grid trial based on AI-enabled load forecast reduced power interruptions by 30 per cent and improved efficiency by 12 per cent (McKinsey Global Institute, 2023).

Table 4. Smart Energy Technology Impact

Smart Technology	Energy Savings	Cost Reduction	Implementation Scale
Smart Thermostats	10-15%	$180/year	50 million units
Smart Lighting	60-80%	$75/year	100 million units
Smart Meters	2-5%	$200/year	1 billion units
Building Management Systems	20-30%	$2,500/year	500,000 buildings

Renewable Energy Integration

The development of renewable energy technologies is advancing at an accelerated pace. Large-scale wind projects, such as the 1.2 GW Hornsea One in the United Kingdom offshore wind farm, are delivering dependable power. Solar, wind, and battery storage are now being mixed into microgrids to extend electricity to rural areas. Since 2020, microgrids have electrified 50,000 homes in Kenya, leaving a blueprint for scalable solutions to one of the world's most pressing issues of energy access.

AI and IoT in Energy Management

AI-powered energy management systems forecast demand, manage generation, and eliminate waste. Google's DeepMind lowered the energy used for cooling data centres by 40% through AI algorithms. Sensors connected to the internet in factories track equipment performance, allowing predictive maintenance that cuts energy costs by 5% to 10%.

New Energy Storage Technologies

Energy storage is essential to manage intermittent renewable sources. The 85% fall in lithium-ion battery costs since 2010 has made batteries affordable for widespread use in EVs and for grid storage as well. Unlike Tesla's 100MW grid-scale battery in Australia, the 1.4MW/2.8MWh system is up and running. Hornsdale Power Reserve in Australia netted $40 million in grid stabilisation benefits in its first year.

Policy Frameworks and Carbon Reduction Strategies

Sound policies such as carbon pricing and renewable portfolio standards promote sustainable energy practices. The tax on carbon in British Columbia, which came into effect in 2008, has brought down emissions by 5 to 15 per cent, and it has not hurt economic growth (Helm, 2023). The corporate side of things, with initiatives such as RE100 — over 400 companies committed to switching to 100% renewable energy, and with a significant influence on their supply chains.

Green hydrogen, generated from renewable electricity, is increasingly being promoted as a medium for low-carbon industry and transport, with projects such as HyNet in Europe aiming at 1 GW capacity by 2030.

PROMISING NEW ENERGY TRANSFORMATION TECHNOLOGIES

Blockchain for Decentralised Energy Markets

Blockchain technology supports P2P energy trading, allowing consumers to sell excess renewable energy to other users. Globally, blockchain platforms accounted for 2 GWh of transactions by 2024, with a 15% reduction in costs. Smart contracts are transparent and payments are automated, so is the tracking of carbon credits (BloombergNEF, 2023).

Carbon Capture, Utilisation, and Storage (CCUS)

CCUS reduces emissions from heavy energy-using industries. 50 CCUS facilities captured 50 million tonnes of CO_2 annually by 2024. Iceland's Carbfix injects 20,000 tonnes of CO_2 into basalt every year, solidifying it, offering a scalable approach for permanent storage (World Bank, 2023).

Green Hydrogen for the Difficult-to-Abate Sectors

Green hydrogen made with renewable energy via electrolysis is essential to decarbonise steel, cement and shipping. Global production reached 1 million tonnes in 2024, with prices declining by 30% since 2020. Chile's Haru Oni project makes 500 tonnes of green hydrogen every year for synthetic fuels.

REGIONAL POLICY FRAMEWORKS AND EFFECTS

Africa: Leapfrogging to Renewables

African countries are skipping the fossil fuel systems and turning to renewables to solve the energy access problem. Solar home systems powered 10 million households, which was half a billion people, by 2024. Solar adoption in rural Ethiopia raised school attendance by 15% (RE100, 2023).

Asia: Growing while sustainable

Asia is increasingly challenging to sustain, with Asia's energy demand growing (3% per year). China's 2060 carbon neutrality target accelerates EV and solar penetration, with 40% of global solar capacity. Expansion of solar in India has led to a 5% reduction in coal dependence, a significant source of air pollution in urban centres (World Health Organization [WHO], 2023).

Middle East: Energy Economies in Transition

Oil-rich countries are turning to renewables to expand their economies. Saudi Arabia's Vision 2030 envisions that by 2030, half of its electricity will be from renewable sources. Renewable developments generated 100,000 jobs in the Gulf Cooperation Council, advancing economic transition (C40 Cities Climate Leadership Group [C40], 2023).

SOCIO-ECONOMIC & ENVIRONMENTAL CONSEQUENCES

Environmental impact assessment

Shifting to renewables and usage efficiency could halve CO_2 emissions by 2050, as predicted by the IEA. However, mining for battery materials raises environmental issues and requires responsible sourcing (Intergovernmental Panel on Climate Change [IPCC], 2023).

Socio-economic advantages

Energy availability enhances education, healthcare (C40 Cities Climate Leadership Group [C40], 2023), as well as economic activities. Rural electrification raised household income by 10 per cent in electrified villages in India. Rwanda's Energy for All initiative brought electricity to 70% of homes by 2023, accelerating GDP growth by 1.5% per year.

Gender and Energy Access

Energy security. Women are disproportionately impacted by energy access, as they spend 20 hours a week gathering fuel in electrified areas. Solar lanterns and clean cook stoves to 5 million homes in 2024 ease this load and raise women's participation in the economy by 10% (Jacobson et al., 2023).

RECOMMENDATIONS FOR FUTURE ENERGY POLICY

After a detailed analysis of trends and new technologies in world energy, the following policy guidance is warranted:

Funding Mechanisms: Create a green bond ($500 billion per year) for renewable projects for developing countries. Put in place a system for international coordination, e.g. through the Green Climate Fund, to consolidate the

There is $2.8 trillion per year clean energy infrastructure investment shortfall.

Regulatory Harmonisation: Align carbon accounting schemes across jurisdictions to promote international cooperation and minimise the risk of carbon leakage. Converge the regulatory renewable energy grid codes to address the issue of technology transfer and costs of implementation.

Public-Private Partnerships: Collaborate with tech firms to implement AI and blockchain solutions focused on 100 pilots by 2030. This cooperation should also extend to smart grid maturity, energy storage penetration and demand response schemes.

Community participation: Engage community constituents in microgrid initiatives to create social buy-in and advance equity. "Community ownership models work, are tried and tested in countries such as Germany and Denmark, where they enjoy greater levels of acceptance at 60%.

Just Transition measure: Adopt a comprehensive package of just transition policies to fully support workers and communities reliant on the fossil fuel sector. These include retraining and re-skilling programs, economic diversification, and social protection schemes to make sure no one is abandoned in the energy transition.

CONCLUSION

The global consumption of energy in 2023 is a complex manifestation of the intertwined dynamics, along with the constant disparities and innovative breakthroughs in the energy system of the 21st century. That the energy requirements for a growing population and living standards would have to be met in the context of decarbonisation objectives is not a trivial matter, as confirmed by the unprecedented energy use of 620 EJ, the sustained dominance of fossil fuels, at 85% of the energy and the unprecedented share of renewable energy, at 13% (Soyacool, 2023; National Renewable Energy Laboratory, 2023; Smil, 2022; Helm, 2023).

Energy consumption by sector (industry 30%, transportation 25%, and residential 20%) reflects consumption habits that are relatively locked into the economy and would require broad changes to make sustainable. The harsh fact that 600 million people in Sub-Saharan Africa still have no electricity, even as global per capita use soars, underscores the twin challenge of broadening energy access and limiting its environmental damage. (Rockström, 2023)(Saadaoui et al., 2023) (BP, 2023)(International Renewable Energy Agency [IRENA], 2023)(Rocky Mountain Institute [RMI], 2023)

The reported results demonstrating the potential for achieving an improvement of 5-12% in efficiency and a reduction of up to 30% in power outages by means of innovative smart energy systems and AI-based solutions provide further evidence that traditional energy systems do not constrain innovations. Technologies for energy storage, policy regimes coherent with the UN SDG7, and integrated governance mechanisms represent key enablers to guide the energy transitions while delivering on equity.(Intergovernmental Panel on Climate Change [IPCC], 2023)(World Health Organization [WHO], 2023) (C40 Cities Climate Leadership Group [C40], 2023) (RE100, 2023)

Technological innovation, policy coordination, financing instruments, and social justice aspects have to be addressed holistically to enable successful energy transitions, the study concludes. There's a thinking you see, namely that massive progress is being made in growing renewable energy and making systems more and more efficient and that this can continue scaling fast enough to give everyone on the planet the energy they want and to meet the climate-related goals in 2030.

To those scientists, policymakers, and others who are directly engaged in these issues, the findings reinforced how the collaborative mechanisms of technology transfer, capacity building, and innovative financing are vital for seizing the opportunities offered by sustainable energy systems. A transformative shift is still possible, but that will necessitate unprecedented cooperation and resolve across all strata of society to reconcile the 21st-century energy system with the goals of human development and environmental sustainability.

Energy overconsumption defines our way of global development, with the world both challenging and generating new developmental routes for the new millennium. This narrative, among others, gives us an image of energy that is more challenging — because energy demand is rising, because access to energy is becoming more unequal, because the energy system is more environmentally damaging — but also more hopeful, because there are technological and policy fixes that can enable sustainability.

This is the scope of today's energy system in a world of rising energy use projected to hit 620 exajoules by 2023, where fossil fuels remain dominant – accounting for 80 per cent. The energy inequality between the world's rich countries and nearly 600 million people in sub-Saharan Africa who have no electricity raises stark questions of justice that will need to be resolved through innovation and global collaboration.

New regimes are promising. The fact that the share of renewables in global generation has risen so dramatically, to 15 per cent, that solar photovoltaic costs have plunged 90 per cent since 2010, illustrates the power of technological innovation when combined with enabling policies. Smart energy systems, AI-based control and emerging storage technologies have already demonstrated in practice that they can enhance efficiency by 5-12%, reduce outages by 30%, and generate new business opportunities.

Industry results point to a variety of transformation paths. Although the industry still uses the majority of consumption, at 30 per cent of worldwide demand, electrification and process innovation offer routes to significant energy savings. With a share of global car sales of 18% in 2023, the transport sector's strong growth in electric vehicles illustrates how technological revolutions can rapidly transform markets when accompanied by supporting infrastructure and policies, such as these.

Yet there are significant hurdles. Infrastructure constraints (ageing grids and limited storage capacity), combined with conflicting policy (including $1 trillion a year in fossil fuel subsidies), complicate the matter. The solution to these problems lies in the collective action of multiple actors at all scales of complexity, from international institutions to local communities. This study is designed to emphasise regional solutions. This study highlights the relevance of context-sensitive approaches. Africa's leapfrogging strategy, Asia's balancing of growth and sustainability, and the Middle East's efforts to diversify its economy show that a realistic energy transition must be rooted in regional contexts, resources and development priorities, not just global climate ambitions.

Looking to the future, the spread of emerging technologies, including blockchain-based energy trading, green hydrogen for hard-to-electrify sectors and super-intelligent AI systems for grid management, will transform the system in ways previously unimagined. Such technological developments, together with an enhanced set of policy instruments and better international cooperation, could help accelerate progress towards SDG 7 and other sustainable development goals.

The Path for the Future Is Ongoing Dedication to Creativity, Justice and Environmental Responsibility. Everyone engaged in energy has the potential to help drive an energy future that is more sustainable and inclusive, one that recognises the boundaries of the planet, by focusing on energy efficiency, accelerating renewable energy deployment, taking advantage of promising new technologies and supporting just transitions in communities that are highly affected. The overlap of what is technically and economically feasible and politically desirable to do represents an extraordinary moment in time to change energy systems globally for the benefit of this and future generations.

REFERENCES

C40 Cities Climate Leadership Group. (2023). *Just Transition towards Environmentally Sustainable Economies and Societies for All.* ILO Publications.

C40 Cities Climate Leadership Group. (2023). *Climate Action Planning Framework.* London: C40 Publications.

Bloomberg, N. E. F. (2023). *New Energy Outlook 2023.* Bloomberg Publications.

BP. (2023). *Statistical Review of World Energy 2023.* BP Publications.

Creutzig, F., Niamir, L., Bai, X., Callaghan, M., Cullen, J., Díaz-José, J., Figueroa, M., Grubler, A., Lamb, W. F., Leip, A., Masanet, E., Mata, É., Mattauch, L., Minx, J. C., Mirasgedis, S., Mulugetta, Y., Nugroho, S. B., Pathak, M., Perkins, P., & Ürge-Vorsatz, D. (2023). Demand-Side Solutions to Climate Change Mitigation Consistent with High Levels of Well-Being. *Nature Climate Change*, *12*(1), 36–42. DOI: 10.1038/s41558-021-01219-y

European Commission. (2019). *The European Green Deal.* EC Publications.

Helm, D. (2023). *Net Zero: How We Stop Causing Climate Change.* Yale University Press.

Intergovernmental Panel on Climate Change. (2023). *Climate Change 2023: Mitigation of Climate Change.* Cambridge University Press.

International Energy Agency. (2023). *Energy Efficiency 2023.* IEA Publications.

International Energy Agency. (2023). *World Energy Outlook 2023.* IEA Publications.

International Finance Corporation. (2023). *Emerging Market Clean Energy Investment Trends.* IFC Publications.

International Renewable Energy Agency. (2023). *Global Energy Transformation: A Roadmap to 2050.* IRENA Publications.

Jacobson, M. Z.. (2023). 100% Clean and Renewable Wind, Water, and Sunlight All-Sector Energy Roadmaps for Countries. *Nature Energy*, *8*(4), 234–247.

Modi, V., ... (2023). Energy Services for the Millennium Development Goals. UN Development Programme.

National Renewable Energy Laboratory. (2023). *2023 Annual Technology Baseline.* NREL Publications.

OECD. (2023). Energy Policies of IEA Countries: Review and Recommendations. Paris: OECD Publications. [National energy policy analysis and best practices] (Rockström, 2023). Rockström, J. (2023). Planetary Boundaries and Human Development. Nature Publications.

RE100. (2023). *Annual Report: Corporate Renewable Energy Procurement*. The Climate Group Publications.

Rocky Mountain Institute. (2023). *Clean Energy Transitions Program*. RMI Publications.

Saadaoui, K., Rhazi, K. S., Mejdoub, Y., & Aboudou, A. (2023). An analysis of the prospects and efficiency of floating and overland photovoltaic systems. *Int. J. Appl. Power Eng*, *12*(2), 144–152.

Smil, V. (2022). *Energy Transitions: Global and National Perspectives*. Praeger Publishers.

Sovacool, B. K. (2023). Energy Justice: Conceptual Insights and Practical Applications. *Applied Energy Journal*, *312*, 118–135.

Technology cost and performance projections, McKinsey Global Institute. (2023).

The Net-Zero Transition: What It Would Cost, What It Could Bring. New York: McKinsey Publications.

United Nations. (2015). *Transforming Our World: The 2030 Agenda for Sustainable Development*. UN Publications.

World Bank. (2023). *State and Trends of Carbon Pricing 2023*. World Bank Publications.

World Health Organization. (2023). *Ambient Air Quality and Health*. WHO Publications.

Compilation of References

Abakah, E. J. A., Tiwari, A. K., Adekoya, O. B., & Oteng-Abayie, E. F. (2023). An analysis of the time-varying causality and dynamic correlation between green bonds and US gas prices. *Technological Forecasting and Social Change, 186*, 122134. DOI: 10.1016/j.techfore.2022.122134

Abbas, M. H., Ali, H., & Musa, M. (2023). Integration between ESG goals and value-based intermediation approach within the paradigm of Islamic finance. *Islamic Banking & Finance Review, 10*(2), 31–54.

Abere, S. S., & Akinbobola, T. O. (2020). External Shocks, Institutional Quality, and Macroeconomic Performance in Nigeria. *SAGE Open, 10*(2), 2158244020919518. Advance online publication. DOI: 10.1177/2158244020919518

Aebi, V., Sabato, G., & Schmid, M. (2012). Risk management, corporate governance, and Bank performance in the financial crisis. *Journal of Banking & Finance, 36*(12), 3213–3226. DOI: 10.1016/j.jbankfin.2011.10.020

Afandy, C., Mirda, R., Fauzi, R. U. A., Hatidja, H., Fachri, M., & Jamaludin, J. (2024). Sustainable Finance and Green Bonds: Bridging the Gap between Profitability and Environmental Responsibility. eScience Humanity Journal, 5(1), 243-252.

Afonso, A., Alves, J., & Monteiro, S. (2024a). Beyond borders: Assessing the influence of Geopolitical tensions on sovereign risk dynamics. *European Journal of Political Economy, 83*, 102550. DOI: 10.1016/j.ejpoleco.2024.102550

Afonso, A., Alves, J., & Monteiro, S. (2024b). Sovereign risk dynamics in the EU: The time varying relevance of fiscal and external (im)balances*. *Journal of International Financial Markets, Institutions and Money, 94*, 102026. DOI: 10.1016/j.intfin.2024.102026

Afonso, A., Furceri, D., & Gomes, P. (2012). Sovereign credit ratings and financial markets linkages: Application to European data. *Journal of International Money and Finance, 31*(3), 606–638. DOI: 10.1016/j.jimonfin.2012.01.016

Afzal, A., Hasnaoui, J., Firdousi, S., & Noor, R. (2024). Climate change and the European banking sector: The effect of green technology adaptation and human capital. *Review of Accounting and Finance, 23*(3), 394–418. DOI: 10.1108/RAF-10-2023-0341

Aggarwal, D., & Mohanty, P. (2018). Do Indian stock market sentiments impact contemporaneous returns? *South Asian Journal of Business Studies, 7*(3), 332–346. *Scopus*. Advance online publication. DOI: 10.1108/SAJBS-06-2018-0064

Agnello, L., Castro, V., Hammoudeh, S., & Sousa, R. M. (2020). Global factors, uncertainty, weather conditions and energy prices: On the drivers of the duration of commodity price cycle phases. *Energy Economics, 90*, 104862. DOI: 10.1016/j.eneco.2020.104862

Agnello, L., Castro, V., & Sousa, R. M. (2018). The Legacy and the Tyranny of Time: Exit and Re-Entry of Sovereigns to International Capital Markets. *Journal of Money, Credit and Banking, 50*(8), 1969–1994. DOI: 10.1111/jmcb.12474

Agnihotri, S., & Chauhan, K. (2022). Modeling tail risk in Indian commodity markets using conditional EVT-VaR and their relation to the stock market. *Investment Management and Financial Innovations, 19*(3), 1–12. *Scopus*. Advance online publication. DOI: 10.21511/imfi.19(3).2022.01

Agoraki, M.-E. K., Kouretas, G. P., & Laopodis, N. T. (2022). Geopolitical risks, uncertainty, and stock market performance. *Economic and Political Studies, 10*(3), 253–265. DOI: 10.1080/20954816.2022.2095749

Aguirre, M., & Ibikunle, G. (2014). Determinants of renewable energy growth: A global sample analysis. *Energy Policy, 69*, 374–384. DOI: 10.1016/j.enpol.2014.02.036

Ahmad, F., Boumaiza, A., Sanfilippo, A., & Al-Fagih, L. (2025). A Detailed Comprehensive Role of Digital Technologies in Green Finance Initiative for Net-Zero Energy Transition. *Advanced Energy and Sustainability Research, 2500066*(10), 2500066. Advance online publication. DOI: 10.1002/aesr.202500066

Ahmad, N., Malik, M. S., & Hussain, M. (2023). Bank efficiency and green finance: Evidence from emerging markets. *Journal of Cleaner Production, 380*, 135029. DOI: 10.1016/j.jclepro.2022.135029

Ahmed, F., Syed, A. A., Kamal, M. A., López-garcía, M. L. N., Ramos-requena, J. P., & Gupta, S. (2021). Assessing the impact of COVID-19 pandemic on the stock and commodity markets performance and sustainability: A comparative analysis of south asian countries. *Sustainability (Switzerland), 13*(10). *Sustainability (Basel), 13*(10), 5669. Advance online publication. DOI: 10.3390/su13105669

Aizenman, J., Edwards, S., & Riera-Crichton, D. (2012). Adjustment patterns to commodity terms of trade shocks: The role of exchange rate and international reserves policies. *Journal of International Money and Finance*, *31*(8), 1990–2016. DOI: 10.1016/j.jimonfin.2012.05.003

Ajzen, I. (1991). The theory of planned behavior. *Organizational Behavior and Human Decision Processes*, *50*(2), 179–211. DOI: 10.1016/0749-5978(91)90020-T

Akmal, I., Ibrahim, F. W., Bambang, S., & Syafiq, M. H. (2025). Rethinking financial deepening in Islamic countries: Political system and economic growth. *International Journal of Islamic and Middle Eastern Finance and Management*, *18*(2), 467–485. DOI: 10.1108/IMEFM-10-2023-0376

Akram, Q. F. (2009). Commodity prices, interest rates and the dollar. *Energy Economics*, *31*(6), 838–851. DOI: 10.1016/j.eneco.2009.05.016

Alan, S., Attanasio, O., & Browning, M. (2009). Estimating Euler equations with noisy data: Two exact GMM estimators. *Journal of Applied Econometrics*, *24*(2), 309–324. DOI: 10.1002/jae.1037

Aldy, J. E., & Stavins, R. N. (2012). The promise and problems of pricing carbon: Theory and experience. *Journal of Environment & Development*, *21*(2), 152–180. DOI: 10.1177/1070496512442508

Alemzero, D. A., Iqbal, N., Iqbal, S., Mohsin, M., Chukwuma, N. J., & Shah, B. A. (2021). Assessing the perceived impact of exploration and production of hydrocarbons on households perspective of environmental regulation in Ghana. *Environmental Science and Pollution Research International*, *28*(5), 5359–5371. DOI: 10.1007/s11356-020-10880-3 PMID: 32960443

Ali Q, Khan MTI, Khan MNI (2018) Dynamics between financial development, tourism, sanitation, renewable energy, trade and total reserves in 19 Asia cooperation dialogue members. J Clean Prod 179:114–131ArticleGoogle Scholar

Ali, M., Jamal, S., Zeb, A., Dil, R., & Ali, O. (2025). The Role of Green Bonds in Advancing Climate Finance Goals in Emerging Economies: A Policy-Oriented Assessment. *Social Science Review Archives*, *3*(3), 1144–1157. DOI: 10.70670/sra.v3i3.946

Almaiman, S., & McLaughlin, P. (2018). Investigating the role of organizational culture in facilitating continuous improvement within Saudi non-profit organizations. Journal of Human Resources Management Research. *Article ID*, *724461*, 1–16. Advance online publication. DOI: 10.5171/2018.724461

Al-Saidi, M., & Elagib, N. A. (2022). Ecological modernization and renewable energy transitions in the Gulf Cooperation Council countries. *Renewable & Sustainable Energy Reviews, 160*, 112278.

Alsaleh, M., & Abdul-Rahim, A. (2019). Financial development and bioenergy consumption in the EU28 region: Evidence from panel auto-regressive distributed lag bound approach. *Resources, 8*(1), 44. DOI: 10.3390/resources8010044 PMID: 33366414

Apau, R., & Sibindi, A. B. (2025). Sustainable Finance and Insurance: Opportunities and Challenges. Sustainable Finance and Insurance in Africa, 9-21.

Apergis, N., Chatziantoniou, I., & Cooray, A. (2020). Monetary policy and commodity markets: Unconventional versus conventional impact and the role of economic uncertainty. *International Review of Financial Analysis, 71*, 101536. DOI: 10.1016/j.irfa.2020.101536

Arifin, A. H., Hudaeni, N., Rasyid, A., Suhardi, S., Suprayitno, D., Redjeki, F., & Purwati, T. (2024). The effectiveness of green bonds in promoting sustainable finance: A meta-analysis. [Journal of Humanities and Social Studies]. *JHSS, 8*(3), 830–839.

Armendáriz, B., & Morduch, J. (2010). *The economics of microfinance* (2nd ed.). MIT Press.

Arner, D. W., Buckley, R. P., Zetzsche, D. A., & Veidt, R. (2020). Sustainability, FinTech and Financial Inclusion. *European Business Organization Law Review, 21*(1), 7–35. DOI: 10.1007/s40804-020-00183-y

Arouri, M., Mhadhbi, M., & Shahrour, M. H. (2025). Dynamic Connectedness and Hedging Effectiveness between Green Bonds, ESG Indices, and Traditional Assets. European Financial Management.

Aslan, C., Bulut, E., Cepni, O., & Yilmaz, M. (2022). Does climate change affect bank lending behavior? *Economics Letters, 220*, 110859. DOI: 10.1016/j.econlet.2022.110859

Asumadu, S. S., Ahmed, M., & Owusu, P. A. (2023). Advancing COP26 climate goals: Leveraging energy innovation, governance readiness, and socio-economic factors for enhanced climate resilience and sustainability. *Journal of Cleaner Production, 431*, 139757. DOI: 10.1016/j.jclepro.2023.139757

Awadai, M. A. M. (2023). The Reality of Strategic Planning in Non-Profit Organizations to Achieve Sustainable Development in the Kingdom of Saudi Arabia. *Migration Letters : An International Journal of Migration Studies, 20*(S3), 1115–1129. DOI: 10.59670/ml.v20iS3.4137

Baek, S., & Kang, M. (2025). Does ESG enhance asset quality and funding cost management in banking diversification? *Finance Research Letters*, *73*, 106542. DOI: 10.1016/j.frl.2024.106542

Bai, M., & Wang, W. (2024). Role of green bonds in developing ESG projects in Eastern provinces of China. *Humanities & Social Sciences Communications*, *11*(1), 1–8. DOI: 10.1057/s41599-024-04353-y

Baldacci, E., Gupta, S., & Mati, A. (2011). Political and Fiscal Risk Determinants of Sovereign Spreads in Emerging Markets. *Review of Development Economics*, *15*(2), 251–263. DOI: 10.1111/j.1467-9361.2011.00606.x

Balli, F., Balli, H. O., Hasan, M., & Gregory-Allen, R. (2022). Geopolitical risk spillovers and its determinants. *The Annals of Regional Science*, *68*(2), 463–500. DOI: 10.1007/s00168-021-01081-y

Balta, N., Fernandez, I. V., & Ruscher, E. (2013). Assessing the impact of uncertainty on consumption and investment. [QREA]. *Q Rep Euro Area*, *12*(2), 7–16.

Banker, R. D., Charnes, A., & Cooper, W. W. (1984). Some models for estimating technical and scale inefficiencies in data envelopment analysis. *Management Science*, *30*(9), 1078–1092. DOI: 10.1287/mnsc.30.9.1078

Barbier, E. B. (2020). Greening the post-pandemic recovery: A sustainable growth strategy. *Environmental and Resource Economics*, *76*(4), 685–703. DOI: 10.1007/s10640-020-00437-w PMID: 32836827

Barbi, F., & da Costa Ferreira, L. (2017). Governing climate change risks: Subnational climate policies in Brazil. *Chinese Political Science Review*, *2*(2), 237–252. DOI: 10.1007/s41111-017-0061-3

Basurto, S., & Sánchez, G. (2022). Technical efficiency in small and medium-sized firms in Mexico: A stochastic frontier analysis. *Estudios Económicos de El Colegio de México*, *37*(1), 103–147.

Batsaikhan, A., Hachinger, S., Kurtz, W., Heller, H., & Frank, A. (2020). Application of Modern Web Technologies to the Citizen Science Project BAYSICS on Climate Research and Science Communication. *Sustainability (Basel)*, *12*(18), 7748. DOI: 10.3390/su12187748

Battese, G. E., & Coelli, T. J. (1995). A model for technical inefficiency effects in a stochastic frontier production function for panel data. *Empirical Economics*, *20*(2), 325–332. DOI: 10.1007/BF01205442

Baur, D. G., & Smales, L. A. (2020). Hedging geopolitical risk with precious metals. *Journal of Banking & Finance, 117*, 105823. DOI: 10.1016/j.jbankfin.2020.105823

Bazán-Navarrete, M., & Bravo-Peña, L. C. (2021). Clean energy certificates market performance in Mexico: An empirical analysis of market liquidity and price formation. *Renewable Energy, 162*, 107–119. DOI: 10.1016/j.renene.2020.08.025

Bazilian, M., & Pielke, R.Jr. (2013). Making energy access meaningful. *Issues in Science and Technology, 29*(4), 74–79.

Bell, J., Battisti, G., & Guin, B. (2023). The greening of lending: Evidence from banks' pricing of energy efficiency before climate-related regulation. *Economics Letters, 230*, 111212. DOI: 10.1016/j.econlet.2023.111212

Ben Flah, I., Samet, K., El Ammari, A., & Terzi, C. (2025). Shock and Volatility Transmissions Across Global Commodity and Stock Markets Spillovers: Empirical Evidence from Africa. *Journal of Risk and Financial Management, 18*(6), 332. DOI: 10.3390/jrfm18060332

Ben-Ner, A., & Van Hoomissen, T. (1991). Nonprofit organizations in the mixed economy: A demand and supply analysis. *Annals of Public and Cooperative Economics, 62*(4), 519–550. DOI: 10.1111/j.1467-8292.1991.tb01366.x

Berger, A. N., & Mester, L. J. (1997). Inside the black box: What explains differences in the efficiencies of financial institutions? *Journal of Banking & Finance, 21*(7), 895–947. DOI: 10.1016/S0378-4266(97)00010-1

Berg, F., Kölbel, J. F., & Rigobon, R. (2019). *Aggregate confusion: The divergence of ESG ratings* [Working paper]. MIT Sloan School of Management.

Bergman, U. M., Hutchison, M. M., & Hougaard Jensen, S. E. (2019). European policy and markets: Did policy initiatives stem the sovereign debt crisis in the euro area? *European Journal of Political Economy, 57*, 3–21. DOI: 10.1016/j.ejpoleco.2018.06.003

Bernard, J., Steinführer, A., Klärner, A., & Keim-Klärner, S. (2023). Regional opportunity structures: A research agenda to link spatial and social inequalities in rural areas. *Progress in Human Geography, 47*(1), 103–123. DOI: 10.1177/03091325221139980

Bhattacharyya, S. C. (2011). *Energy economics: Concepts, issues, markets and governance*. Springer Science & Business Media. DOI: 10.1007/978-0-85729-268-1

Birindelli, G., Bonanno, G., Dell'Atti, S., & Lannuzzi, A. (2022). Climate Change Commitment, credit risk and the country's environmental performance: Empirical evidence from a sample of international banks. *Business Strategy and the Environment, 31*(4), 1641–1655. DOI: 10.1002/bse.2974

Birindelli, G., Dell'Atti, S., Di Tommaso, C., Lannuzzi, A., & Pacelli, V. (2024). The impact of banks' climate engagement on systemic risk. does committing a little or a lot make a difference?'. *Research in International Business and Finance, 70B*, 102392. DOI: 10.1016/j.ribaf.2024.102392

Bloomberg, N. E. F. (2023). *New Energy Outlook 2023*. Bloomberg Publications.

Bouri, E., De Boyrie, M. E., & Pavlova, I. (2017). Volatility transmission from commodity markets to sovereign CDS spreads in emerging and frontier countries. *International Review of Financial Analysis, 49*, 155–165. DOI: 10.1016/j.irfa.2016.11.001

BP. (2023). *Statistical Review of World Energy 2023*. BP Publications.

Breitenstein, M., Nguyen, D. K., & Walther, T. (2021). Environmental hazards and risk management in the financial sector: A Systematic Literature Review. *Journal of Economic Surveys, 35*(2), 512–538. DOI: 10.1111/joes.12411

Brockman, P., Dang, T. L., & Pham, T. P. (2024). Stock price synchronicity and stock liquidity: International evidence. *Journal of Empirical Finance, 79*, 101541. DOI: 10.1016/j.jempfin.2024.101541

Brüggemann, T., & Lueg, R. (2025). Bank Responses to Physical and Transition Risks in Lending: A Diagnostic Framework from a Systematic Literature Review. *Business Strategy and the Environment, 0*, 1–18. DOI: 10.1002/bse.70176

Brundtland, G. H. (1987). *Our Common Future*. Oxford University Press.

Bruno, B., & Lombini, S. (2023). Climate transition risk and bank lending. *Journal of Financial Research, 46*(S1), 59–106. DOI: 10.1111/jfir.12360

Bulkeley, H., & Betsill, M. (2021). Cities and climate change governance. Routledge Handbook of Environmental Governance, 231–245.

Bulkeley, H., Broto, V., & Edwards, G. (2021). *Governing climate change: The new politics of urban sustainability*. Routledge.

Busch, T., & Lewandowski, S. (2020). The EU Taxonomy for sustainable activities: Implications for investor decision-making and capital allocation [Working paper]. University of Hamburg. https://www.wiso.uni-hamburg.de/fachbereich-vwl/professuren/busch/publikationen/eu-taxonomy-investor-decisions.pdf

C40 Cities Climate Leadership Group. (2023). *Climate Action Planning Framework.* London: C40 Publications.

C40 Cities Climate Leadership Group. (2023). *Just Transition towards Environmentally Sustainable Economies and Societies for All.* ILO Publications.

Cafiso, G., Missale, A., & Rivolta, G. (2025). The credit channel of the sovereign spread: A Bayesian SVAR analysis. *Economic Modelling, 144*, 106984. DOI: 10.1016/j.econmod.2024.106984

Caldara, D., & Iacoviello, M. (2022). Measuring Geopolitical Risk. *The American Economic Review, 112*(4), 1194–1225. DOI: 10.1257/aer.20191823

Campiglio, E. (2015). Beyond carbon pricing: The role of banking and monetary policy in financing the transition to a low-carbon economy. *Ecological Economics, 121*, 220–230. DOI: 10.1016/j.ecolecon.2015.03.020

Campiglio, E., Dafermos, Y., Monnin, P., Ryan-Collins, J., Schotten, G., & Tanaka, M. (2018). Climate change challenges for central banks and financial regulators. *Nature Climate Change, 8*(6), 462–468. DOI: 10.1038/s41558-018-0175-0

Cao, S., & Dong, W. (2025). Production networks and the macroeconomic impacts of commodity price shocks. *The Canadian Journal of Economics. Revue Canadienne d'Economique, 58*(2), 690–715. DOI: 10.1111/caje.70007

Carlsson, B., & Stankiewicz, R. (1991). On the nature, function and composition of technological systems. *Journal of Evolutionary Economics, 1*(2), 93–118. DOI: 10.1007/BF01224915

Charan, J. K., Chandra, S., Vaissnave, V., Ragupathi, T., Satyanarayana, P., Selvakumar, P., & Manjunath, T. C. (2025). Impact of Generative AI on Industries and Organizations. In Zhao, J., Kumar, V., Katina, P., & Richards, J. (Eds.), *Humans and Generative AI Tools for Collaborative Intelligence* (pp. 509–528). IGI Global Scientific Publishing., DOI: 10.4018/979-8-3693-8332-2.ch022

Chaudhry, N., & Kumari, D. (2024). How do banks price carbon risk? evidence from India. *Pacific-Basin Finance Journal, 84*, 102304. DOI: 10.1016/j.pacfin.2024.102304

Chauhan, S. S., Suri, P., Twala, B., Priyadarshi, N., & Ali, F. (2025). *Exploring the relationship between macroeconomic indicators and sectoral indices of Indian stock market* (No. 14:180). F1000Research. DOI: 10.12688/f1000research.160668.2

Chava, S., Malakar, B., & Singh, M. (2024). Impact of corporate subsidies on borrowing costs of local governments: Evidence from municipal bonds. *Review of Finance*, *28*(1), 117–161. DOI: 10.1093/rof/rfad021

Chen, A. A., Stephens, A. J., Koon, R. K., Ashtine, M., & Koon, K. M. K. (2020). Pathways to climate change mitigation and stable energy by 100% renewable for a small island: Jamaica as an example. *Renewable & Sustainable Energy Reviews*, *121*, 109671. DOI: 10.1016/j.rser.2019.109671

Chen, J., Li, Z., Dong, Y., Song, M., Shahbaz, M., & Xie, Q. (2020). Coupling coordination between carbon emissions and the eco-environment in China. *Journal of Cleaner Production*, *276*, 123848. DOI: 10.1016/j.jclepro.2020.123848

Chen, S., Xu, Q., Zeng, J., & Zhao, Y. (2025). Ark amidst floods: Evidence from personal credit at a Chinese bank. *International Review of Financial Analysis*, *106*, 104514. DOI: 10.1016/j.irfa.2025.104514

Climate Bonds Initiative. (2021). *Green bonds market outlook*. Climate Bonds Initiative. https://www.climatebonds.net

Climate Transparency. (2025, April). Assessment of financial instruments for Mexico's energy transition: Cases of FOTEASE and FIDE. https://www.climate-transparency.org/wp-content/uploads/2025/04/Assessment-of-financial-instruments-for-Mexicos-energy-transition_-cases-of-FOTEASE-and-FIDE.pdf

Coelli, D., Rao, T. P., O'Donnell, C., & Battese, G. (2005). *An Introduction to Efficiency and Productivity Analysis* (2nd ed.). Springer.

Conlon, T., Ding, R., Huan, X., & Zhang, Z. (2024). Climate risk and financial stability: Evidence from syndicated lending. *European Journal of Finance*, *30*(17), 2001–2031. DOI: 10.1080/1351847X.2024.2343111

Creutzig, F., Niamir, L., Bai, X., Callaghan, M., Cullen, J., Díaz-José, J., Figueroa, M., Grubler, A., Lamb, W. F., Leip, A., Masanet, E., Mata, É., Mattauch, L., Minx, J. C., Mirasgedis, S., Mulugetta, Y., Nugroho, S. B., Pathak, M., Perkins, P., & Ürge-Vorsatz, D. (2023). Demand-Side Solutions to Climate Change Mitigation Consistent with High Levels of Well-Being. *Nature Climate Change*, *12*(1), 36–42. DOI: 10.1038/s41558-021-01219-y

Cucinelli, D., Nieri, L., & Piserà, S. (2024). Climate and environment risks and opportunities in the banking industry: The role of Risk Management. *International Journal of Finance & Economics*, *0*, 1–17. DOI: 10.1002/ijfe.3083

D'Orazio, P. (2023). The politics of climate finance and policy initiatives to promote sustainable finance and address ESG issues. In *Sustainable finance and ESG: Risk, management, regulations, and implications for financial institutions* (pp. 145–171). Springer International Publishing. DOI: 10.1007/978-3-031-24283-0_7

Datta, S., & Hatekar, N. (2025). Range Volatility Spillover across Sectoral Stock Indices during COVID 19 Pandemic: Evidence from Indian Stock Market. *Journal of Applied Economic Sciences, 20*(1), 79–100. *Scopus*. Advance online publication. DOI: 10.57017/jaes.v20.1(87).06

de Bandt, O., Kuntz, L., Pankratz, N., Pegoraro, F., Solheim, H., Sutton, G., Takeyama, A., & Xia, F. D. (2025). The effects of climate change-related risks on banks: A literature review. *Journal of Economic Surveys*, *39*(4), 1553–1594. DOI: 10.1111/joes.12665

de Novellis, G., Pedrazzoli, A., Pennetta, D., & Venturelli, V. (2025). Greenwashing, greenhushing, and the path to Green Banking. *Global Finance Journal*, *67*, 101147. DOI: 10.1016/j.gfj.2025.101147

Degryse, H., Goncharenko, R., Theunisz, C., & Vadasz, T. (2023). When green meets green. *Journal of Corporate Finance*, *78*, 102355. DOI: 10.1016/j.jcorpfin.2023.102355

Del Sarto, N., & Ozili, P. K. (2025). FinTech and financial inclusion in emerging markets: A bibliometric analysis and future research agenda. *International Journal of Emerging Markets*, *20*(13), 270–290. DOI: 10.1108/IJOEM-08-2024-1428

Delis, M. D., Greiff, K., Losifidi, M., & Ongena, S. (2024). Being stranded with fossil fuel reserves? climate policy risk and the pricing of Bank Loans. *Financial Markets, Institutions &amp. Instruments*, *33*(3), 239–265.

Delmas, M. A., & Montes-Sancho, M. J. (2010). Voluntary agreements to improve environmental quality: Symbolic and substantive cooperation. *Strategic Management Journal*, *31*(6), 575–601. DOI: 10.1002/smj.826

Demetriades, E., & Politsidis, P. N. (2025). Bank lending to fossil fuel firms. *Journal of Financial Stability*, *76*, 101349. DOI: 10.1016/j.jfs.2024.101349

Demirgüç-Kunt, A., Peria, M., Soledad, M., & Thierry, T. (2020). The global financial crisis and the capital structure of firms: Was the impact more severe among SMEs and non-listed firms? *Journal of Corporate Finance*, *60*(C), 101514. DOI: 10.1016/j.jcorpfin.2019.101514

Demski, C., Capstick, S., Pidgeon, N., Sposato, R. G., & Spence, A. (2017). Experience of extreme weather affects climate change mitigation and adaptation responses. *Climatic Change*, *140*(2), 149–164. DOI: 10.1007/s10584-016-1837-4 PMID: 32355377

Dewan, P., & Dharni, K. (2022). Herding and spillover effects in the Indian commodity futures market. *Journal of Agribusiness in Developing and Emerging Economies*, *13*(5), 748–761. DOI: 10.1108/JADEE-11-2021-0288

Dieckelmann, D., Larkou, C., McQuade, P., Pancaro, C., & Rößler, D. (2025). Geopolitical risk and euro area bank CDS spreads and stock prices: Evidence from a new index. *Economics Letters*, *254*, 112461. DOI: 10.1016/j.econlet.2025.112461

DiMaggio, P. J., & Powell, W. W. (1983). The Iron Cage Revisited: Institutional isomorphism and collective rationality in organisational fields. *American Sociological Review*, *48*(2), 147. DOI: 10.2307/2095101

Doan, A., Tran, A., Tran, H., & Nguyen, C. (2025). A global perspective on climate adaptation, ownership, and Bank Performance. *International Journal of the Economics of Business*, •••, 1–43. DOI: 10.1080/13571516.2025.2479754

Doelman, J. C., Stehfest, E., van Vuuren, D. P., Tabeau, A., Hof, A. F., Braakhekke, M. C., Gernaat, D. E. H. J., van den Berg, M., van Zeist, W.-J., Daioglou, V., van Meijl, H., & Lucas, P. L. (2020). Afforestation for climate change mitigation: Potentials, risks and trade-offs. *Global Change Biology*, *26*(3), 1576–1591. DOI: 10.1111/gcb.14887 PMID: 31655005

Doytch, N., & Narayan, S. (2016). Does FDI influence renewable energy consumption? An analysis of sectoral FDI impact on renewable and non-renewable industrial energy consumption. *Energy Economics*, *54*, 291–301. DOI: 10.1016/j.eneco.2015.12.010

Duanmu, J., Li, Y., Lin, M., & Tahsin, S. (2022). Natural disaster risk and residential mortgage lending standards. *Journal of Real Estate Research*, *44*(1), 106–130. DOI: 10.1080/08965803.2021.2013613

Eckstein, D., Künzel, V., & Schäfer, L. (2022). *Global Climate Risk Index 2022: Who suffers most from extreme weather events? Weather-related loss events in 2020 and 2001–2020.* Germanwatch. https://germanwatch.org/en/cri

Economics Observatory. (2025, April 24). *How could green energy policy boost Mexico's economy?* https://economicsobservatory.com/how-could-green-energy-policy-boost-mexicos-economy

Ehlers, T., Packer, F., & de Greiff, K. (2022). The Pricing of Carbon Risk in Syndicated loans: Which Risks Are Priced and why? *Journal of Banking & Finance, 136*, 106180. DOI: 10.1016/j.jbankfin.2021.106180

Ember. (2025, May 29). Renewables point the way to Mexico's energy security. https://ember-energy.org/latest-insights/renewables-point-the-way-to-mexicos-energy-security/

Eren, B. M., Taspinar, N., & Gokmenoglu, K. K. (2019). The impact of financial development and economic growth on renewable energy consumption: Empirical analysis of India. *The Science of the Total Environment, 663*, 189–197. DOI: 10.1016/j.scitotenv.2019.01.323 PMID: 30711585

Erhemjamts, O., Huang, K., & Tehranian, H. (2024). Climate risk, ESG performance, and ESG sentiment in US Commercial Banks. *Global Finance Journal, 59*, 100924. DOI: 10.1016/j.gfj.2023.100924

European Commission. (2019). *The European Green Deal*. EC Publications.

European Commission. (2021). *The European Green Deal: 2021 Progress Report*. European Union.

Fan, D., Breslin, D., Callahan, J. L., & Iszatt-White, M. (2022). Advancing literature review methodology through rigour, generativity, scope and transparency. *International Journal of Management Reviews, 24*(2), 171–180. DOI: 10.1111/ijmr.12291

Feng, T., Xu, J., Zhou, Z., & Luo, Y. (2025). How green credit policy affects commercial banks' credit risk? *Journal of Cases on Information Technology, 26*(1), 1–21. DOI: 10.4018/JCIT.333858

Feng, Y., Pan, Y., Sun, C., & Niu, J. (2024). Assessing the effect of green credit on risk-taking of commercial banks in China: Further analysis on the two-way Granger causality. *Journal of Cleaner Production, 437*, 140698. DOI: 10.1016/j.jclepro.2024.140698

Fernández, A. de J., & Watson, J. (2022). Mexico's renewable energy innovation system: Geothermal and solar photovoltaics case study. *Journal of Cleaner Production, 345*, 131264. DOI: 10.1016/j.jclepro.2022.131264

Fernandez-Perez, A., Gómez-Puig, M., & Sosvilla-Rivero, S. (2025). Examining the transmission of credit and liquidity risks: A network analysis for EMU sovereign debt markets. *The North American Journal of Economics and Finance, 77*, 102407. DOI: 10.1016/j.najef.2025.102407

Flammer, C. (2021). Green bonds: Effectiveness and implications for corporate strategy and public policy. *Journal of Financial Economics, 142*(2), 475–498.

Flori, A., Pammolli, F., & Spelta, A. (2021). Commodity prices co-movements and financial stability: A multidimensional visibility nexus with climate conditions. *Journal of Financial Stability, 54*, 100876. DOI: 10.1016/j.jfs.2021.100876

Folke, C. (2021). Resilience (Republished). *Ecology and Society, 26*(4), 11–25.

Friede, G., Busch, T., & Bassen, A. (2015). ESG and financial performance: aggregated evidence from more than 2000 empirical studies. Journal of sustainable finance & investment, 5(4), 210-233.

Friedman, A. L., & Miles, S. (2002). Developing stakeholder theory. *Journal of Management Studies, 39*(1), 1–21. DOI: 10.1111/1467-6486.00280

Friedrich, T. J., Velte, P., & Wulf, I. (2023). Corporate climate reporting of European banks: Are these institutions compliant with climate issues? *Business Strategy and the Environment, 32*(6), 2817–2834. DOI: 10.1002/bse.3272

Fu, C., Lu, L., & Pirabi, M. (2023). Advancing green finance: A review of sustainable development. *Digital Economy and Sustainable Development, 1*(1), 20. DOI: 10.1007/s44265-023-00020-3

Gaganis, C., Pasiouras, F., Tasiou, M., & Zopounidis, C. (2023). *Sustainable Finance and ESG. Risk, Management, Regulations, and Implications for Financial Institutions*. Springer International Publishing., DOI: 10.1007/978-3-031-24283-0

Gaies B, Goutte S, Guesmi K (2019a) Banking crises in developing countries–what crucial role of exchange rate stability and external liabilities? Financ Res Lett 31

Gaies, B., Kaabia, O., Ayadi, R., Guesmi, K., & Abid, I. (2019b). Financial development and energy consumption: Is the MENA region different? *Energy Policy, 135*, 111000. DOI: 10.1016/j.enpol.2019.111000

Galletta, S., Goodell, J., Mazzù, S., & Paltrinieri, A. (2023). Bank reputation and operational risk: The IMPACT OF ESG. *Finance Research Letters, 51*, 103494. DOI: 10.1016/j.frl.2022.103494

Galletta, S., & Mazzù, S. (2022). ESG controversies and Bank Risk Taking. *Business Strategy and the Environment, 32*(1), 274–288. DOI: 10.1002/bse.3129

Galletta, S., Mazzù, S., & Naciti, V. (2021). Banks' business strategy and Environmental Effectiveness: The monitoring role of the Board of Directors and the managerial incentives. *Business Strategy and the Environment, 30*(5), 2656–2670. DOI: 10.1002/bse.2769

Gambacorta, L., Polizzi, S., Reghezza, A., & Scannella, E. (2024). Do banks practice what they preach? Brown lending and environmental disclosure in the Euro Area. [Preprint]. *Journal of Financial Services Research.* Advance online publication. DOI: 10.1007/s10693-024-00435-9

Geels, F. W. (2002). Technological transitions as evolutionary reconfiguration processes: A multi-level perspective and a case-study. *Research Policy*, *31*(8-9), 1257–1274. DOI: 10.1016/S0048-7333(02)00062-8

Geels, F. W., & Schot, J. (2010). The dynamics of transitions: A socio-technical perspective. In *Transitions to Sustainable Development* (pp. 11–104). Routledge.

Geetha, S., Suwetha, C., Sangavi, A., & Shabana Nazrin, S. (2024). Time Series Financial Market Forecasting Based on Machine Learning. *2024 2nd International Conference on Artificial Intelligence and Machine Learning Applications Theme: Healthcare and Internet of Things (AIMLA)*, 1–5. DOI: 10.1109/AIMLA59606.2024.10531346

Ge, W., Qi, Z., Wu, Z., & Yu, L. (2025). Abnormal temperatures, climate risk disclosures and bank loan pricing: International evidence. *British Journal of Management*, *36*(2), 726–744. DOI: 10.1111/1467-8551.12867

Ghosh, I., Datta Chaudhuri, T., Alfaro-Cortés, E., Gámez Martínez, M., & García Rubio, N. (2021). Estimating the relative effects of raw material prices, sectoral outlook and market sentiment on stock prices. *Resources Policy*, *73*, 102158. DOI: 10.1016/j.resourpol.2021.102158

Giuzio M, Krušec D, Levels A, Melo AS, Mikkonen K, Radulova P (2019) Climate change and financial stability. Financ Stabil Rev 1

Global Reporting Initiative (GRI). (2020). *Global sustainability reporting standards*. GRI. https://www.globalreporting.org

Goel, R., Sidana, N., Bhattacharjee, A., Singh, T., & Kaur, M. (2025). Shifting sands: How major events shape gold futures in the Indian commodity market. *Cogent Economics & Finance*, *13*(1), 2483867. DOI: 10.1080/23322039.2025.2483867

Goldthau, A., & Sovacool, B. K. (2012). The uniqueness and diversity of energy security. In *Handbook of global energy policy* (pp. 13–29). John Wiley & Sons.

Gonzalez, E. D., Sijariya, R., Singh, A. K., & Garg, V. (2025). Integrating Climate Change, Social Responsibility and Electronic Financial Inclusion: A Pathway to Sustainable Development. In *Climate Change and Social Responsibility* (pp. 147–164). Emerald Publishing Limited. https://www.emerald.com/insight/content/doi/10.1108/S2043-052320250000025008/full/html

González-Mahecha, R. E., Lecuyer, O., Hallack, M., Bazilian, M., & Vogt-Schilb, A. (2019). Committed emissions from existing and planned power plants in Mexico: A preliminary assessment. *Environmental Research Letters*, *14*(5), 054019. DOI: 10.1088/1748-9326/ab10c2

Guru Basava Aradhya, S., Chithambar Gupta, V., Selvakumar, P., Raman, M. S., Singh, J., Manjunath, T. C., & Sharma, M. (2025). Health Crises and Digitalization: Digital Health. In Yıldırım, D., Yıldırım, S., & Kandpal, V. (Eds.), *Sustainable Digitalization Strategies in Business and Healthcare* (pp. 171–198). IGI Global Scientific Publishing., DOI: 10.4018/979-8-3373-5087-5.ch007

Hallegatte, S., Bangalore, M., & Vogt-Schilb, A. (2016). Assessing socioeconomic resilience to floods in 90 countries. *World Bank Policy Research Working Paper*, (7663).

Harmanpreet, S. P., Chandra, S., Kalshetti, P., Satyanarayana, P., Sharma, M., & Manjunath, T. C. (2025). Digital Transformation in the Public Sector Entities. In A. Santos Ferreira & C. Lourenço dos Santos (Eds.), Enhancing Public Sector Accountability and Services Through Digital Innovation (pp. 353-380). IGI Global Scientific Publishing. DOI: 10.4018/979-8-3693-9251-5.ch011

Hassan, M. K., Hasan, R., Bataineh, H., & Haque, M. (2025). Bank Board Diversity and Environmental Risk Management: Does critical mass matter in project finance decisions? *Managerial Finance*, *51*(6), 879–895. DOI: 10.1108/MF-03-2024-0168

Hassine, M. B., & Harrathi, N. (2017). The causal links between economic growth, renewable energy, financial development and foreign trade in gulf cooperation council countries. *Int J Energy Econ Policy*, *7*(2), 76–85.

Helm, D. (2023). *Net Zero: How We Stop Causing Climate Change*. Yale University Press.

Hilal, M. Sumbal, & Imad, M. (2025). An Impact of Sustainable Finance Initiatives on Financial Markets and Economic Stability. In Emerging Disruptive Technologies for Society 5.0 in Developing Countries: Challenges and Applications (pp. 89-96). Cham: Springer Nature Switzerland.

Ho, K., & Wong, A. (2023). Effect of climate-related risk on the costs of bank loans: Evidence from syndicated loan markets in emerging economies. *Emerging Markets Review*, *55*, 100977. DOI: 10.1016/j.ememar.2022.100977

Hrazdil, K., Anginer, D., Li, J., & Zhang, R. (2024). Climate reputation and bank loan contracting. *Journal of Business Ethics*, *192*(4), 875–896. DOI: 10.1007/s10551-023-05517-7

Huang, F., Zhu, H., & Wu, S. (2025). Green Credit, monetary policy, and bank risk-taking. *Finance Research Letters, 79*, 107289. DOI: 10.1016/j.frl.2025.107289

Huang, J., Wang, X., Liu, H., & Iqbal, S. (2021). Financial consideration of energy and environmental nexus with energy poverty: Promoting financial development in G7 economies. *Frontiers in Energy Research, 9*, 666. DOI: 10.3389/fenrg.2021.777796

Hughes, T. P. (1983). *Networks of power: Electrification in Western society, 1880-1930*. Johns Hopkins University Press.

Hussain, M., Jaweed, M., Murtaza, M., & Ali, M. (2025). Green Bonds and Sustainable Investment Strategies: Evaluating Risk-Return Profiles, Market Growth, and the Role of Climate-Conscious Portfolios in Sustainable Finance. *Advance Journal of Econometrics and Finance, 3*(2), 39–53. DOI: 10.63075/5nggp275

Hu, Z., & Borjigin, S. (2025). Climate Information Disclosure Quality and systemic risk in the U.S. Banking Industry. *Journal of Financial Stability, 79*, 101420. DOI: 10.1016/j.jfs.2025.101420

IMF. (2024). *Annual Report 2024*. International Monetary Fund.

Inoune, M., Iwasaki, M., Otani, T., Sasazuki, S., Noda, M., & Tsugane, S.Japan Public Health Center–Based Prospective Study Group. (2006). Diabetes mellitus and the risk of cancer: Results from a large-scale population-based cohort study in Japan. *Archives of Internal Medicine, 166*(17), 1871–1877. DOI: 10.1001/archinte.166.17.1871 PMID: 17000944

Intergovernmental Panel on Climate Change (IPCC). (2023). *Sixth Assessment Report: Climate Change 2023*. United Nations.

Intergovernmental Panel on Climate Change. (2023). *Climate Change 2023: Mitigation of Climate Change*. Cambridge University Press.

International Energy Agency (IEA). (2022). *World Energy Investment Report 2022*. IEA Publications.

International Energy Agency (IEA). (2022). *World Energy Outlook 2022*. IEA Publications.

International Energy Agency. (2023). *Energy Efficiency 2023*. IEA Publications.

International Energy Agency. (2023). *World Energy Outlook 2023*. IEA Publications.

International Finance Corporation (IFC). (2019). *Financing a sustainable future: The role of green finance in emerging markets*. IFC. https://www.ifc.org

International Finance Corporation. (2023). *Emerging Market Clean Energy Investment Trends*. IFC Publications.

International Financial Reporting Standards (IFRS). (2021). *IFRS standards and sustainability reporting*. IFRS Foundation. https://www.ifrs.org

International Islamic Financial Market (IIFM). (2023, August). IIFM Sukuk Report 2023 (12th edition): A comprehensive study of the global Sukuk market. Manama, Kingdom of Bahrain.

International Labor Organization (ILO). (2022). *World Employment and Social Outlook: The Role of Green Jobs*. ILO.

International Renewable Energy Agency (IRENA). (2022). *Innovation Outlook: Renewable Power-to-Hydrogen*. IRENA.

International Renewable Energy Agency (IRENA). (2022). *World Energy Transitions Outlook 2022*. IRENA.

International Renewable Energy Agency. (2023). *Global Energy Transformation: A Roadmap to 2050*. IRENA Publications.

International Trade Administration. (2023, November 5). *Mexico - Renewable Energy*. https://www.trade.gov/country-commercial-guides/mexico-renewable-energy

Iqbal S, Bilal AR (2021) Energy financing in COVID-19: how public supports can benefit?. China Finance Rev Int

Iqbal, S., & Bilal, A. R. (2021). Investment performance: Emotional beasts are dragging into the darkness of the castle. *Global Business Review*, •••, 09721509211044309.

Iqbal, S., Bilal, A. R., Nurunnabi, M., Iqbal, W., Alfakhri, Y., & Iqbal, N. (2021a). It is time to control the worst: Testing COVID-19 outbreak, energy consumption and CO 2 emission. *Environmental Science and Pollution Research International*, *28*(15), 19008–19020. DOI: 10.1007/s11356-020-11462-z PMID: 33184786

Islam, E., & Singh, M. (2025). Geographic diversification, climate risk, and bank lending: Evidence from farm loans. *Journal of Financial Intermediation*, *63*, 101152. DOI: 10.1016/j.jfi.2025.101152

Islamic Financial Services Board (IFSB). (2024). Islamic Financial Services Industry Stability Report 2024. Kuala Lumpur, Malaysia.

Islam, M. J., Moreira, F., & Douch, M. (2025). Does banks' environmental engagement impact funding costs? *Journal of International Financial Markets, Institutions and Money*, *103*, 102184. DOI: 10.1016/j.intfin.2025.102184

Ivanov, I. T., Kruttli, M. S., & Watugala, S. W. (2024). Banking on Carbon: Corporate Lending and Cap-and-Trade Policy. *Review of Financial Studies, 37*(5), 1640–1684. DOI: 10.1093/rfs/hhad085

Ivanovski, K., & Marinucci, N. (2021). Policy uncertainty and renewable energy: Exploring the implications for global energy transitions, energy security, and environmental risk management. *Energy Research & Social Science, 82*, 102415. DOI: 10.1016/j.erss.2021.102415

Jacobson, M. Z.. (2023). 100% Clean and Renewable Wind, Water, and Sunlight All-Sector Energy Roadmaps for Countries. *Nature Energy, 8*(4), 234–247.

Jain, A., & Biswal, P. C. (2016). Dynamic linkages among oil price, gold price, exchange rate, and stock market in India. *Resources Policy, 49*, 179–185. *Scopus.* Advance online publication. DOI: 10.1016/j.resourpol.2016.06.001

Jena, S. K., Lahiani, A., Tiwari, A. K., & Roubaud, D. (2021). Uncovering the complex asymmetric relationship between trading activity and commodity futures price: Evidenced from QNARDL study. *Resources Policy, 74*, 102277. DOI: 10.1016/j.resourpol.2021.102277

Jha, K. K., & Katiyar, K. (2024). *Impact of Russia-Ukraine war on Oil & Gas, Commodities and Metal Sector of Indian Economy*. 2024 IEEE International Conference on Interdisciplinary Approaches in Technology and Management for Social Innovation, IATMSI 2024. Scopus. DOI: 10.1109/IATMSI60426.2024.10503500

Jha, S., & Dangwal, R. C. (2025). Fintech services and financial inclusion: A systematic literature review of developing nations. *Journal of Science and Technology Policy Management, 16*(7), 1167–1198. DOI: 10.1108/JSTPM-03-2023-0034

Ji, Q., Li, J., & Sun, X. (2019). Measuring the interdependence between investor sentiment and crude oil returns: New evidence from the CFTC's disaggregated reports. *Finance Research Letters, 30*, 420–425. DOI: 10.1016/j.frl.2019.02.005

Ji, Q., & Zhang, D. (2019). How much does financial development contribute to renewable energy growth and upgrading of energy structure in China? *Energy Policy, 128*, 11–124. DOI: 10.1016/j.enpol.2018.12.047

Joshi, D., Paul, S., Singh, P., Mane, N. S., & Yadav, M. (2024). Sustainable Finance: The Role of Green Bonds in Driving ESG Investments. Library of Progress-Library Science, Information Technology & Computer, 44(3).

Kahia, M., Jebli, M. B., & Belloumi, M. (2019). Analysis of the impact of renewable energy consumption and economic growth on carbon dioxide emissions in 12 MENA countries. *Clean Technologies and Environmental Policy, 21*(4), 871–885. DOI: 10.1007/s10098-019-01676-2

Kahneman, D., & Tversky, A. (1979). Prospect theory: An analysis of decision under risk. *Econometrica, 47*(2), 263–291. DOI: 10.2307/1914185

Karmakar, M., & Inani, S. (2019). Information share and its predictability in the Indian stock market. *Journal of Futures Markets, 39*(10), 1322–1343. *Scopus*. Advance online publication. DOI: 10.1002/fut.22041

Kaur, G., & Dhiman, B. (2024). A short on the current scenario of fluctuations in commodity market in India. *BIO Web of Conferences, 86*, 01056. DOI: 10.1051/bioconf/20248601056

Kaur, H., & Chaudhary, A. (2024). Volatility Transmission Role of Indian Equity and Commodity Markets. In *Deep Learning Tools for Predicting Stock Market Movements* (pp. 429–443). Scopus., DOI: 10.1002/9781394214334.ch18

Kaur, R., Kumar, P., Radulescu, M., Mohd, S., & Dascalu, N. (2025). Spillover Nexus among Green Cryptocurrency, Sectoral Renewable Energy Equity Stock and Agricultural Commodity: Implications for Portfolio Diversification. *Economics, 19*(1), 20250138. Advance online publication. DOI: 10.1515/econ-2025-0138

Kaushik, S., & Nagarkar, J. (2023). Seasonality index in different asset classes in India. *AIP Conference Proceedings, 2523*(1), 030019. DOI: 10.1063/5.0110536

Keswani, S., & Wadhwa, B. (2019). Evaluating the impact of macroeconomic variable on Indian stock market. *International Journal of Engineering and Advanced Technology, 8*(6), 4427–4434.

Khamitdkhanovich, N. A., Rajendran, M., Nailevich, I. R., & Krishnakumari, S. (2025, June). Green finance and ESG investing: Driving sustainable growth through innovation. In AIP Conference Proceedings (Vol. 3306, No. 1, p. 030025). AIP Publishing LLC.

Khan, M. A., Zulkhibri, M., & Zaman, K. (2022). The role of Islamic finance in sustainable development goals: Empirical evidence from sukuk markets. *Resources Policy, 77*, 102665. DOI: 10.1016/j.resourpol.2022.102665

Khan, S. (2020). Toward a theory of professionalization of the Saudi nonprofit sector. *Journal of Nonprofit Education & Leadership, 10*(1). Advance online publication. DOI: 10.18666/JNEL-2020-V10-I1-9472

Khan, T., & Badjie, F. (2022). Islamic blended finance for circular economy impactful SMEs to achieve SDGs. *The Singapore Economic Review, 67*(01), 219–244. DOI: 10.1142/S0217590820420060

Kim, H., Batten, J. A., & Ryu, D. (2020). Financial crisis, bank diversification, and financial stability: OECD countries. *International Review of Economics & Finance, 65*, 94–104. DOI: 10.1016/j.iref.2019.08.009

King Khalid Foundation. (2020). Nonprofit sector in Saudi Arabia: Progress and prospects under. *Vision (Basel), 2030*, •••. https://kkf.org.sa/en/research/nonprofit-sector-vision-2030

Kingdom of Saudi Arabia. (2016). Vision 2030: Kingdom of Saudi Arabia. Riyadh: Council of Economic and Development Affairs. https://www.vision2030.gov.sa

Kirithiga, S., Naresh, G., & Thiyagarajan, S. (2018). Spillover between commodity and equity benchmarking indices. *Benchmarking, 25*(7), 2512–2530. *Scopus*. Advance online publication. DOI: 10.1108/BIJ-06-2017-0143

Kohli, K., Kumar, B., Kohli, P., & Kashyap, M. (2024). Examining the factors affecting mobile banking adoption among users in the Himalayan region of Uttarakhand: An extension of the technology acceptance model through structural equation modelling. *International Journal of Electronic Banking, 4*(3), 294–317. DOI: 10.1504/IJEBANK.2024.139221

Koparde, S. A., Selvakumar, P., Jha, R. P., Deepthi Reddy, K., Kuchipudi, J. D., Manjunath, T. C., & Sharma, M. (2026). Generative AI in Predicting and Preventing Foodborne Illnesses. In Madaan, G., Singh, A., Chahal, B., David, A., & Singh, G. (Eds.), *Generative AI in Food Systems: Predictive Demand, Smart Supply Chains, and Sustainable Service Futures* (pp. 205–232). IGI Global Scientific Publishing., DOI: 10.4018/979-8-3373-5182-7.ch010

Kopp, E., Kaffenberger, L., & Wilson, C. (2017). *Cyber risk, market failures, and financial stability*. International Monetary Fund. DOI: 10.5089/9781484313787.001

Koval, V., Laktionova, O., Atstāja, D., Grasis, J., Lomachynska, I., & Shchur, R. (2022). Green Financial Instruments of Cleaner Production Technologies. *Sustainability (Basel), 14*(17), 10536. DOI: 10.3390/su141710536

Kraaijenbrink, J., Spender, J. C., & Groen, A. J. (2019). The resource-based view: A review and assessment of its critiques. *Journal of Management, 36*(1), 349–372. DOI: 10.1177/0149206309350775

Krueger, P., Sautner, Z., & Starks, L. T. (2020). *The importance of climate risks for institutional investors* [Working paper]. European Corporate Governance Institute. DOI: 10.1093/rfs/hhz137

Kumar, K. K., Stalin, G. A., & Rajeswari, B. (2024). An Empirical Study: Long Run and Short Run Impact of Macroeconomic Variables on Indian Stock Market Prices. In *Studies in Systems, Decision and Control* (Vol. 223, pp. 273–281). Scopus. DOI: 10.1007/978-3-031-51997-0_23

Kumar, N., & Goel, D. (2024). Macroeconomic Determinants of Price Dynamics in the Indian Stock Market. *Indian Journal of Finance*, 25–41. DOI: 10.17010/ijf/2024/v18i9/174457

Kumar, P., Taneja, S., & Ozen, E. (2025). Exploring the influence of green bonds on sustainable development through low-carbon financing mobilization. *International Journal of Law and Management*, *67*(2), 249–270. DOI: 10.1108/IJLMA-01-2024-0030

Kunz, J., & Heitz, M. (2021). Banks' risk culture and management control systems: A Systematic Literature Review. *Journal of Management Control*, *32*(4), 439–493. DOI: 10.1007/s00187-021-00325-4

Lasco, R. D., Delfino, R. J. P., & Espaldon, M. L. O. (2014). Agroforestry systems: Helping smallholders adapt to climate risks while mitigating climate change. *Wiley Interdisciplinary Reviews: Climate Change*, *5*(6), 825–833. DOI: 10.1002/wcc.301

Liang, J., & Ullah, I. (2024). Analysis of crude oil and gold price volatility and their correlation during socio-economic crises. *Resources Policy*, *98*, 105311. DOI: 10.1016/j.resourpol.2024.105311

Lim, C. Y., Woods, M., Humphrey, C., & Seow, J. (2017). The paradoxes of risk management in the banking sector. *The British Accounting Review*, *49*(1), 75–90. DOI: 10.1016/j.bar.2016.09.002

Lindfors, L. (2024). *The role of user-centric design in global platform economy value creation*. https://aaltodoc.aalto.fi/items/e111add3-f142-4c11-a488-1d05c58dcd7c

Lin, H., Shen, H., Hsieh, Y., & Hsiao, H. (2025). Corporate carbon emissions and the costs of bank borrowing: Evidence from an emerging market. *Corporate Social Responsibility and Environmental Management*, *0*(6), 1–17. DOI: 10.1002/csr.70087

Liu, C., Tian, M., & Huang, B. (2025). Volatility spillover dynamics between fintech and traditional financial industries and their rich determinants: New evidence from Chinese listed institutions. *International Review of Financial Analysis, 101. International Review of Financial Analysis, 101*, 104034. Advance online publication. DOI: 10.1016/j.irfa.2025.104034

Liu, J.-L., Ma, C.-Q., Ren, Y.-S., & Zhao, X.-W. (2020). Do real output and renewable energy consumption affect CO2 emissions? Evidence for selected BRICS countries. *Energies, 13*(4), 960. DOI: 10.3390/en13040960

Liu, Z., Li, J., & Sun, H. (2024). Climate transition risk and bank risk-taking: The role of Digital Transformation. *Finance Research Letters, 61*, 105028. DOI: 10.1016/j.frl.2024.105028

Lockett, A., Thompson, S., & Morgenstern, U. (2009). The development of the resource-based view of the firm: A critical appraisal. *International Journal of Management Reviews, 11*(1), 9–28. DOI: 10.1111/j.1468-2370.2008.00252.x

Lusardi, A., & Mitchell, O. S. (2014). The economic importance of financial literacy: Theory and evidence. *Journal of Economic Literature, 52*(1), 5–44. DOI: 10.1257/jel.52.1.5 PMID: 28579637

Malik, M. S., Rehman, A. U., & Hashmi, S. H. (2021). Barriers to green finance in Islamic banking: An empirical study. *Environmental Science and Pollution Research International, 28*, 15298–15310. DOI: 10.1007/s11356-021-13588-7

Mamadiyarov, Z. (2024). Comparative Analysis of Digital Banking Adoption in Central Asia: Opportunities and Barriers. *International Multi-Disciplinary Journal of Education, 2*(8), 452–458.

Mannucci, P. M., & Franchini, M. (2017). Health effects of ambient air pollution in developing countries. *International Journal of Environmental Research and Public Health, 14*(9), 1048. DOI: 10.3390/ijerph14091048 PMID: 28895888

Markard, J., Raven, R., & Truffer, B. (2012). Sustainability transitions: An emerging field of research and its prospects. *Research Policy, 41*(6), 955–967. DOI: 10.1016/j.respol.2012.02.013

Maurya, P. K., Bansal, R., & Mishra, A. K. (2025). *Sectoral connectedness, volatility spillover and hedging opportunities: Evidence from Indian stock market*. Eurasian Economic Review., DOI: 10.1007/s40822-025-00324-z

Meadowcroft, J. (2022). Governance, transitions, and the green state. *Environmental Politics, 31*(4), 575–594.

Memon, B. A., & Nusratova, G. (2025). Do Financial Markets and Safe-Haven Assets Affect CBDCs? Examining the Nexus between CBDC, Stock Index, Metal Commodity Futures, Oil Price, and Volatility. *Journal of Central Banking Theory and Practice*, *14*(2), 151–167. DOI: 10.2478/jcbtp-2025-0017

Mexico Business News. (n.d.). *Mexico's Energy Transition Progress Slows Down*: WEF. https://mexicobusiness.news/energy/news/mexicos-energy-transition-progress-slows-down-wef-0

Miniaoui, H., & Bellalah, M. (2025). Green Growth Opportunities and Sustainable Finance in the Gulf Cooperation Council Region: Strengthening the Response to Global Shocks (éditeurs). Gulf Studies, Springer. DOI: 10.1007/978-981-96-8287-4

Mio, C., Panfilo, S., & Blundo, B. (2020). Sustainable development goals and the strategic role of Business: A Systematic Literature Review. *Business Strategy and the Environment*, *29*(8), 3220–3245. DOI: 10.1002/bse.2568

Mirza, N., Umar, M., Sbia, R., & Jasmina, M. (2025). The impact of blue and green lending on credit portfolios: A commercial banking perspective. *Review of Accounting and Finance*, *24*(2), 238–253. DOI: 10.1108/RAF-11-2023-0389

Mishra, A., & Kumar, R. P. (2023). Threshold Cointegration and Price Transmission in Commodity Marketsof India. *Finance: Theory and Practice*, *27*(4), 66–79. *Scopus*. Advance online publication. DOI: 10.26794/2587-5671-2023-27-4-66-79

Mishra, B. R., Selvakumar, P., Kaur, R., Kalshetti, P., Mishra, K. K., Sharma, M., & Manjunath, T. C. (2025). Innovation and Digital Transformation in the Public Sector Entities. In Santos Ferreira, A., & Lourenço dos Santos, C. (Eds.), *Enhancing Public Sector Accountability and Services Through Digital Innovation* (pp. 409–436). IGI Global Scientific Publishing., DOI: 10.4018/979-8-3693-9251-5.ch013

Mishra, S. P., Kumar, R., & Rout, J. (2023). An overview of green, social, sustainability, and sustainability-linked (GSSS) bonds. *Journal of Social and Economic Development*, *25*(S1, Suppl 1), 127–145. DOI: 10.1007/s40847-023-00275-8

Mishra, U., & Dash, A. S. (2025). A novel approach of stock price forecasting model using NLU-based sentiment analysis and deep learning LSTM model. *Neural Computing and Applications. Neural Computing & Applications*, *37*(22), 18449–18478. Advance online publication. DOI: 10.1007/s00521-025-11395-z

Mi, Z., Guan, D., Liu, Z., Liu, J., Viguié, V., Fromer, N., & Wang, Y. (2019). Cities: The core of climate change mitigation. *Journal of Cleaner Production*, *207*, 582–589. DOI: 10.1016/j.jclepro.2018.10.034

Modi, V., ... (2023). Energy Services for the Millennium Development Goals. UN Development Programme.

Mohammed, R. Y. M. (2024). Sustainability Leadership in Saudi Non-profit Organizations: A Qualitative Insight (Governance). *Public Administration Review, 13*(1), 13–27. DOI: 10.5539/par.v13n1p13

Mookerjee, J., Bhuriya, M. K., & Josphin, R. (n.d.). *Digital Banking and Financial Inclusion in Rural Economies*. Retrieved September 17, 2025, from https://core.ac.uk/download/pdf/640070834.pdf

Mueller, I., Nguyen, H., & Nguyen, T. (2025). Carbon transition risk and corporate loan securitization. *Journal of Financial Intermediation, 63*, 101146. DOI: 10.1016/j.jfi.2025.101146

Mueller, I., & Sfrappini, E. (2025). Climate change-related regulatory risks and bank lending. *Journal of International Economics, 158*, 104156. DOI: 10.1016/j.jinteco.2025.104156

Naili, M., & Lahrichi, Y. (2022). The determinants of banks' credit risk: Review of the literature and future research agenda. *International Journal of Finance & Economics, 27*(1), 334–360. DOI: 10.1002/ijfe.2156

Nandru, P., Byram, A., & Rentala, S. (2015). Factors influencing financial inclusion through banking services. [JCRM]. *Journal of Contemporary Research in Management, 10*(4). http://jcrm.psgim.ac.in/index.php/jcrm/article/download/448/302

Naruetharadhol, P., Ketkaew, C., Hongkanchanapong, N., Thaniswannasri, P., Uengkusolmongkol, T., Prasomthong, S., & Gebsombut, N. (2021). Factors Affecting Sustainable Intention to Use Mobile Banking Services. *SAGE Open, 11*(3), 21582440211029925. DOI: 10.1177/21582440211029925

Naseer, M. M., & Bagh, T. (2024). Green finance and climate challenges: Paving the way for economic, social, and governance readiness and sustainable development goals. In *The Palgrave Handbook of Green Finance for Sustainable Development* (pp. 123–153). Springer International Publishing. DOI: 10.1007/978-3-031-65756-6_5

National Renewable Energy Laboratory. (2023). *2023 Annual Technology Baseline*. NREL Publications.

Nautiyal, N., & Kandpal, V. (2025). Nonlinearity Between Economic Indicators and Indian Capital Market. *FIIB Business Review, 14*(1), 43–57. DOI: 10.1177/23197145211035483

Ndlovu, X., Schaling, E., & Alagidede, P. (2014). Modelling the rand and commodity prices: A Granger causality and cointegration analysis. *Suid-Afrikaanse Tydskrif vir Ekonomiese en Bestuurswetenskappe, 17*(5), 5. Advance online publication. DOI: 10.4102/sajems.v17i5.334

Negi, P., & Jaiswal, A. (2024). *Sustainable bonds as a sustainable financial instrument: thematic insights and future prospects*. Qualitative Research in Financial Markets.

Neuhoff, K., Fankhauser, S., Guerin, E., Hourcade, J. C., Jackson, H., Rajan, R., & Ward, J. (2009). *Structuring international financial support to support domestic climate change mitigation in developing countries*. Climate Strategies.

NewClimate Institute. (2024, November). *Understanding finance needs for a just transition of the Mexican power sector*. https://newclimate.org/sites/default/files/2024-11/Report_Understanding%20finance%20needs%20for%20a%20just%20transition%20of%20the%20Mexican%20power%20sector_nov2024.pdf

Newell, P., & Mulvaney, D. (2023). The political economy of just transitions: Toward inclusive and sustainable energy transitions. *Energy Research & Social Science, 99*, 103081.

Nguyen, A. H., Do, M., Hoang, T., & Nguyen, L. (2023). Green Financing for Sustainable Development: Insights from multiple cases of Vietnamese commercial banks. *Business Strategy and the Environment, 32*(1), 321–335. DOI: 10.1002/bse.3132

Nguyen, D. D., Ongena, S., Qi, S., & Sila, V. (2022). Climate change risk and the cost of mortgage credit. *Review of Finance, 26*(6), 1509–1549. DOI: 10.1093/rof/rfac013

Nguyen, H. C., & Duong, H. K. (2025). An international empirical study of the impact of green bonds on sustainable development goals. *Review of Behavioral Finance, 17*(3), 499–523. DOI: 10.1108/RBF-08-2024-0243

Nirupama, E., Seenivasan, S., Bhardwaj, K., Karunakaran, N. B., Basava Aradhya, S. G., Selvakumar, P., & Manjunath, T. C. (2025). Transforming Digital Supply Chains With Gamification and Metaverse Integration. In Malik, G., Singh, D., & Bansal, R. (Eds.), *Addressing Practical Problems Through the Metaverse and Game-Inspired Mechanics* (pp. 123–152). IGI Global Scientific Publishing., DOI: 10.4018/979-8-3693-8886-0.ch006

North, D. C. (1990). *Institutions, institutional change and economic performance*. Cambridge University Press. DOI: 10.1017/CBO9780511808678

O'Connor, N., & Madden, M. G. (2005, December). A neural network approach to predicting stock exchange movements using external factors. In International Conference on Innovative Techniques and Applications of Artificial Intelligence (pp. 64-77). London: Springer London.

OECD. (2023). *Climate Finance Provided and Mobilised by Developed Countries in 2013-2021: Aggregate Trends and Opportunities for Scaling Up Adaptation and Mobilised Private Finance. Climate Finance and the USD 100 Billion Goal series.* OECD Publishing.

OECD. (2023). Energy Policies of IEA Countries: Review and Recommendations. Paris: OECD Publications. [National energy policy analysis and best practices] (Rockström, 2023). Rockström, J. (2023). Planetary Boundaries and Human Development. Nature Publications.

OECD. (2024). *Governing for Sustainable Prosperity in the Middle East and North Africa, OECD Public Governance Reviews.* OECD Publishing., DOI: 10.1787/d0da1d30-

Ogunbode, C. A., Böhm, G., Capstick, S. B., Demski, C., Spence, A., & Tausch, N. (2019). The resilience paradox: Flooding experience, coping and climate change mitigation intentions. *Climate Policy*, *19*(6), 703–715. DOI: 10.1080/14693062.2018.1560242

Ohler, A., & Fetters, I. (2014). The causal relationship between renewable electricity generation and GDP growth: A study of energy sources. *Energy Economics*, *43*, 125–139. DOI: 10.1016/j.eneco.2014.02.009

Oh, N., & Kim, K. N. (2023). Struggle to survive: Case of flood risk on US community banks. *Agricultural Finance Review*, *83*(4/5), 617–634. DOI: 10.1108/AFR-02-2023-0018

Omri, A., & Nguyen, D. K. (2014). On the determinants of renewable energy consumption: International evidence. *Energy*, *72*, 554–560. DOI: 10.1016/j.energy.2014.05.081

Organisation for Economic Co-operation and Development (OECD). (2021). *Green Growth and Sustainable Development Report 2021*. OECD.

Organisation for Economic Co-operation and Development (OECD). (2022). *Policy Coherence for Sustainable Energy Transition*. OECD.

Organization for Economic Co-operation and Development. (2020). *OECD/INFE 2020 international survey of adult financial literacy*. OECD Publishing.

Ostrom, E. (2010). Polycentric systems for coping with collective action and global environmental change. *Global Environmental Change*, *20*(4), 550–557. DOI: 10.1016/j.gloenvcha.2010.07.004

Paramati, S. R., Ummalla, M., & Apergis, N. (2016). The effect of foreign direct investment and stock market growth on clean energy use across a panel of emerging market economies. *Energy Economics*, *56*, 29–41. DOI: 10.1016/j.eneco.2016.02.008

Park, E., Boyer, C. N., & Neill, C. L. (2025). A Markov regime-switching event response model: Beef price spread response to processing capacity shocks. *Empirical Economics*, *68*(3), 1039–1071. DOI: 10.1007/s00181-024-02677-x

Park, S., & Gupta, S. (2012). Comparison of SML and GMM estimators for the random coefficient logit model using aggregate data. *Empirical Economics*, *43*(3), 1353–1372. DOI: 10.1007/s00181-011-0519-3

Pastpipatkul, P., Sriboonchitta, S., & Subsai, P. (2024). Relationships Among Financial and Commodity Markets on Economic Growth: New Evidence from Bayesian Estimation of Seemingly Unrelated Time Series Equations. In V. Kreinovich, W. Yamaka, & S. Leurcharusmee (Eds.), *Applications of Optimal Transport to Economics and Related Topics* (pp. 605–621). Springer Nature Switzerland. DOI: 10.1007/978-3-031-67770-0_41

Pata, U. K. (2018). Renewable energy consumption, urbanization, financial development, income and CO2 emissions in Turkey: Testing EKC hypothesis with structural breaks. *Journal of Cleaner Production*, *187*, 770–779. DOI: 10.1016/j.jclepro.2018.03.236

Patel, A. S., Rao, V. K., & Radhakrishnan, M. K. (2023). Impact of mobile banking platforms PayTM and Google Pay on financial inclusion in rural and Semi-Urban areas in India. *Journal of Finance and Accounting*, *7*(5), 113–122. DOI: 10.53819/81018102t4205

Peng, J., Patwary, A. K., Rasul, F., Koley, B., Nicomedes, C. J. C., Dutta, A., & Rahman, M. H. (2018). An empirical investigation of financial literacy and behavioral biases on investment decision: Evidence from Chittagong Stock Exchange. *Cogent Economics & Finance*, *6*(1), 1455723. DOI: 10.1080/23322039.2018.1455723

Periasamy, P., & Satish, R. (2015). A study on relationship between Indian commodity market and Indian stock market with special reference to exchanges in India—An analytical framework. [Scopus.]. *International Journal of Applied Business and Economic Research*, *13*(8), 6323–6334.

Pesaran, M. H., Shin, Y., & Smith, R. J. (2001). Bounds testing approaches to the analysis of level relationships. *Journal of Applied Econometrics, 16*(3), 289–326. *Scopus.* Advance online publication. DOI: 10.1002/jae.616

Pfeffer, J., & Salancik, G. R. (1978). *The external control of organizations: A resource dependence perspective.* Harper & Row.

Pfenninger, S., Hawkes, A., & Keirstead, J. (2014). Energy systems modeling for twenty-first century energy challenges. *Renewable & Sustainable Energy Reviews, 33*, 74–86. DOI: 10.1016/j.rser.2014.02.003

Phan, D. H. B., Iyke, B. N., Sharma, S. S., & Affandi, Y. (2021). Economic policy uncertainty and financial stability–is there a relation? *Economic Modelling, 94*, 1018–1029. DOI: 10.1016/j.econmod.2020.02.042

Piper, D. L. A. (2016, January). *Mexico's new Energy Transition Law: key points.* https://www.dlapiper.com/en-ar/insights/publications/2016/01/mexico-new-energy-transition-law

Pranchana, R., Sudhamathi, S., & Benneet, S. (2025). Evaluating the Impact of Sectoral Indices on Stock Market Performance in the National Stock Exchange. *Indian Journal of Information Sources and Services, 15*(1), 238–243. DOI: 10.51983/ijiss-2025.IJISS.15.1.30

Putrevu, J., & Mertzanis, C. (2024). The adoption of digital payments in emerging economies: Challenges and policy responses. *Digital Policy. Regulation & Governance, 26*(5), 476–500.

Qamruzzaman, M., & Jianguo, W. (2020). The asymmetric relationship between financial development, trade openness, foreign capital flows, and renewable energy consumption: Fresh evidence from panel NARDL investigation. *Renewable Energy, 159*, 827–842. DOI: 10.1016/j.renene.2020.06.069

Qiubin, H., & Mingting, K. (2024). Does climate policy uncertainty impair or improve corporate investment efficiency? *Economics Letters, 244*, 112015. DOI: 10.1016/j.econlet.2024.112015

Rafindadi, A. A., & Ozturk, I. (2017). Impacts of renewable energy consumption on the German economic growth: Evidence from combined cointegration test. *Renewable & Sustainable Energy Reviews, 75*, 1130–1141. DOI: 10.1016/j.rser.2016.11.093

Raman, M. S., Koppa, K. B., Chithambar Gupta, V., Bhat, V. A., Guru Basava Aradhya, S., & Selvakumar, P. (2025). Digital Transformation and Sustainability Artificial Intelligence and Sustainable Development. In Yıldırım, D., Yıldırım, S., & Kandpal, V. (Eds.), *Effects of Digitalization and Circular Economy on Sustainable Policy and Climate Change Prevention* (pp. 69–96). IGI Global Scientific Publishing., DOI: 10.4018/979-8-3693-9909-5.ch003

Ratnovski, L. (2013). Liquidity and transparency in Bank Risk Management. *Journal of Financial Intermediation*, *22*(3), 422–439. DOI: 10.1016/j.jfi.2013.01.002

Raza, S. A., Shah, N., Qureshi, M. A., Qaiser, S., Ali, R., & Ahmed, F. (2020). Nonlinear threshold effect of financial development on renewable energy consumption: Evidence from panel smooth transition regression approach. *Environmental Science and Pollution Research International*, *27*(25), 32034–32047. DOI: 10.1007/s11356-020-09520-7 PMID: 32506406

RE100. (2023). *Annual Report: Corporate Renewable Energy Procurement.* The Climate Group Publications.

Reghezza, A., Altunbas, Y., Marques-Ibanez, D., Rodriguez d'Acri, C., & Spaggiari, M. (2022). Do banks fuel climate change? *Journal of Financial Stability*, *62*, 101049. DOI: 10.1016/j.jfs.2022.101049

REN21. (2024). Renewables 2024 Global Status Report. Paris: REN21 Secretariat.

Rocky Mountain Institute. (2023). *Clean Energy Transitions Program.* RMI Publications.

Rodrigues, J. P., & Vinnarasi, B. (2023). *A Study of Financialization of Commodity Markets in India* (Vijayan V. & Senthil Kumar T.S., Eds.; Vol. 399). Scopus. DOI: 10.1051/e3sconf/202339907011

Rodríguez-Monroy, C., Mármol-Acitores, G., & Nilsson-Cifuentes, G. (2018). Electricity generation in Mexico: Current status, future prospects, and economic implications of clean energy reforms. *Renewable & Sustainable Energy Reviews*, *92*, 1–17. DOI: 10.1016/j.rser.2018.04.077

Roy, R. P., & Sinha Roy, S. (2017). Financial contagion and volatility spillover: An exploration into Indian commodity derivative market. *Economic Modelling, 67*, 368–380. *Scopus.* Advance online publication. DOI: 10.1016/j.econmod.2017.02.019

Saadaoui, K., Rhazi, K. S., Mejdoub, Y., & Aboudou, A. (2023). An analysis of the prospects and efficiency of floating and overland photovoltaic systems. *Int. J. Appl. Power Eng*, *12*(2), 144–152.

Sahadudheen, I., & Kumar, P. K. S. (2023). Time-Varying Relationship and Volatility Spillovers Among Oil, Gold, Forex and Stock Markets in Indian Context: The Juxtaposition of Global Economic Crisis and COVID-19 Pandemic. *Indian Economic Journal, 71*(4), 748–767. *Scopus.* Advance online publication. DOI: 10.1177/00194662231166168

Sahu, P., Şanli, O., Janjua, L. R., & Rao, N. M. (2024). Evolutionary Perspective of Green Bond Financing under the Shadow of ESG Readiness. In *Green Bonds and Sustainable Finance* (pp. 104–119). Routledge. DOI: 10.4324/9781032686844-7

Salamon, L. M., & Anheier, H. K. (1997). *Defining the nonprofit sector: A cross-national analysis.* Manchester University Press.

Sánchez-Peña, J., García-Rendón, J., & López-Feldman, A. (2020). Renewable energy policies and economic growth in Mexican states. *Energy Journal, 41*(2), 47–72. DOI: 10.5547/01956574.41.2.jsan

Sarangi, P. K., Singh, E., Kaushal, A., Sharma, B., Dutta, M., & Dubey, V. P. (2025). *Analysing the Fluctuations in Commodity Prices and Forecasting the Future Directions Using Machine Learning Techniques. 259*, 1827–1836. *Scopus.* Advance online publication. DOI: 10.1016/j.procs.2025.04.138

Sarfaraz, J. (2017). Unified theory of acceptance and use of technology (UTAUT) model-mobile banking. *Journal of Internet Banking and Commerce*, *22*(3), 1–20.

Sarwar, S., Khalfaoui, R., Waheed, R., & Dastgerdi, H. G. (2019). Volatility spillovers and hedging: Evidence from Asian oil-importing countries. *Resources Policy, 61*, 479–488. *Scopus.* Advance online publication. DOI: 10.1016/j.resourpol.2018.04.010

Saudi Green Initiative. (2023). *Annual Report 2023.* Ministry of Energy.

Sealey, C.Jr, & Lindley, J. (1977). Inputs, Outputs, and the Theory of Production and Cost of Depository Financial Institutions. *The Journal of Finance*, *32*(4), 1251–1266. DOI: 10.1111/j.1540-6261.1977.tb03324.x

Sehrawat, M., & Giri, A. K. (2017). A Sectoral Analysis of the Role of Stock Market Development on Economic Growth: Empirical Evidence from Indian Economy. *Global Business Review, 18*(4), 911–923. *Scopus.* Advance online publication. DOI: 10.1177/0972150917692242

Seth, S., & Pathan, M. I. (2025). Green Sukuk: Contextual Spectrum and Alignments. *Journal of Marketing & Social Research*, *2*(2), 540–548.

Shaban, A. (2024). Digitalization and Exclusion—Digital Divides and Development. In Shaban, A. (Ed.), *Digital Geographies—Theory, Space, and Communities* (pp. 255–496). Springer Nature Singapore., DOI: 10.1007/978-981-97-4734-4_4

Shaikh, A. A., Alamoudi, H., Alharthi, M., & Glavee-Geo, R. (2023). Advances in mobile financial services: A review of the literature and future research directions. *International Journal of Bank Marketing*, *41*(1), 1–33. DOI: 10.1108/IJBM-06-2021-0230

Shalini, V., & Prasanna, K. (2016). Impact of the financial crisis on Indian commodity markets: Structural breaks and volatility dynamics. *Energy Economics, 53*, 40–57. *Scopus.* Advance online publication. DOI: 10.1016/j.eneco.2015.02.011

Shamsher, S. (2021). Financialisation of commodities – Empirical evidence from the Indian financial market. *IIMB Management Review, 33*(1), 38–49. *Scopus.* Advance online publication. DOI: 10.1016/j.iimb.2021.03.001

Sharma, A., & Piplani, N. (2017). Digital banking in India: A review of trends, opportunities and challenges. [IRJMST]. *International Research Journal of Management Science & Technology*, *8*(1), 168–180.

Sharma, C. (2020). Exchange rate volatility and exports from India: A commodity-level panel data analysis. *Journal of Financial Economic Policy, 12*(1), 23–44. *Scopus.* Advance online publication. DOI: 10.1108/JFEP-11-2018-0157

Sharma, S. S., & Thuraisamy, K. (2013). Oil price uncertainty and sovereign risk: Evidence from Asian economies. *Journal of Asian Economics*, *28*, 51–57. DOI: 10.1016/j.asieco.2013.06.001

Shikimi, M. (2025). Firms' climate risks and bank lending: Evidence from the COVID-19 crisis. *Finance Research Letters*, *73*, 106606. DOI: 10.1016/j.frl.2024.106606

Siddiqui, S., & Roy, P. (2019). Predicting volatility and dynamic relation between stock market, exchange rate and select commodities. *Acta Universitatis Agriculturae et Silviculturae Mendelianae Brunensis, 67*(6), 1597–1611. *Scopus.* Advance online publication. DOI: 10.11118/actaun201967061597

Sidhu, A., & Katoch, R. (2021). Do international gold prices and NSE nifty 50 move together? *Advances in Mathematics: Scientific Journal, 10*(1), 497–506. *Scopus.* Advance online publication. DOI: 10.37418/amsj.10.1.49

Sikka, V., & Bhayana, P. (2024). Barriers to Comprehensive Financial Inclusion Across the Globe: From Sociocultural Norms to Systemic Challenges. In Mansour, N., Baral, S., & Garg, V. (Eds.), *E-Financial Strategies for Advancing Sustainable Development* (pp. 89–126). Springer Nature Switzerland., DOI: 10.1007/978-3-031-67523-2_7

Silva, M., & Cerqueira, P. A. (2019). Assessing the determinants of household electricity prices in the EU: A system-GMM panel data approach. *Renewable & Sustainable Energy Reviews, 73*, 1131–1137. DOI: 10.1016/j.rser.2017.02.016

Singh Chauhan, S., Suri, P., Alam, F., Hani, U., Johri, A., & Ali, F. (2025). A causality investigation into stock prices and macroeconomic indicators in the Indian stock market. *F1000Research, 13*. Scopus. DOI: 10.12688/f1000research.157041.3

Singhal, S., & Ghosh, S. (2016). Returns and volatility linkages between international crude oil price, metal and other stock indices in India: Evidence from VAR-DCC-GARCH models. *Resources Policy, 50*, 276–288. *Scopus.* Advance online publication. DOI: 10.1016/j.resourpol.2016.10.001

Singhania, M., Chadha, G., & Prasad, R. (2024). Sustainable finance research: Review and agenda. *International Journal of Finance & Economics, 29*(4), 4010–4045. DOI: 10.1002/ijfe.2854

Singh, G., & Padmakumari, L. (2020). Stock market reaction to inflation announcement in the Indian stock market: A sectoral analysis. *Cogent Economics and Finance, 8*(1). *Cogent Economics & Finance, 8*(1), 1723827. Advance online publication. DOI: 10.1080/23322039.2020.1723827

Sivanand, R., Selvakumar, P., & Sharma, N. (2025). Nanotechnology Applications in Cancer Drug Delivery Systems: A Review. *Journal of Environmental Nanotechnology, 14*(2), 458–470. DOI: 10.13074/jent.2025.06.2511221

Smil, V. (2022). *Energy Transitions: Global and National Perspectives.* Praeger Publishers.

Snyder, H. (2019). Literature review as a research methodology: An overview and guidelines. *Journal of Business Research, 104*, 333–339. DOI: 10.1016/j.jbusres.2019.07.039

Somu, B. (2025). *The Future of Financial IT: Agentic Artificial Intelligence and Intelligent Infrastructure in Modern Banking*. Deep Science Publishing. https://books.google.com/books?hl=en&lr=&id=VeplEQAAQBAJ&oi=fnd&pg=PA1&dq=Agentic+AI,+working+in+tandem+with+digital+asset+innovation,+introduces+adaptive+intelligence+into+these+markets-+capable+of+autonomously+managing+tokenized+securities,+monitoring+carbon+credit+compliance,+and+optimizing+risk+strategies+for+next-generation+green+instruments&ots=Fxwl2XSwul&sig=5YqN-BLKmdmw46YaPfa7FhctPWo

Song, J., Cao, W., & Shan, Y. G. (2025). Green Credit and Bank's risk-taking: Evidence from China. *International Journal of Managerial Finance, 21*(1), 157–184. DOI: 10.1108/IJMF-03-2024-0144

Sovacool, B. K. (2023). Energy Justice: Conceptual Insights and Practical Applications. *Applied Energy Journal, 312*, 118–135.

Sovacool, B. K., & Griffiths, S. (2020). The cultural barriers to a low-carbon future: A review of energy justice. *Energy Policy, 138*, 111227.

Sovacool, B. K., Hook, A., Martiskainen, M., & Baker, L. (2021). The justice and equity implications of the clean energy transition. *Nature Energy, 6*(7), 728–739.

Sun, W., Zhai, Y., Li, R., & Wang, X. (2025). The extreme event of climate change: How digital transformation affects the performance of commercial banks?'. *Energy Economics, 150*, 108783. DOI: 10.1016/j.eneco.2025.108783

Swart, R. (2019). Assessing physical climate risks for investments: A risky promise. *Climate Services, 14*, 15–18. DOI: 10.1016/j.cliser.2019.04.001

Technology cost and performance projections, McKinsey Global Institute. (2023).

Tekin, B. (2025). Do economic uncertainty and political risk steer CDS dynamics? An analysis of the Türkiye CDS. *International Journal of Islamic and Middle Eastern Finance and Management, 18*(2), 249–270. DOI: 10.1108/IMEFM-05-2024-0215

The Impact of Economic Events on Stock Market Returns: Evidence from India | Asian Economic and Financial Review. (n.d.). Retrieved July 21, 2025, from https://archive.aessweb.com/index.php/5002/article/view/2011

The Net-Zero Transition: What It Would Cost, What It Could Bring. New York: McKinsey Publications.

Tiwari, A. K., Nasreen, S., Shahbaz, M., & Hammoudeh, S. (2020). Time-frequency causality and connectedness between international prices of energy, food, industry, agriculture and metals. *Energy Economics, 85*, 104529. DOI: 10.1016/j.eneco.2019.104529

Topcu, M., & Tugcu, C. T. (2020). The impact of renewable energy consumption on income inequality: Evidence from developed countries. *Renewable Energy, 151*, 1134–1140. DOI: 10.1016/j.renene.2019.11.103

Trinh, V. Q., Elnahass, M., Salama, A., & Izzeldin, M. (2020). Board busyness, performance and financial stability: Does bank type matter? *European Journal of Finance*, *26*(7–8), 774–801. DOI: 10.1080/1351847X.2019.1636842

Tugcu, C. T., Ozturk, I., & Aslan, A. (2012). Renewable and non-renewable energy consumption and economic growth relationship revisited: Evidence from G7 countries. *Energy Economics, 34*(6), 1942–1950. DOI: 10.1016/j.eneco.2012.08.021

Umar, Z., Nasreen, S., Solarin, S. A., & Tiwari, A. K. (2019). Exploring the time and frequency domain connectedness of oil prices and metal prices. *Resources Policy, 64*, 101516. DOI: 10.1016/j.resourpol.2019.101516

UN-Habitat. (2022). *Climate Action in Cities: Integrated Approaches for Urban Resilience*. UN-Habitat.

United Nations Development Program (UNDP). (2022). *Energy Governance for Sustainable Development*. UNDP.

United Nations Environment Program (UNEP). (2023). *Adaptation Gap Report 2023*. UNEP.

United Nations Environment Programme (UNEP). (2016). The green finance playbook: A guide to unlocking private capital for a sustainable future. https://www.unepfi.org/wordpress/wp-content/uploads/2017/05/UNEPFI-2016-ANNUAL-OVERVIEW.pdf

United Nations Environment Programme Finance Initiative (UNEP FI). (2020). Financial sector transformation for sustainable development. UNEP FI. https://www.unepfi.org

United Nations Framework Convention on Climate Change (UNFCCC). (2023). *Climate Finance Delivery Plan 2023–2026*. UNFCCC Secretariat.

United Nations. (2015). *Transforming Our World: The 2030 Agenda for Sustainable Development*. UN Publications.

United Nations. (2022). *Sustainable Development Goals Progress Report 2022*. UN Publications.

Vairasigamani, P., & Amilan, S. (2025). Unveiling the dynamic linkages and Hedging between Indian Sectoral Indices and Oil, Gold and Cryptocurrency in times of crisis episodes. *Revista Mexicana de Economía y Finanzas Nueva Época REMEF*, *20*(3), 3. Advance online publication. DOI: 10.21919/remef.v20i3.1211

Valerievna, T. O., & Igorevna, G. I. (2024). Sustainable financing as a factor influencing eco-friendly business transformation: Existing challenges and potential solutions. *The American Journal of Interdisciplinary Innovations and Research*, *6*(10), 8–20. DOI: 10.37547/tajiir/Volume06Issue10-02

van Zanden, J. L. (2023). Examining the relationship of information and communication technology and financial access in Africa. *Journal of Business and Economic Options*, *6*(3), 26–36.

Vannoy, S. A., & Palvia, P. (2010). The social influence model of technology adoption. *Communications of the ACM*, *53*(6), 149–153. DOI: 10.1145/1743546.1743585

Vashishth, T. K., Sharma, V., Kumar, B., & Sharma, K. K. (2024). Cloud-Based Data Management for Behavior Analytics in Business and Finance Sectors. In Data-Driven Modelling and Predictive Analytics in Business and Finance (pp. 133-155). Auerbach Publications. DOI: 10.1201/9781032618845-9

Vashishth, T. K., Sharma, V., Kumar, B., Chaudhary, S., Panwar, R., & Sharma, K. K. (2024, May). Guarding Finances: The Role of Image Processing in Unmasking Counterfeit Currency. In 2024 2nd International Conference on Advancement in Computation & Computer Technologies (InCACCT) (pp. 668-671). IEEE.

Vashishth, T. K., Sharma, V., Sharma, K. K., Kumar, B., Chaudhary, S., & Kaushik, V. (2025). The Role of Blockchain in Disrupting Traditional Banking with Decentralized Finance. Decentralized Finance and Blockchain: A Game Changing Duo, 143-174.

Vashishth, T. K., Kumar, B., Alam, M. S., Sharma, V., & Chaudhary, S. (2024). The Impact of Artificial Intelligence on the Future of Computing: A Comparative Study. In *Applications of Block Chain technology and Artificial Intelligence: Lead-ins in Banking, Finance, and Capital Market* (pp. 1–17). Springer International Publishing. DOI: 10.1007/978-3-031-47324-1_1

Vasylieva, T., Lyulyov, O., Bilan, Y., & Streimikiene, D. (2019). Sustainable economic development and greenhouse gas emissions: The dynamic impact of renewable energy consumption, GDP, and corruption. *Energies*, *12*(17), 3289. DOI: 10.3390/en12173289

Veningston, K., Ronalda, M., Shaikh, T. A., Gazi, F., & Bhagat, S. (2025). Modeling and Evaluating Stock Market and Commodity Price Fluctuation Using Time Series Sequence Learning. *SN Computer Science*, *6*(5), 435. Advance online publication. DOI: 10.1007/s42979-025-03931-4

VijayaVenkataRaman, S., Iniyan, S., & Goic, R.VijayaVenkataRaman S. (2012). A review of climate change, mitigation and adaptation. *Renewable & Sustainable Energy Reviews*, *16*(1), 878–897. DOI: 10.1016/j.rser.2011.09.009

Viscusi, W. K., & Zeckhauser, R. J. (2006). The perception and valuation of the risks of climate change: A rational and behavioral blend. *Climatic Change*, *77*(1), 151–177. DOI: 10.1007/s10584-006-9075-9

Vishnuvardhan, B., Manjula, B., & Lakshman Naik, R. (2020). A Study of Digital Banking: Security Issues and Challenges. In K. S. Raju, A. Govardhan, B. P. Rani, R. Sridevi, & M. R. Murty (Eds.), *Proceedings of the Third International Conference on Computational Intelligence and Informatics* (Vol. 1090, pp. 163–185). Springer Singapore. DOI: 10.1007/978-981-15-1480-7_14

Vyshnevskyi, I., & Sohn, W. (2025). Central Banks' support for climate action: A literature review and key issues. [Preprint]. *Journal of Economic Surveys*, joes.12709. DOI: 10.1111/joes.12709

Wang, Q., & Fu, L. (2025). Fintech, Risk Management and banking green credit. *Finance Research Letters*, *83*, 107686. DOI: 10.1016/j.frl.2025.107686

Wang, Y., Wang, R., & Mason, T. A. (2025). Exploring the impact of green bonds on sustainable development and financial markets: An ARDL-based causality approach. *Finance Research Letters*, *85*, 107943. DOI: 10.1016/j.frl.2025.107943

Warale, P. N., Limbore, N. V., Anute, N., Ahluwalia, G. K., Raut, A., & Selvakumar, P. (2025). AI-Enhanced Customer Engagement and Experience: Chatbots, Virtual Assistants, and AI-Powered Customer Interactions. In Hussain, Z., Khan, A., Uzir, M., Sharipudin, M., & Shaheen, A. (Eds.), *Impacts of AI-Generated Content on Brand Reputation* (pp. 335–362). IGI Global Scientific Publishing., DOI: 10.4018/979-8-3373-4327-3.ch014

Warren, R., VanDerWal, J., Price, J., Welbergen, J. A., Atkinson, I., Ramirez-Villegas, J., Osborn, T. J., Jarvis, A., Shoo, L. P., Williams, S. E., & Lowe, J. (2013). Quantifying the benefit of early climate change mitigation in avoiding biodiversity loss. *Nature Climate Change, 3*(7), 678–682. DOI: 10.1038/nclimate1887

Weber, C. (2017). What drives financial literacy and investment behavior? Evidence from survey data. *Applied Economics, 49*(48), 4861–4878. DOI: 10.1080/00036846.2017.1296551

Weisbrod, B. A. (1998). *To profit or not to profit: The commercial transformation of the nonprofit sector*. Cambridge University Press., DOI: 10.1017/CBO9780511625947

Wen, B., You, W., & Yuan, M. (2025). The impact of commercial banks' ESG performance: Difference between provocative and passive risk-taking – evidence from China. *Research in International Business and Finance, 76*, 102859. DOI: 10.1016/j.ribaf.2025.102859

Windmeijer, F. (2005). A finite sample correction for the variance of linear efficient two-step GMM estimators. *Journal of Econometrics, 126*(1), 25–51. DOI: 10.1016/j.jeconom.2004.02.005

World Bank Group. (2018). *Green finance for emerging markets: Leveraging finance for sustainable development*. World Bank. https://www.worldbank.org

World Bank. (2020). *Mexico renewable energy development*. World Bank Group. https://openknowledge.worldbank.org/handle/10986/34587

World Bank. (2023). *Energy Sector Directions Paper: Powering a Green Future*. World Bank.

World Bank. (2023). *Financing the Energy Transition: Global Outlook 2023*. World Bank.

World Bank. (2023). Results and Performance of the World Bank Group 2023. Washington, DC: Independent Evaluation Group. World Bank. http://hdl.handle.net/10986/40886

World Bank. (2023). *State and Trends of Carbon Pricing 2023*. World Bank Publications.

World Economic Forum. (2025, August 29). *Global Risks Report 2025*. https://www.weforum.org/publications/global-risks-report-2025/

World Health Organization (WHO). (2021). *Air Pollution and Public Health Report 2021*. WHO.

World Health Organization. (2023). *Ambient Air Quality and Health.* WHO Publications.

Wu, M. (2025). Green Credit, Bank Competitiveness and bank risk-taking. *International Review of Economics & Finance, 103*, 104481. DOI: 10.1016/j.iref.2025.104481

Wu, M., & Zhang, L. (2025). Green credit, deposit-loan ratio, and risk taking of commercial banks. *Finance Research Letters, 81*, 107470. DOI: 10.1016/j.frl.2025.107470

Xiong, J., Zhang, Y., & Mao, Z. (2025). A Study on the Impact of Green Bonds on Corporate ESG Performance. GBP Proceedings Series, 3, 159-165. DOI: 10.70088/6yb5je33

Yang, X., Wang, X., Cao, J., Zhao, L., & Huang, C. (2024). Cross-regional connectedness of financial market: Measurement and determinants. *The North American Journal of Economics and Finance, 72*, 102157. DOI: 10.1016/j.najef.2024.102157

Yildirim, E., Saraç, Ş., & Aslan, A. (2012). Energy consumption and economic growth in the USA: Evidence from renewable energy. *Renewable & Sustainable Energy Reviews, 16*(9), 6770–6774. DOI: 10.1016/j.rser.2012.09.004

Yin, H. (2019). Bank globalization and financial stability: International evidence. *Research in International Business and Finance, 49*, 207–224. DOI: 10.1016/j.ribaf.2019.03.009

Yu, H., Lu, R., Yang, H., & Zhang, B. (2024). Wealthy individual investors and stock markets' tail risk. *PLoS One, 19*(5), e0282173. DOI: 10.1371/journal.pone.0282173

Yunus, M. (2008). Creating a world without poverty: Social business and the future of capitalism, Global Urban Development, 4(2), 1-19. https://www.globalurban.org/GUDMag08Vol4Iss2/Yunus.pdf

Zagorchev, A., Vasconcellos, G., & Bae, Y. (2011). Financial development, technology, growth and performance: Evidence from the accession to the EU. *Journal of International Financial Markets, Institutions and Money, 21*(5), 743–759. DOI: 10.1016/j.intfin.2011.05.005

Zairis, G., Liargovas, P., & Apostolopoulos, N. (2024). Sustainable finance and ESG importance: A systematic literature review and research agenda. *Sustainability (Basel), 16*(7), 2878. DOI: 10.3390/su16072878

Zehri, C., Alsadan, A., & Saleh Ben Ammar, L. (2025). Asymmetric impacts of geopolitical risks on energy Trade: Divergent vulnerabilities in emerging vs. advanced economies. *Journal of Economic Asymmetries, 32*, e00427. DOI: 10.1016/j.jeca.2025.e00427

Zhanfeng L, & Zhiying Y (2009). The relationship between consumption, investment, export and economic growth in China [J]. Statist Res 2.

Zhang, B., Wang, B., & Wang, Z. (2017). Role of renewable energy and non-renewable energy consumption on EKC: Evidence from Pakistan. *Journal of Cleaner Production*, *156*, 855–864. DOI: 10.1016/j.jclepro.2017.03.203

Zhang, R. (2025). Green Finance and ESG Investment Strategies under Climate Hussain Management. *Economics and Management Innovation*, *2*(1), 140–148. DOI: 10.71222/00d71j98

Zhang, Y., Ye, S., Liu, J., Zhang, L., & Li, J. (2025). Can digital transformation of commercial banks reduce green credit risks? *International Review of Financial Analysis*, *99*, 103934. DOI: 10.1016/j.irfa.2025.103934

Zhao, C., Cai, J., & Yang, S. (2025). A hybrid stock prediction method based on periodic/non-periodic features analyses. *EPJ Data Science*, *14*(1), 1. Advance online publication. DOI: 10.1140/epjds/s13688-024-00517-7

Zhao, P., Pan, J., Yue, Q., & Zhang, J. (2021). Pricing of financial derivatives based on the Tsallis statistical theory. *Chaos, Solitons, and Fractals*, *142*, 110463. DOI: 10.1016/j.chaos.2020.110463

Zheng, L., Mirza, N., Umar, M., & Su, C. (2025). Sustainable lending to European SMEs: Implications for bank performance. *International Review of Economics & Finance*, *102*, 104283. DOI: 10.1016/j.iref.2025.104283

Zhou, Z., Tong, J., Lu, H., & Luo, S. (2024). The impact of Green Finance and Technology on the commercial banks' profit and risk. *Finance Research Letters*, *66*, 105715. DOI: 10.1016/j.frl.2024.105715

Zhu, J. (2014). "Data Envelopment Analysis," International Series in Operations Research & Management Science, in: Quantitative Models for Performance Evaluation and Benchmarking, edition 3, chapter 1, pages 1-9, Springer.

Zhu, B., & Zhao, Y. (2022). Carbon risk and the cost of bank loans: Evidence from China. *Technological Forecasting and Social Change*, *180*, 121741. DOI: 10.1016/j.techfore.2022.121741

About the Contributors

Abdelkader Mohamed Sghaier Derbali is an Assistant Professor in Finance at the Department of Finance and Accounting, Higher Institute of Informatics and Management of Kairouan, Kairouan University, Kairouan, Tunisia and at the Department of Administrative and Financial Sciences, Applied College, Taibah University, Saudi Arabia. He is one of the Editorial Board Members in the Journal of Energy Markets, International Business Review, Cogent Economic & Finance, International Review of Applied Economics, African Journal of Accounting, Auditing and Finance, and African Journal of Economic and Sustainable Development. His research interests include Risk Management, Systemic risk, International finance, Capital markets and institutions, Banking and market microstructure and Islamic Finance. He has published articles, among others, in Research in International Business and Management, Cogent Business & Management, Cogent Economic & Finance, Journal of Chinese Governance, The Chinese Economy, Journal of Energy Markets, and International Journal of Trade and Global Markets.

***

Kanagaraj A. R. is a professor in the Department of Corporate Secretaryship at Dr. N.G.P. Arts and Science College, Coimbatore, Tamil Nadu. With an teaching career spanning over 27 years, he has made significant contributions to the field of corporate education. Dr. Kanagaraj has published more than 45 research articles in reputed journals, reflecting his deep academic insight and commitment to knowledge dissemination. He is a qualified NET (National Eligibility Test) holder and has served as a member of the Board of Studies in various colleges, playing a vital role in curriculum development and academic governance.

Vanshita Arora is a student of Masters in Statistics with research aspirations for my higher studies (PHd). She has been working on several research projects involving GDP forecasting.

Sachin Chaudhary is member of International Association of Engineers, The Society of Digital Information and Wireless Communications, Global Professors Welfare Association, International Association of Academic plus Corporate (IAAC), Computer Science Teachers Association and Internet Society. His research interest includes Cloud Computing, Artificial Intelligence, Machine Learning and Operations Research.

Amir Ahmad Dar is the Faculty in the Department of Mathematics at Lovely Professional University Punjab. The area of interest: Financial mathematics, financial economics, statistics and business mathematics.

Avinash Hanmant Ghadage is an academic and Associate Professor at D.Y. Patil Vidyapeeth's Global Business School and Research Centre, located in Tathawade, Pune.

Shashank M Hiremath is an academician & researcher with 2 decades of experience. Corporate trainer, NSE certified financial analyst and investment advisor. Handling subjects like Financial Derivatives, Corporate Taxation, Security Analysis and Portfolio Management, Advanced Financial Management, Merchant Banking, Business Research Methods, etc. Attended and organized various international and national conferences. Guest speaker at various institutions.

Suhashini J. is an academic associated with the Department of Management Studies at GTN Arts College, located in Dindigul, Tamil Nadu.

Lamia Jamel is a PhD in Economics at the Faculty of Economic Sciences and Management of Sousse in University of Sousse, Tunisia and is an assistant Professor at the Department of Finance and Economics, College of Business Administration, Taibah University, Medinah, 42353, Saudi Arabia. His research interests include Economic analysis, financial economics, quantitative finance, financial development, and energy commodities.

Sukriti Kakkar is pursuing an M.Sc. in Statistics & Data Analytics and holds a B.Sc. in Economics with Mathematics. Her research focuses on predictive modeling, probability distributions, and time series analysis. She has worked on projects including GDP Prediction, Applications of Probability in real life and student lifestyle analytics using Python, R, and Power BI. Sukriti aims to apply statistical methods to real-world decision making.

Erhan Kilincarslan is a Reader in Accounting and Finance at the School of Business, Education and Law, University of Huddersfield. Erhan's primary research interest centres on the fields of corporate finance, empirical finance, corporate governance, ownership structure and corporate social responsibility. He has a growing interest in the other fundamental properties and applications of accounting, finance, and management in various areas such as, but not limited to, tourism, banking, REITs, energy, and green stocks. Erhan's research appears in different reputable international journals, including ABS 4*/3* and ABDCA*/A journals.

Jiafan Li is a Lecturer in Accounting at the University of Huddersfield. She started her academic career in 2022. She is interested in accountability, SMEs, small accounting practices and access to finance, and sustainability. She is also interested in multidisciplinary research in accounting with management theories and other disciplines in business. She has several publications in reputable international journals and presented her work in the UK and EU conferences and workshops.

Gayathri M. is a Assistant Professor in the Department of Data Science and Business Systems at SRM University, Chennai. She has over 8 years of teaching experience, with academic interests in data analytics, information visualization, database systems, and artificial intelligence. She is a member of IEEE, ACM, and a life member of ISTE, along with several other international professional associations. She continues to contribute actively to research and teaching in computer science and engineering.

T. C. Manjunath was born on 6th February 1967. He did his B.E. from RV College of Engg. in 1989 (B'lore Univ), M.E. in Robotics & Automation from Govt.'s LD College of Engg., (Gujarat Univ., Ahmedabad) in 1992 with FCD-First Rank & PhD in Systems & Control Engg. from IIT Bombay in the year 2007, got best research scholar award for the entire IITB. He is currently working as the Dean Research (R & D) and as Professor in Dept. of Computer Science & Engg. with specialization in IoT, Cyber Security, Blockchain Technology in Rajarajeswari College of Engineering, Bengaluru, Karnataka. He has got a teaching (academic), research & administrative experience of more than 35+ years (10 years as Principal & 8 years as HoD, 1 year as Dean Research & rest as faculty) and has published more than 1500+ research papers in various national & international journals & conferences, with 175+ patents, 100+ book-chapters, 25 awards, 100s of Keynote talks and also published 25 text books in various field of science, engineering & technology, completed projects worth Rs. 9 Crores till date & guided 10 PhD scholars in VTU. He is a member of 30 Professional Society Chapters and is a fellow of IE, IETE, AMIEE, Sr. Member of IEEE and a chartered engineer.

P. Selvakumar is currently an Associate Professor in the Department of Science and Humanities at Nehru Institute of Technology, Coimbatore, Tamil Nadu, India. Prior to joining Nehru Institute of Technology, he held positions as an Assistant Professor in the Department of Science and Humanities various engineering colleges in Coimbatore, Tamil Nadu, India, spanning from 2011 to 2023. Dr.P. Selvakumar's academic and research interests lie primarily in the field of phytochemistry. He has made significant contributions to his field, with an impressive record that includes 30 journal publications, 12 patents, and 20 conference papers. Additionally, he has organized 5 seminars and conferences, further demonstrating his commitment to academic and scientific exchange. Throughout his career, Dr. P.Selvakumar has been recognized for his dedication and excellence in teaching, having been honored with several awards, including the Best Teacher Awards. He holds a prominent position in the academic community, contributing actively to research.

Prabakaran is an Assistant Professor at Sri Balaji University, Pune, where he is recognized for his dedication to academic excellence and effective teaching practices. he brings a practical and student-focused approach to education. Mr. Prabakaran is actively involved in academic research, curriculum development, and faculty development initiatives, contributing to the university's mission of shaping industry-ready professionals. His commitment to continuous learning, interactive pedagogy, and student mentorship makes him a valued member of the academic community at Sri Balaji University.

Priyanka Rana is an active academician and researcher. Working as an associate professor in School of Commerce and Management, IIMT University, Meerut, U.P, INDIA

Rajveer Rawlin received his Ph.D. from Bharathiar University, Tamil Nadu, India, and his MBA in finance from the Cardiff Metropolitan University, Wales, UK. He teaches in the Department of Business & Management at CHRIST University. His research interests include areas of Capital Markets, Investment Analysis, and Portfolio Management, and he has more than 25 years of experience in the above, covering the US and Indian Markets. He is the author of the popular financial blog Ahead of the Curve. He reviews for Emerald and Sage.

Riyazuddin is a finance scholar and educator with over 16 years of teaching and academic service. He holds a Ph.D. in Finance, UGC-NET qualification, and an MBA in Finance. Currently serving as Associate Professor in the School of

Business Management at Maharishi University of Information Technology, Noida, he has previously worked with leading institutes such as NIET Greater Noida and Dr. K.N. Modi Institute. His research contributions include ABDC, Scopus, and peer-reviewed journals, along with authored books and chapters. Bilingual in Hindi and English, he specializes in Accounting, Finance, and Business Management.

Guru Basava Aradhya S. is a academician with 18 years of experience in management education. He currently serves as the Director in Padmashree Institute of Management and Sciences, Bengaluru. He holds a Ph.D. in Management and has mentored five Ph.D. scholars. His expertise includes HR, marketing, and business analytics. He has published in SCOPUS, ABDC, and UGC CARE journals. He has organized numerous international conferences and contributed to NAAC and NBA accreditations. His industry collaborations include faculty exchange programs and consultancy projects.

Vikas Sharma is an active academician and researcher in the field of computer science with 6 years of experience. He earned M. Tech degree specialized in Computer Science. He is involved in academic development and scholarly activities. He is member of International Association of Engineers, The Society of Digital Information and Wireless Communications, Global Professors Welfare Association, International Association of Academic plus Corporate (IAAC), Computer Science Teachers Association and Internet Society. His research interest includes Cloud Computing, Artificial Intelligence, Machine Learning; published more than 55 research articles with 4 books and 90 book chapters in edited books. He is contributing as member of editorial and reviewers' boards in conferences and various computer journals and edited books published by Elsevier, CRC Press, Taylor and Francis, Springer, IGI global, Wiley Publication, IET Publication, De Gruyter, River Publication etc.

Mercy Toni is a Lecturer in the Department of Economics and Finance, College of Economics, Management and Information Systems, University of Nizwa since 2008. She has over 32 years of experience in the field of teaching Economics, Statistics and other Economics and Business-related courses. Her field of interest is Micro & Macro Economics and has been working on areas such as Co-operative Learning, Money and Banking, and Population Growth. She has also attended the State Council of Education Research and Training in 2006 from India and regularly participates in Teacher's Cluster and Guild events. Dr. Mercy has completed her Ph.D. in Economics from Bharathiar University, Coimbatore, India, M.A (Economics) from M.G. University, Kottayam, India, B.Ed. (Social Science) from M.G. University, Kottayam, Kerala, B.A (Economics) from M.G. University,

Kottayam, Kerala, DCE and DCA. She also has 3 years of industrial experience in the field of Finance, Accounting & Management and 3 Years of experience in the field of teaching Communicative & General English.

Tarun Kumar Vashishth is an active academician and researcher in the field of Computer Science with 22 years of experience. He earned Ph.D. Mathematics degree specialized in Operations Research; served several academic positions such as Dy. Director, HoD, NAAC Coordinator, Academic Coordinator, Member Secretary of Department Research Committee, Assistant Center Superintendent and Head Examiner in University Examinations. He is involved in academic development and scholarly activities. He is member of International Association of Engineers, The Society of Digital Information and Wireless Communications, Global Professors Welfare Association, International Association of Academic plus Corporate (IAAC), Computer Science Teachers Association and Internet Society . His research interest includes Cloud Computing, Artificial Intelligence, Machine Learning and Operations Research; published more than 50 research articles & conference papers with 2 book and 80 book chapters. He is contributing as member of editorial and reviewer's boards in conferences and various computer journals and edited books published by Elsevier, CRC Press, Taylor and Francis, Springer, IGI global, Wiley Publication, IET Publication, De Gruyter, River Publication etc.

Index

B

C

D

E

F

G

I

M

P

R

S

V

www.ingramcontent.com/pod-product-compliance
Lightning Source LLC
LaVergne TN
LVHW062329300926
849161LV00007B/1626
9798337367668